LIVERPOOL
JOHN MOORES UNIVERSITY
AVRIL ROBARTS LRC
TITHEBARN STREET
LIVERPOOL L2 2ER
TEL. 0151 231 4022

# Forensic Biology

# Forensic Biology

## Richard Li

CRC Press is an imprint of the
Taylor & Francis Group, an **informa** business

CRC Press
Taylor & Francis Group
6000 Broken Sound Parkway NW, Suite 300
Boca Raton, FL 33487-2742


Printed in the United States of America on acid-free paper
10 9 8 7 6 5 4 3 2 1

International Standard Book Number-13: 978-1-4200-4343-3 (Hardcover)



**Library of Congress Cataloging-in-Publication Data**

Li, Richard.
Forensic biology / Richard Li.
p. cm.
Includes bibliographical references and index.
ISBN 978-1-4200-4343-3 (alk. paper)
1. Forensic biology. 2. Forensic genetics. 3. Forensic serology. I. Title.

QH313.5.F67L5 2008
363.25'62--dc22 2008000782

**Visit the Taylor & Francis Web site at**
**http://www.taylorandfrancis.com**

**and the CRC Press Web site at**
**http://www.crcpress.com**

# Contents

# VI Forensic Issues

# Preface

This text is aimed toward an undergraduate audience and was written specifically to provide a general understanding of forensic biology. Hopefully, it will be utilized by students, particularly those interested in forensic biology, to further enhance their education and training.

This text defines forensic biology as analyses performed in the forensic biology sections of forensic laboratories and thus focuses on forensic serology and forensic DNA analysis. The intent is to provide students with a scientific grounding in the area of forensic biology by offering an introduction to methods and techniques employed by forensic biology laboratories.

The aim of this book is to emphasize the basic science and its application to forensic science in an effort to make the principles more understandable. In addition, it introduces the language of forensic biology, thus enabling students to become comfortable with its use and it provides clear explanations of the principles of forensic analysis. The text also acknowledges the benefits and limitations that apply to forensic biology techniques. I hope that this text will assist students in becoming more knowledgeable about the field of forensic biology and the wealth of available information. The 24 chapters are designed to be covered in a single-semester course. To convey a general understanding of the concepts of forensic biology, it is necessary to include explanations of various techniques that are employed in the field. The techniques introduced in this text are accompanied by brief background descriptions and discussions of basic principles and techniques. Schematic illustrations are included where necessary.

Many people contributed to this book. The editorial and production staffs at Taylor & Francis Group should be acknowledged for their contributions. I would like to thank Becky Masterman, senior acquisitions editor, for excellent suggestions and support; Kari Budyk, project coordinator, for helping to keep the project on track; and Prudy Board, project editor; and the production staff for their patience and thorough reviews.

I must also acknowledge my students at Indiana University–Purdue University Indianapolis (IUPUI). Szehong Chang, Margaret de la Rosa, Christy Gelback, Lydia Higgins, Kenneth Polezoes, and Azusa Tanahara spent hours assisting with the completion of this book by handling tasks such as proofreading, reference collection, and editing. Peer reviewers provided excellent comments, suggestions, and even a few criticisms. They include Dr. Diane

Leland (IU Medical School), Dr. Deo Sapna (IUPUI), Dr. Sudhir Sinha (ReliaGene Technologies), Carl Sobieralski (Indiana State Police Laboratory), Dr. Mark Timken (California Department of Justice), and Dr. John Watson (IUPUI). Without their help, this textbook would not be as comprehensive and well organized.

Finally, my family members should be acknowledged for their patience and understanding. Countless hours, days, and months were spent to research and write this book. I would like to thank them for their patience during the many hours I spent sitting in front of my computer.

I hope my readers will find this text full of useful information, presented in a way that is easily understood. I continue to be open to suggestions and criticisms for the future.

**Richard C. Li, Ph.D.**
*Indianapolis, Indiana, USA*

# Biological Evidence

# I

# Forensic Biology: A Subdiscipline of Forensic Science

# 1

## 1.1 Forensic Laboratory Services

### 1.1.1 Common Disciplines

Forensic laboraties provide scientific testing to investigative agencies. To date, nine disciplines are commonly practiced in most municipal, county, and state forensic laboratories in the United States:

- Crime scene investigation
- Latent print examination
- *Forensic biology*
- Controlled substance analysis
- Postmortem toxicology
- Questioned document examination
- Firearm, toolmark, and other impression evidence examination
- Explosive and fire debris examination
- Transfer (trace) evidence examination

Table 1.1 describes services normally provided by a forensic laboratory. Examples of these analyses are shown in Figure 1.1 through Figure 1.8.

### 1.1.2 Other Forensic Science Services

A number of specialized forensic science services beyond those provided by forensic laboratories are routinely available to law enforcement personnel. For example, forensic services related to biological evidence and involving more specialized analysis are available. These services are important aids to a criminal investigation and require the expertise of individuals who have highly specialized skills.

#### *1.1.2.1 Forensic Pathology*

When a death is deemed suspicious or unexplained, medical examiners frequently perform autopsies to determine the exact cause (Figure 1.9 through Figure 1.11). The manner of death is classified into one of five categories: (1) natural, (2) homicide, (3) suicide, (4) accident, or (5) undetermined, based

**Table 1.1 Common Services Provided by U.S. Forensic Laboratories**

| Service | Function | Method |
|---|---|---|
| Crime scene investigation | Evidence recognition, documentation, collection, and preservation | Crime scene responses and related endeavors are diverse and vary with case and type of evidence |
| Latent print examination | Analysis of friction-ridge detail in fingerprints<br>Activities include visualization, recording, comparison, storage, and recovery of latent prints | Alternate light sources, physical (powder), and chemical enhancements<br>Direct lifts, photography, and digital imaging<br>Use of an Automated Fingerprint Identification System (AFIS) database |
| Forensic biology | Identification of biological fluids (blood, semen, and saliva)<br>DNA profiling for individualization | Serological and biochemical methods<br>Polymerase chain reaction (PCR)-based methods<br>Automated electrophoresis platforms<br>Use of Combined DNA Index System (CODIS) |
| Controlled substance analysis | Identification and quantification of drugs present in submitted evidence | Microscopic, chemical, chromatographic, and spectroscopic methodologies<br>Gas chromatography–mass spectrometry or infrared spectrophotometry |
| Post mortem toxicology | Determination of concentrations of substances and their metabolites in biological fluids or tissues | Immunoassays and chemical methods<br>Confirmatory techniques such as gas and liquid chromatography–mass spectrometry |
| Questioned document examination | Investigation of forgeries, tracings, disguised handwritings, computer manipulation of images, and recovery of altered documents<br>Analysis of papers, inks, toners, word processors, typewriters, copiers, and printers | Macroscopic and microscopic comparisons<br>Chromatographic and spectroscopic methods |
| Firearm and toolmark examination | Identification of firearms, tools, and other implements (expertise achieved predominantly through experience) | Microscopic comparisons of questioned and authenticated impressions<br>Comparison of striae on recovered bullets<br>Use of National Integrated Ballistics Information Network (NIBIN) |

**Table 1.1 Common Services Provided by U.S. Forensic Laboratories** (continued)

| Service | Function | Method |
|---|---|---|
| Explosive and fire debris examination | Identification, recovery, and detection of bulk explosives, residues, debris, and accelerants | Microscopic, spectroscopic, and chromatographic methods<br>Gas chromatography–mass spectrometry may be needed to adequately characterize sample |
| Trace evidence examination | Analysis of transferred evidence such as hairs, fibers, soil, paints, and glass | Microscopic analysis of evidence with gas chromatograph mass spectrometers, FTIR microscopes, scanning electron microscopes, basic and advanced microscopy, and capillary electrophoresis |

*Source:* Adapted from Office of Justice Programs, National Institute of Justice, U.S. Department of Justice, 1999. Forensic science: review of status and needs.

**Figure 1.1** Crime scene investigation.

on the circumstances. Additionally, a medical examiner participating in a criminal investigation is often responsible for estimating the time of death.

### *1.1.2.2 Forensic Anthropology*

Forensic anthropology is the identification and examination of human skeletal remains (Figure 1.12). Skeletal remains can reveal a number of individual characteristics that can be useful in attempting to identify an individual. An examination of bones may reveal an individual's origin, sex, approximate age, race, and presence of a skeletal injury. A forensic anthropologist may also assist in creating facial reconstructions to aid in the identification of skeletal remains or may be called upon to help collect and organize bone fragments in the course of identifying victims of mass disasters such as plane crashes.

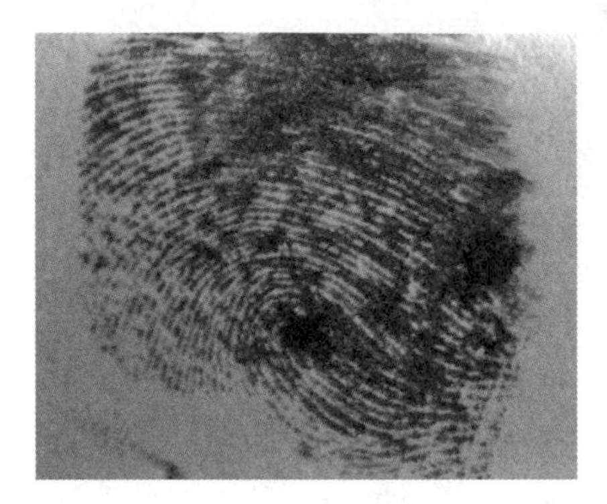

Figure 1.2 Physically enhanced fingerprint.

Figure 1.3 Automated electrophoresis instruments used for forensic DNA profiling.

### *1.1.2.3 Forensic Entomology*

The study of insects in relation to a criminal investigation is known as forensic entomology. This forensic discipline is valuable for estimating the time of death when the circumstances surrounding the crime are otherwise unknown. The stages of development of certain insect species present in or on a body can be identified and allow a forensic entomologist to approximate how long the body was left exposed (Figure 1.13 and Figure 1.14).

### *1.1.2.4 Forensic Odontology*

Practitioners of forensic odontology participate in the identification of victims whose bodies are left in an unrecognizable state. The characteristics of

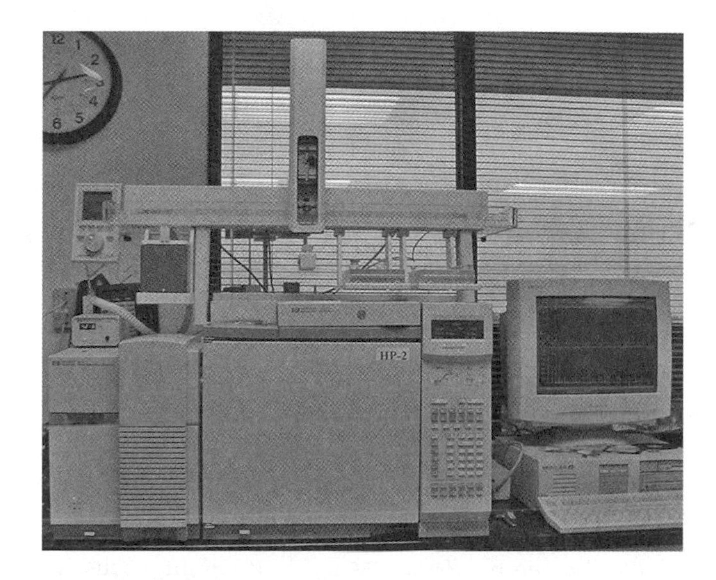


**Figure 1.4** Gas chromatography–mass spectrometry used for controlled substance analysis.

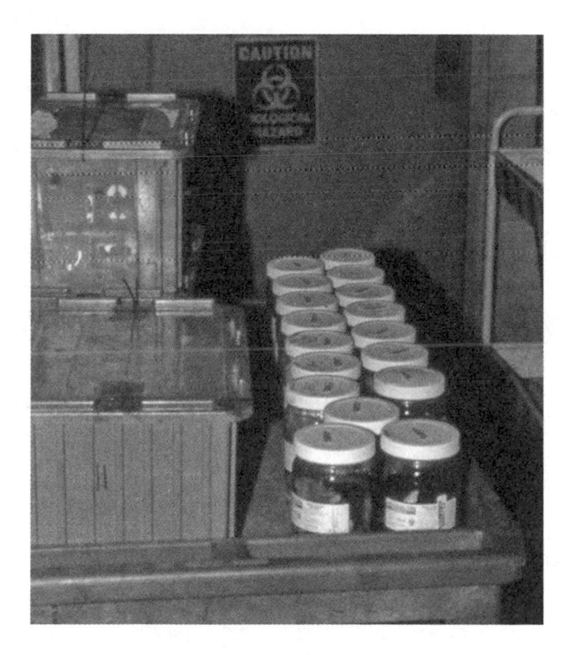


**Figure 1.5** Tissue samples for forensic toxicological analysis.

teeth, their alignment, and the overall structure of the mouth provide evidence that can identify a specific person. Dental records such as x-rays and dental casts allow a forensic odontologist to compare a set of dental remains

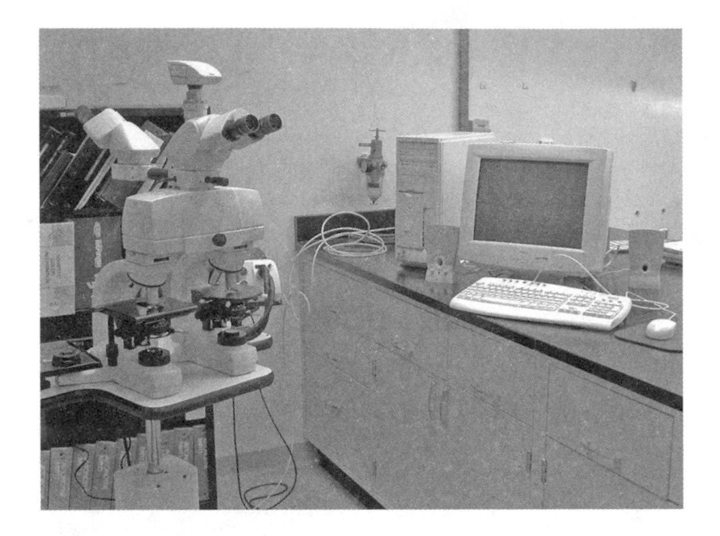

**Figure 1.6** A comparison microscope is used for simultaneous comparison of items of firearms evidence. (*Source*: James, S. H. and J. J. Nordby. *Forensic Science: An Introduction to Scientific and Investigative Techniques*, 2nd ed., Boca Raton: CRC Press, 2005. With permission.)

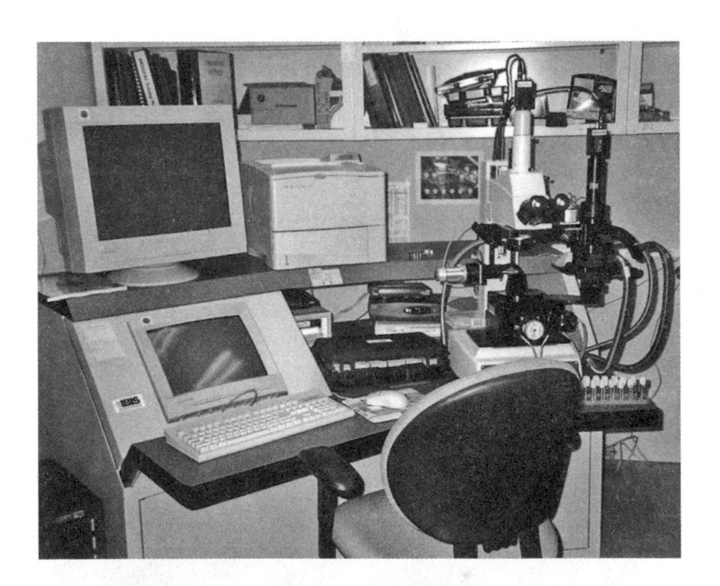

**Figure 1.7** A workstation linked to the firearm database.

and an alleged victim. Another application of forensic odontology in a criminal investigation is bite mark analysis. A forensic odontologist can analyze the marks left on a victim and compare them with tooth structures of a suspect to make a comparison.

**Figure 1.8** Scanning electron microscope used in gunshot residue analysis.

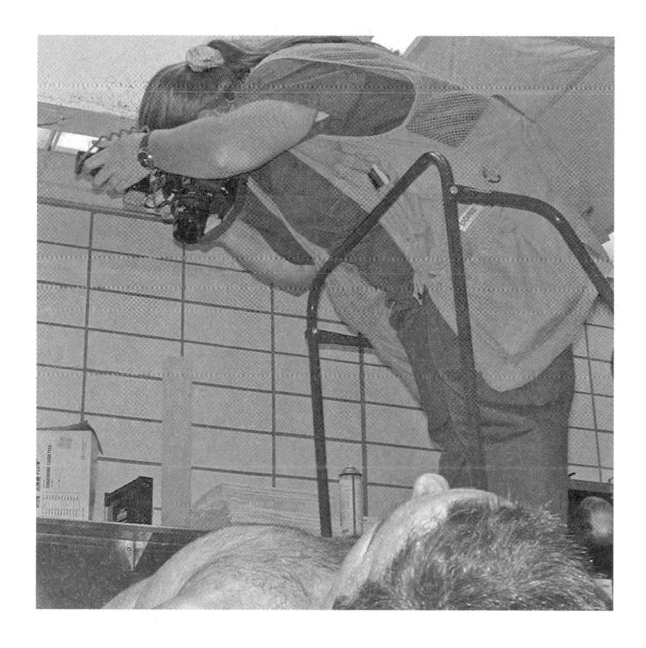

**Figure 1.9** Photographic documentation prior to autopsy.

## 1.2 A Brief History of Development of Forensic Biology

The developmental history of modern forensic biology spans three stages: (1) antigen polymorphism, (2) protein polymorphism, and (3) DNA polymorphism. Figure 1.15 illustrates this history.

**Figure 1.10** View of forensic pathology facility.

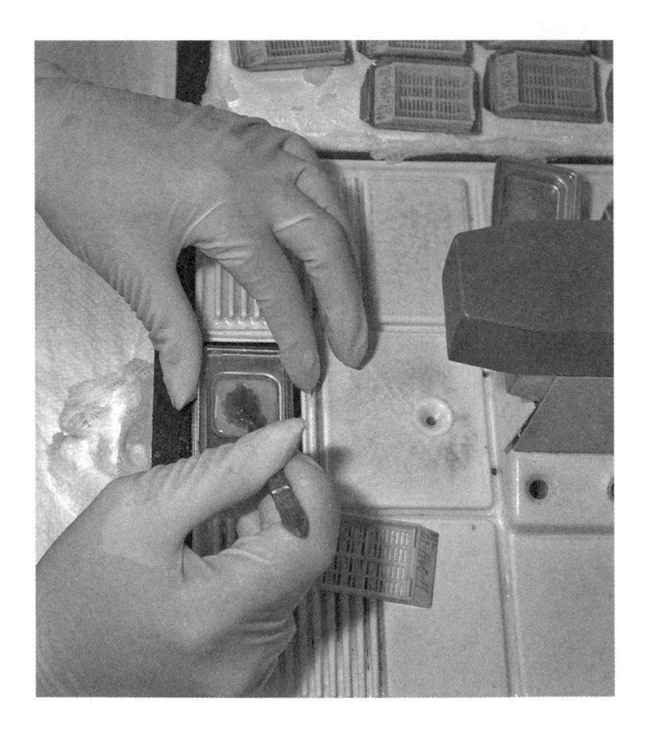

**Figure 1.11** Preparing specimens for histological section for forensic pathological examination.

**Figure 1.12** Human skeleton recovered by forensic anthropologists.

**Figure 1.13** Insects found on dead animal.

### 1.2.1 Antigen Polymorphism

The human ABO blood groups were discovered in 1900 by Karl Landsteiner in a study of the causes of blood transfusion reactions. Landsteiner's discovery made blood transfusions feasible and he received the Nobel Prize in

**Figure 1.14** Blow fly specimen. This insect is commonly encountered at crime scenes.

1930 when he revealed the four groups of human blood cells designated A, B, AB, and O. Subsequent studies found that the blood types were inherited and the frequencies with which the four types appeared in specific human populations were found to differ. This led to the discovery of the first antigen polymorphic marker for use in human identification.

By the 1960s, a dozen blood group systems had been characterized and 29 systems are currently known. Forensic laboratories employ blood group systems in a discipline known as forensic serology. While it is possible to exclude a suspect through the use of blood group typing, the evidence for inclusion of a suspect is weak due to the high probability of a coincidental match between two unrelated persons.

### 1.2.2 Protein Polymorphism

Because of the limitations of antigen polymorphism, protein polymorphism was introduced for forensic identification. Initially, a few polymorphisms in serum proteins and erythrocyte enzymes were reported. By the 1980s, however, approximately a hundred protein polymorphisms had been discovered. A few systems were commonly used in forensic laboratories, including the polymorphisms of erythrocyte enzymes, serum proteins, and hemoglobins. Blood groups and protein polymorphism analysis were combined more effectively in criminal investigations and lowered the probability of a match

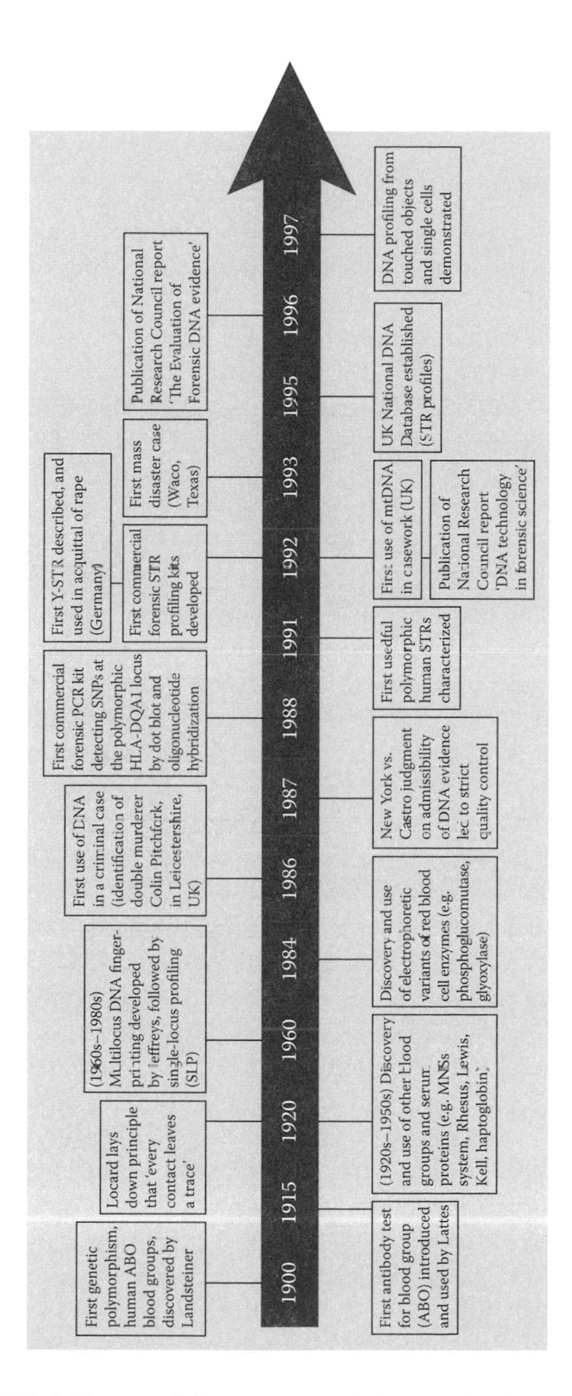


**Figure 1.15** Brief history of development of forensic biology. (*Source*: Jobling, M. A. and P. Gill, 2004. Encoded evidence: DNA in forensic analysis. *Nat Rev Genet* 5:739. With permission.)

between two unrelated individuals. However, more powerful methods were still sought.

### 1.2.3 DNA Polymorphism

In 1984, Sir Alec Jeffreys developed a DNA profiling technique using a variable number tandem repeat (VNTR) technique involving multi-locus profiling and later followed by single-locus profiling. This technique led to the solving of a double murder committed in Leicestershire (U.K.) in the 1980s. The case was the first to apply DNA evidence to a criminal investigation. During the investigation, DNA profiling not only identified the true perpetrator but also excluded an innocent suspect. This case demonstrated the great potential of DNA profiling in forensic investigations.

DNA polymorphism offered a number of advantages compared to earlier systems. The most important is the ability of the technique to reveal far greater individual variability in DNA than can be revealed by antigen and protein polymorphic markers. The probability of two unrelated individuals having the same DNA profile is very low. The great variability of DNA polymorphisms has made it possible to offer strong support for concluding that DNA from a suspect and from a crime scene came from the same source. This technique was subsequently implemented in forensic laboratories worldwide.

In the mid 1980s, Kary Mullis and coworkers developed the polymerase chain reaction (PCR) technique that amplifies a tiny quantity of DNA. Mullis's invention had a powerful impact on molecular biology and earned him a Nobel Prize in 1993. Since the introduction of PCR, additional techniques have been developed for forensic DNA testing purposes.

The application of PCR-based assays makes forensic DNA analysis possible when only minute quantities of DNA can be recovered from a crime scene, for example, from hairs and cigarette butts. These assays have greatly increased the sensitivity of forensic DNA testing. The first forensic application of a PCR-based assay employing single nucleotide polymorphisms (SNPs) at the HLA-DQA1 locus (formerly called DQα) occurred in 1986. One major disadvantage of the assay was the high probability (approximately 1 in 4000) of a match between two unrelated persons. Amplified fragment length polymorphism (AFLP) at the D1S80 locus has also been implemented for forensic laboratories. The D1S80 locus is a small size VNTR marker that can be amplified by PCR. The HLA-DQA1 and AFLP assays were used for some years until the introduction of short tandem repeat (STR) assays.

In the late 1990s, forensic laboratories started employing STR loci. STRs have a number of advantages compared to VNTRs. For example, STRs can be amplified by PCR because of their smaller size, which greatly increases the sensitivity of the assay. Furthermore, STR markers are highly variable. With the application of multiple STR loci, the probability of a match between two unrelated persons becomes extremely low. As a result of DNA testing, perpetrators have been identified and wrongly convicted innocent people have been exonerated.

In 1995, the United Kingdom established the first national DNA database for criminal investigations. By the end of 1998, several other countries, including the United States, created their own national DNA databases. The U.S. has selected 13 STR loci for the Combined DNA Index System (CODIS). These national DNA databases play important roles in solving criminal cases.

Another technique known as mitochondrial DNA (mtDNA) profiling has also been used for forensic testing. mtDNA is a maternally inherited genetic material and is therefore particularly useful for human identification. Each cell contains multiple copies of mtDNA. Thus, mtDNA typing can be useful for analysis when nuclear DNA is severely degraded or present in very limited amounts, such as in cases involving shed hairs.

Alternatively, polymorphic markers at the Y chromosome have also been employed for forensic DNA testing. Y chromosomal markers are paternally inherited so they can be used for paternity testing. These markers are also very useful in analyzing DNA from multiple contributors in sexual assault cases.

## Bibliography

Brettell, T. A., K. Inman, N. Rudin, and R. Saferstein. 1999. Forensic science. *Anal Chem* 71 (12):235R.

Carey, L., and L. Mitnik. 2002. Trends in DNA forensic analysis. *Electrophoresis* 23 (10):1386.

DeForest, P., R. Gaensslen, and H. C. Lee. 1983. *Forensic Science: An Introduction to Criminalistics.* New York: McGraw-Hill.

Deng, Y. J. et al. 2005. Preliminary DNA identification for the tsunami victims in Thailand. *Genomics Proteomics Bioinformatics* 3 (3):143.

Gill, P. 2002. Role of short tandem repeat DNA in forensic casework in the U.K.: Past, present, and future perspectives. *Biotechniques* 32 (2):366.

Gill, P. et al. 1991. Databases, quality control and interpretation of DNA profiling in the Home Office Forensic Science Service. *Electrophoresis* 12 (2–3):204.

Gill, P. et al. 1994. Identification of the remains of the Romanov family by DNA analysis. *Nat Genet* 6 (2):130.

Gill, P., A. J. Jeffreys, and D. J. Werrett. 1985. Forensic application of DNA 'fingerprints.' *Nature* 318 (604 6):577.

Holland, M. M. et al. 1993. Mitochondrial DNA sequence analysis of human skeletal remains: Identification of remains from the Vietnam War. *J Forensic Sci* 38 (3):542.

Howlett, R. 1989. DNA forensics and the FBI. *Nature* 341 (6239):182.

Jeffreys, A. J. 2005. Genetic fingerprinting. *Nat Med* 11 (10):1035.

Jobling, M. A., and P. Gill. 2004. Encoded evidence: DNA in forensic analysis. *Nat Rev Genet* 5 (10):739.

Lee, H. C. et al. 1994. DNA typing in forensic science. I. Theory and background. *Am J Forensic Med Pathol* 15 (4):269.
Monckton, D. G., and A. J. Jeffreys. 1993. DNA profiling. *Curr Opin Biotechnol* 4 (6):660.
Office of Justice Programs, National Institute of Justice, U.S. Department of Justice. 1999. Forensic science: review of status and needs.
_____ . 2000. The future of forensic DNA testing: predictions of the research and development working group.
_____ . 2002. Using DNA to solve cold cases.
Primorac, D., and M. S. Schanfield. 2000. Application of forensic DNA testing in the legal system. *Croat Med J* 41 (1):32.
Saferstein, R. 2004. *Criminalistics: An Introduction to Forensic Science*. New York: Prentice-Hall.
Tracey, M. 2001. Short tandem repeat-based identification of individuals and parents. *Croat Med J* 42 (3):6.
Varsha, S. 2006. DNA fingerprinting in the criminal justice system: An overview. *DNA Cell Biol* 25 (3):181.
Walsh, S. J. 2004. Recent advances in forensic genetics. *Expert Rev Mol Diagn* 4 (1):31.
Wenk, R. E. 2004. Testing for parentage and kinship. *Curr Opin Hematol* 11 (5):357.
Whitehead, P. 1993. A historical review of the characterization of blood and secretion stains in the forensic science laboratory. I. Bloodstains. *Forensic Sci Rev* 5 (1):36.
Williamson, R., and R. Duncan. 2002. DNA testing for all. *Nature* 418 (6898):585.
Willott, G. M. 1975. The role of the forensic biologist in cases of sexual assault. *J Forensic Sci Soc* 15 (4):269.

## Study Questions

1. A blood sample from a controlled substances overdose case should be submitted to:
   (a) A forensic toxicology section of a laboratory
   (b) A controlled substances section of a laboratory

2. Blood group typing can be useful to:
   (a) Exclude a suspect
   (b) Identify a suspect with a high degree of certainty

3. Which of the following was the first forensic DNA technique developed?
   (a) STR
   (b) VNTR
   (c) AFLP
   (d) SNP

4. The polymorphic markers on Y chromosomes can be useful for:
   (a) Paternity testing
   (b) Identifying a victim by comparing samples of maternal relative DNA

5. The polymorphic markers on mitochondrial DNA can be useful for:
   (a) Paternity testing
   (b) Identifying a victim by comparing samples of maternal relative DNA

# Sources of DNA Evidence

# 2

## 2.1 Introduction to Cells

DNA evidence analysis is now widely accepted as a standard forensic technique for the investigation of a wide spectrum of crimes. DNA helps police link offenders to crime scenes by matching DNA profiles. The technique can also be used to eliminate suspects. The DNA evidence routinely encountered at crime scenes can often be categorized into several groups or types. Table 2.1 lists sources of DNA frequently found on personal items. Figure 2.1 and Figure 2.2 illustrate representative types of evidence processed and their success rates.

DNA is located in cells which are the building blocks of the human body. The two types are sex cells (sperm and oocytes) and somatic cells (all other types), which share common characteristics as described below.

### 2.1.1 Cell Membrane

All cells have membranes, also known as plasma membranes, that constitute their outer boundaries (Figure 2.3). The functions of cell membranes include exchanges with the environment, signal transduction, and structural support. The cell membrane is a phospholipid bilayer containing lipids, proteins, and carbohydrates. Membrane proteins can act as enzymes, receptors, or channels. Many cells also have carbohydrate-rich molecules including proteoglycans, glycoproteins, and glycolipids on their membrane surfaces. Many of these molecules act as cell surface antigens.

### 2.1.2 Cytoplasm

The cytoplasm contains the cytosol fluid in which organelles are suspended. The cytosol contains the cytoskeleton, microvilli, centrioles, cilia, ribosomes, and proteasomes. Some organelles such as the ***endoplasmic reticulum,*** the ***Golgi apparatus, lysosomes, peroxisomes***, and ***mitochondria*** are surrounded by phospholipid membranes that isolate them from the cytosol (Figure 2.4). The mitochondria are responsible for energy production through aerobic metabolism by producing molecules containing high energy bonds such as adenosine triphosphate (ATP).

Two types of endoplasmic reticulum (ER) can exist within a cell: the smooth endoplasmic reticulum (SER) is involved in lipid synthesis; the rough

**Table 2.1 Common Items of Evidence**

| Evidence | Possible DNA Location | Source of DNA |
|---|---|---|
| Baseball bat | Handle | Skin cells, sweat, blood, tissue |
| Hat, bandana, mask | Inside surfaces | Sweat, hair, skin cells, dandruff, saliva |
| Eyeglasses | Nose, ear piece, lens | Sweat, skin cells |
| Facial tissue, cotton swab | Surface | Mucus, blood, sweat, semen, ear wax |
| Dirty laundry | Surface | Blood, sweat, semen, saliva |
| Toothpick | Surface | Saliva |
| Used cigarette | Butt (filter area) | Saliva |
| Used stamp, envelope seal | Moistened area | Saliva |
| Tape or ligature | Inside or outside surface | Skin cells, sweat, saliva |
| Bottle, can, glass | Mouthpiece, rim, outer surface | Saliva, sweat, skin cells |
| Used condom | Inside surface, outside surface | Semen, vaginal cells, rectal cells |
| Bed linen | Surface | Sweat, hair, semen, saliva, blood |
| Through-and-through bullet | Outside surface | Blood, tissue |
| Bite mark | Skin surface | Saliva |
| Fingernail, partial fingernail | Scrapings | Blood, sweat, tissue, skin cells |

*Source:* Using DNA to Solve Cold Cases (July, 2002), U.S. Department of Justice, Office of Justice Programs.

endoplasmic reticulum (RER) contains ribosomes on its outer surface and forms transport vesicles. The Golgi apparatus is responsible for the production of secretory vesicles and new membrane components, and also for the packaging of lysosomes (vesicles containing digestive enzymes for the degradation of injured cells). Lastly, peroxisomes carry enzymes that neutralize potentially harmful free radicals.

### 2.1.3 Nucleus

The ***nucleus*** (Figure 2.4) is surrounded by a nuclear envelope and contains chromosomes and a nucleolus which is a dense structure due to its high RNA content. The chromosomal DNA contains genes that encode for specific proteins. The genetic code is read as an array of triplet codes, a sequence of three bases that specifies the identity of a single amino acid. As gene expression is activated, transcription occurs in which messenger RNA (mRNA) is produced from a DNA template. After transcription, the mRNA is transported

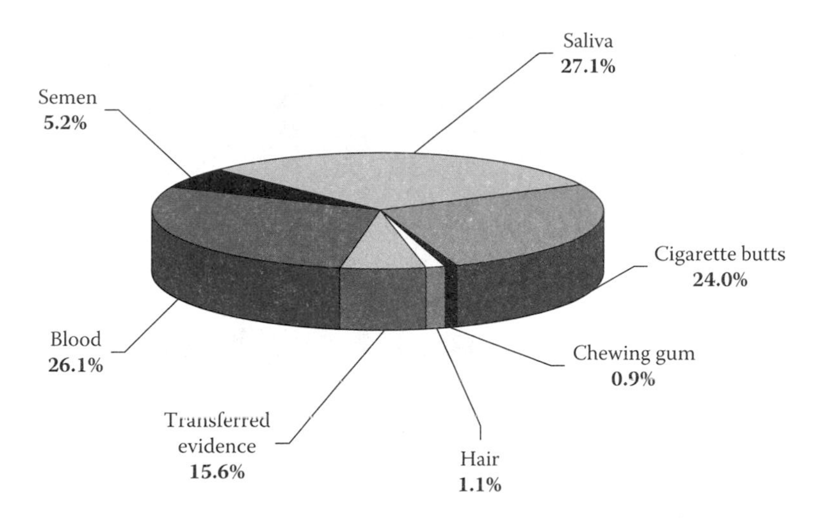


**Figure 2.1** Representative types of evidence samples processed. Data compiled from third quarter 2005 (July–September) results for all police forces in England and Wales. (*Source*: Adapted from Bond, J. W. 2007. Value of DNA evidence in detecting crime. *J Forensic Sci* 52 (1):128.)

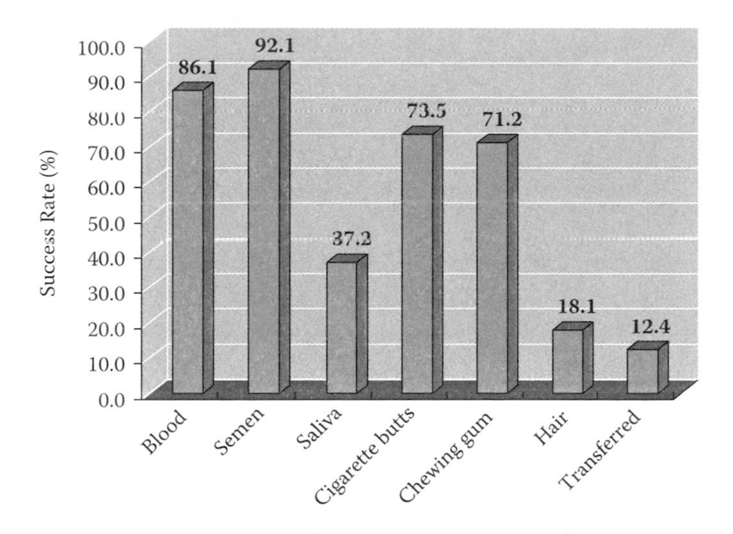


**Figure 2.2** Success rate of obtaining profiles suitable for submitting on DNA database from samples processed. Data is from the third quarter 2005 (July–September) results for all police forces in England and Wales. (*Source*: Adapted from Bond, J. W. 2007. Value of DNA evidence in detecting crime. *J Forensic Sci* 52 (1):128.)

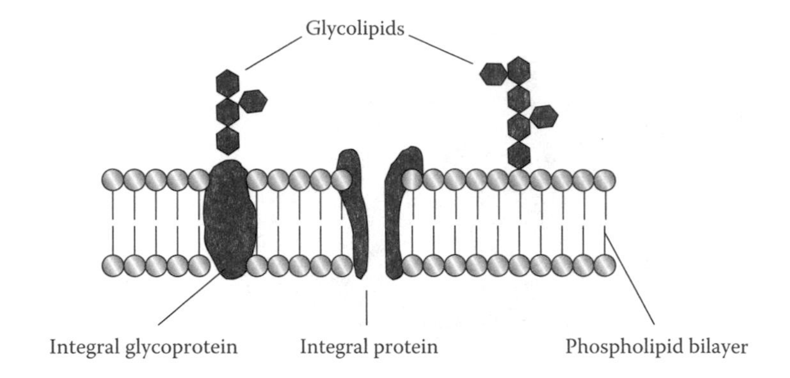


**Figure 2.3** Cell membrane.

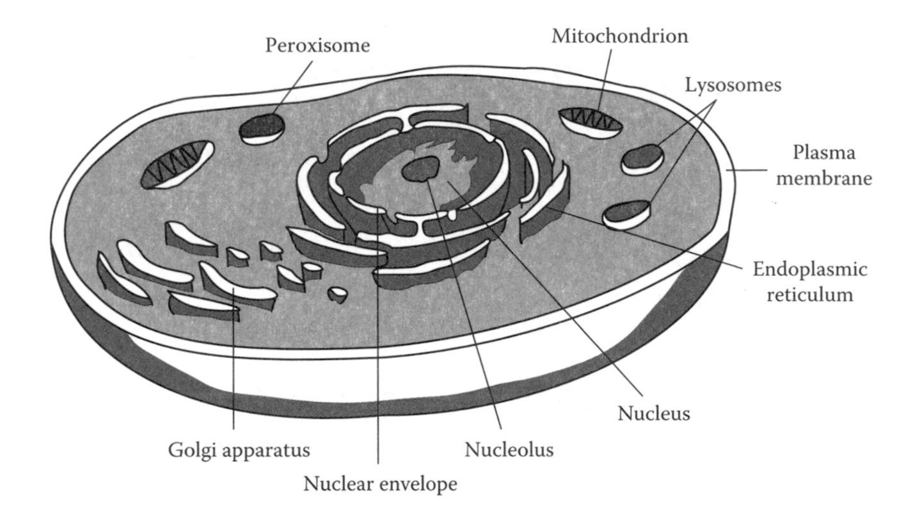


**Figure 2.4** Cross-section view of cell.

from the nucleus to the cytoplasm. The proteins are synthesized in a process known as translation, in which amino acids are assembled based on the codons derived from the triplet code of the DNA contained in the sequence of the mRNA strand. Various components including the ribosomal complex are involved in translation.

## 2.2 Examples of Biological Evidence

### 2.2.1 Blood

Blood evidence is one of the most common types of biological evidence found at crime scenes. Normal blood volume for human beings is about 8% of total body weight. The cellular portion of the blood consists of ***erythrocytes*** (also known as red blood cells), ***leucocytes*** (also known as white blood cells), and ***platelets***. Because mature human erythrocytes and platelets do not have nuclei, they are not useful sources for nuclear DNA. From blood samples (Figure 2.5, Figure 2.6, and Figure 2.7), the nuclear DNA for forensic DNA profiling is primarily isolated from the leucocytes which have nuclei.

Interestingly, the first attempt to isolate DNA from human blood involved the use of human leucocytes by Swiss physician Friedrich Miescher in 1869. This study led to the discovery of the DNA molecule that he referred to as "nuclein." Over the years, various methods for isolating DNA from blood samples were developed. However, many protocols still use the basic principles Miescher developed more than 130 years ago.

**Figure 2.5** Blood samples to be processed for submission to DNA database.

**Figure 2.6** Blood samples are air dried on blood cards for storage.

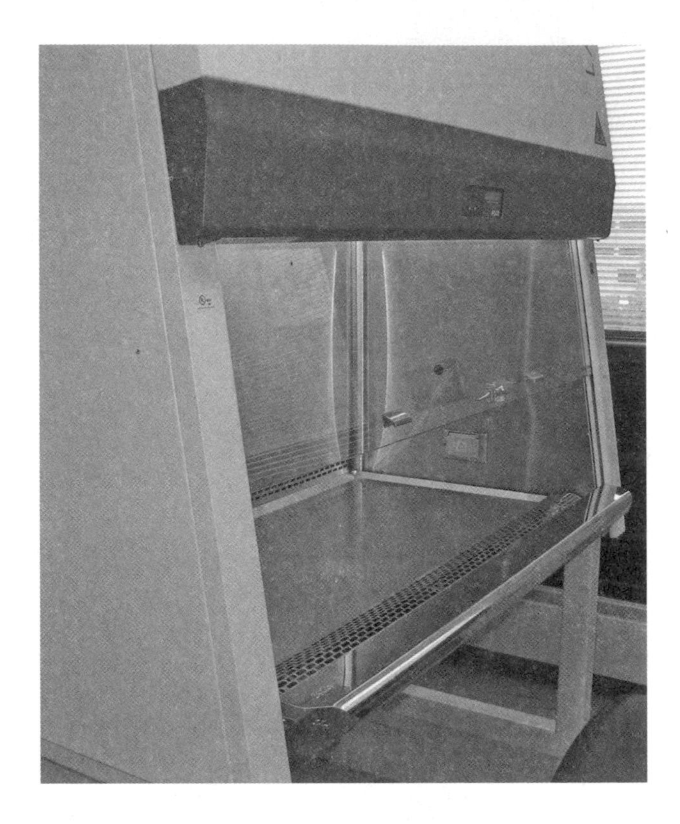

**Figure 2.7** Biosafety hood for extraction of DNA from biological evidence.

## 2.2.2 Hair

### *2.2.2.1 Biology of Hair*

Human hair consists of a ***root*** and a ***shaft*** (Figure 2.8). The root is the portion that anchors the hair to the skin and the shaft is the part that can be seen above the skin surface. The center or core of the hair is called the ***medulla***, which is surrounded by a ***cortex***. The ***cuticle*** consists of overlapping layers of dead, flattened, and keratinized cells that protect the hair.

Hairs are produced in ***follicles***. Each hair follicle is located deep in the dermis and opens onto the surface of the epidermis. A ***papilla*** is located at the base of the hair root and contains nerves and capillaries that supply nutrients. The papilla is surrounded by a ***hair bulb*** that produces the hair.

Hairs grow for a few years and are shed according to a growth cycle. A human head hair grows at a rate of approximately 0.3 mm per day. The growing phase of hair is called the ***anagen*** phase. As a hair grows, it is pushed toward the surface of the skin and becomes longer. By the time a hair approaches the skin surface (which corresponds to the start of the hair shaft),

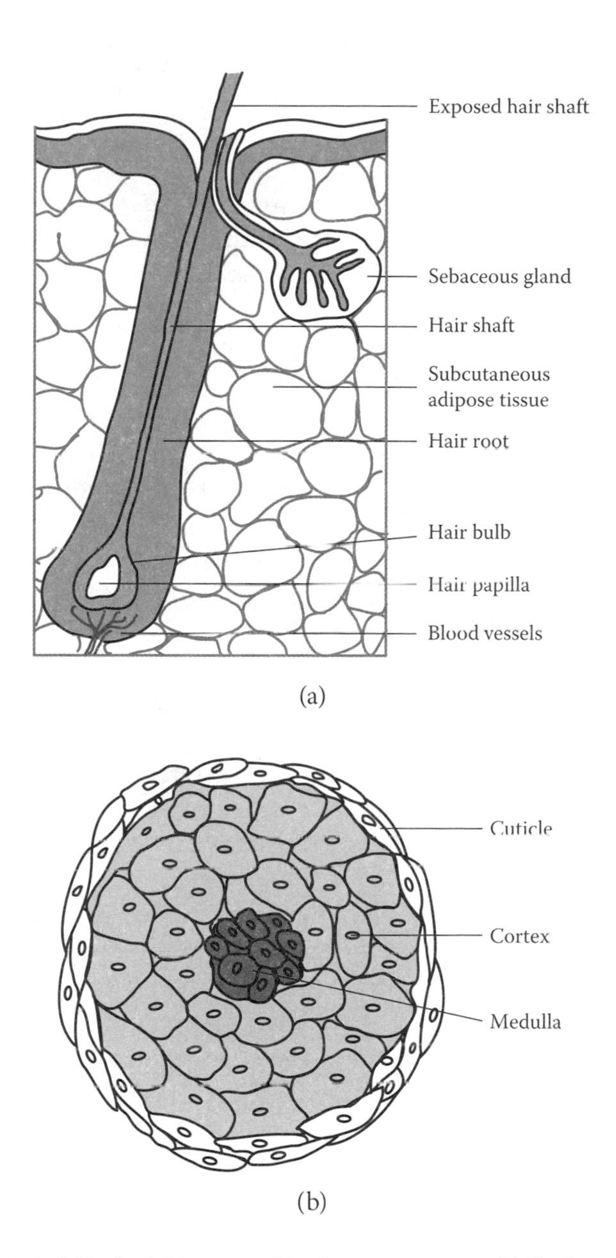


**Figure 2.8** Hair follicle. (a) Longitudinal section view of follicle with accessory structures. (b) Cross-section view of hair shaft.

cell death occurs at the medulla, cortex, and cuticle. At the end of the anagen phase, cells enter the ***catagen*** phase and the follicle, also known as a club hair, becomes inactive. The follicle becomes smaller and becomes detached from the hair root. Eventually, the cells enter the ***telogen*** phase and growth cycle ends. When another cycle begins, the follicle produces a new hair and the telogenic hair is pushed to the surface and shed. An average adult loses approximately 50 head hairs daily.

#### 2.2.2.2 *Hair as Source of DNA Evidence*

Hairs, including head and pubic hairs, frequently constitute biological evidence found at crime scenes and their identification can be of great forensic importance. Formerly, the principal methods employed in forensic hair analysis were limited to morphological analysis and comparisons. Since then, protein polymorphisms provide some potential for identifying individuals from single hairs. However, human hairs contain DNA as a DNA source and may be used for forensic analysis. The development of the polymerase chain reaction (PCR) amplification technique made it possible to analyze very small quantities of DNA in hair and the use of hair as evidence of identification has become more significant.

A cell contains up to several thousand copies of mtDNA. As a result, mtDNA can successfully be isolated from both hair roots and hair shafts. The sequence polymorphism analysis of mtDNA from hair has been carried out by many laboratories. However, mtDNA is maternally inherited and thus cannot be used to perform paternity testing. Additionally, the mtDNA profiling results are not as discriminating as nuclear DNA profiling. Furthermore, mtDNA analysis is labor intensive. Therefore, the typing of nuclear DNA from hair would be preferable for forensic DNA analysis. Additionally, nuclear DNA profiles can be searched against existing DNA profiles in the DNA database.

Nuclear DNA analysis is usually accomplished by using freshly plucked hair roots (Figure 2.8) because cells at the root region contain more abundant nuclear DNA. In comparison, naturally shed hairs or hair shafts contain little nuclear DNA. Thus, current methods use multiple hairs with roots to increase the possibility that some of the hairs will have cells that contain nuclear DNA. Currently, the isolation of DNA from intact roots of hair fibers is routinely used in nuclear DNA analysis. Unfortunately, most human hairs recovered from crime scenes were shed naturally (in the telogen phase) and contain little nuclear DNA. Additionally, shed hairs found at crime scenes may be derived from different individuals. Therefore, the ability to perform forensic DNA analysis of a single shed hair would be highly desirable.

Several research groups have carried out detection of nuclear DNA from a hair shaft. However, nuclear DNA isolation from hair shafts is still far less reliable because hair shafts contain low levels of nuclear DNA. Most DNA in hair is located in the roots and surrounding cells, which may contain as much as 0.5 µg of DNA. The level of DNA in a hair shaft is considerably less when compared to the DNA found in the root ends of freshly plucked hairs. In addition, variations in levels of DNA isolated from hair shafts were

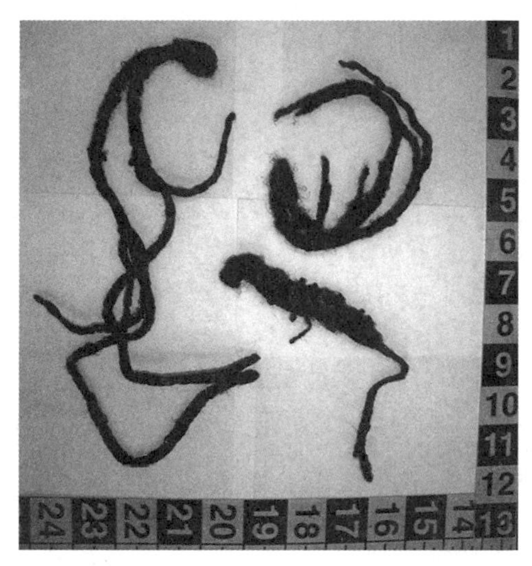

**Figure 2.9** Pulled dreadlocks recovered from a crime scene.

observed in comparisons between different hairs from the same head and hairs from different individuals.

## 2.2.3 Bone

### *2.2.3.1 Biology of Bone*

Bones may be categorized as ***compact*** or ***spongy*** bone (Figure 2.10). Bone tissue, like other connective tissues, contains specialized cells and a matrix. The composition of the matrix is similar in both types of bones. The matrix of bone tissue consists of extracellular protein fibers with the deposition of calcium salts around the protein fibers. The matrix of bone is very dense and strong.

Although ***osteocytes*** are the most abundant cells in bone, they cannot divide once they mature. They have two major functions (1) maintenance of the protein and mineral contents of the surrounding matrix and (2) repair of damaged bone. The repairing of bones is regulated by several types of cells found in bones. A bone contains small numbers of osteoprogenitor cells. These cells divide to produce daughter cells which differentiate into ***osteoblasts*** that produce new bone matrix. ***Osteoclasts*** are responsible for dissolving and recycling the bone matrix. They are giant cells containing 50 or more nuclei.

Human remains undergo a series of changes during decomposition. The rate of degradation of human remains varies greatly with environmental conditions (climate, bacterial growth, and presence of insect and animal scavengers). After a time, soft tissues may be lost while more stable bone tissues remain. Forensic scientists are called upon to attempt to identify human skeletal remains in a variety of situations, including mass fatality incidents,

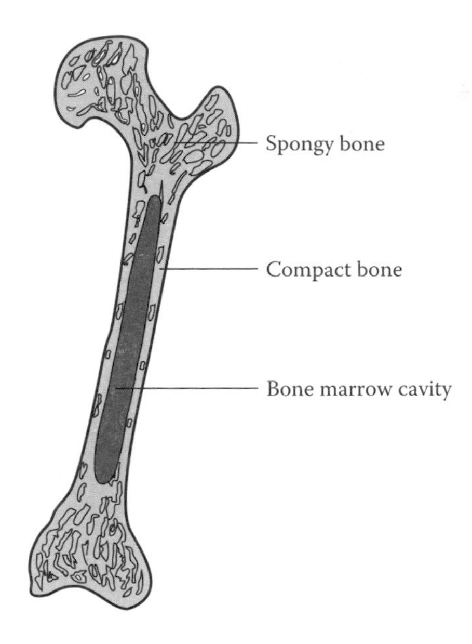


**Figure 2.10** Structure of long bone.

analysis of skeletal remains of military personnel recovered from wars, fires, and explosions, and criminal cases involving skeletal remains.

A number of methods are used to identify human remains. The most common involve the identification of facial characteristics, recognition of individualizing scars, marks, and other special body features, matching dentition with premortem dental x-rays, and fingerprint comparisons. In some situations, these methods cannot be used because of extensive putrefaction or destruction of the remains. The mass fatality terrorist attack on the World Trade Center in 2001 serves as a prime example of a situation where common identification techniques were not useful. Large quantities of compromised human skeletal fragments were recovered at the fatality site. In these cases, DNA typing is a powerful tool for identifying human remains.

### *2.2.3.2 Bone as Source of DNA Evidence*

Most DNA in compact bone is located in the osteocytes. It has been estimated that there are 20,000 to 26,000 osteocytes per $mm^3$ of calcified bone matrix. Therefore, microgram quantities of DNA could potentially be extracted from a gram of bone. Thus, compact bone tissue should contain sufficient DNA for analysis. However, the skeletal fragments recovered from mass fatality incidents are often subjected to a series of decomposition changes. High humidity and temperatures are factors that affect the degradation rate of DNA. Decomposition may be expected to significantly degrade both nuclear and mitochondrial DNA. The identification of partial DNA profiles and complete failures to obtain DNA profiles were reported after samples from the World Trade Center case were analyzed.

One of the greatest challenges in attempting to identify victims of mass fatality incidents is the analysis of DNA from bone. The bone is difficult to process for DNA extraction and quantities of samples recovered may be too small to properly isolate the DNA. To obtain an adequate quality and quantity of a DNA template, strategies to improve the yield of DNA isolation are needed.

A bone sample must be initially cleaned prior to isolating DNA from bone fragments. Due to the potential for commingled remains, adhering inhibitors, and bacterial contamination, the outer surfaces of bone fragments must be cleaned by using a method such as sanding. However, to avoid cross-contamination of samples, the bone dust generated by sanding must be cleaned and removed. Additionally, special safety protection equipment and procedures are necessary to protect analysts from exposure to blood-borne pathogens.

The bone samples can be ground to aid in DNA extraction (Figure 2.11). The osteocytes containing DNA are embedded in a calcified matrix which is a barrier for access to the DNA in the osteocytes during the extraction process. The matrix must be removed in order to improve the yield of DNA. A decalcification method was developed to dissolve calcium ions during the extraction procedure to remove the matrix. Alternatively, the application of

**Figure 2.11** Freezer/mill for grinding of bone samples prior to DNA extraction.

proteinase can be used to digest the matrix barrier. Proteinase digests the bone matrix and thus increases the yield of DNA harvested.

## 2.2.4 Teeth

### *2.2.4.1 Biology of Teeth*

During embryonic development, two sets of teeth begin to form. The first to appear are the ***deciduous teeth*** or ***primary teeth***. Most children have 20 deciduous teeth that are later replaced by 32 teeth known as the ***secondary dentition*** or ***permanent dentition***.

The bulk of each tooth consists of a mineralized matrix similar to the matrix of bone (Figure 2.12). This material, called ***dentin***, differs from bone in that it does not contain cells. Instead, cytoplasmic processes extend into the dentin from cells in the interior chamber known as the central ***pulp cavity***. The dentin of the crown is covered by a layer of enamel that contains calcium phosphate in a crystalline form. The pulp cavity receives blood vessels and nerves through the ***root canal***, a narrow tunnel located at the ***root*** of the tooth. Incisor and cuspid teeth have single roots. Bicuspids have one or two roots. Molars typically have three or more roots.

### *2.2.4.2 Teeth as Sources of DNA Evidence*

The characteristics of teeth, their alignment, and the overall structure of the mouth provide information for identifying a person. The use of dental

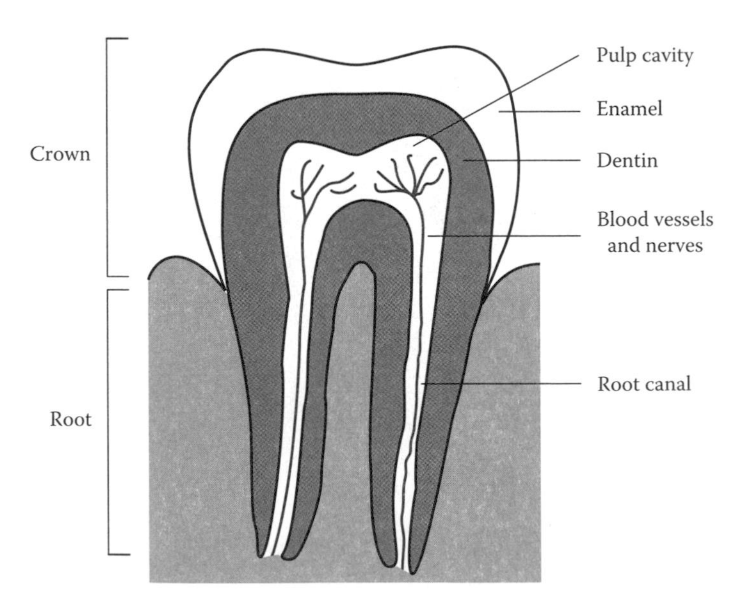


**Figure 2.12** Section of adult tooth.

records such as x-rays and dental casts and even a photograph of a smiling person can allow a set of dental remains to be connected to a victim. Another application of forensic odontology to criminal investigations is bite mark analysis. A forensic odontologist can analyze the marks left on a victim and the tooth structure of a suspect to make a comparison. When no dental comparison is possible due to unavailable dental records or intact teeth, DNA testing can be carried out. Dental pulp tissue is a useful source for DNA isolation.

## Bibliography


Abaz, J. et al. 2002. Comparison of the variables affecting the recovery of DNA from common drinking containers. *Forensic Sci Int* 126 (3):233.

Alvarez Garcia, A. et al. 1996. Effect of environmental factors on PCR DNA analysis from dental pulp. *Int J Legal Med* 109 (3):125.

Anderson, T. D. et al. 1999. A validation study for the extraction and analysis of DNA from human nail material and its application to forensic casework. *J Forensic Sci* 44 (5):1053.

Anzai-Kanto, E. et al. 2005. DNA extraction from human saliva deposited on skin and its use in forensic identification procedures. *Pesqui Odontol Bras* 19 (3):216.

Barbaro, A., P. Cormaci, and A. Barbaro. 2004. Detection of STRs from body fluid collected on IsoCode paper-based devices. *Forensic Sci Int* 146 Suppl:S127.

____. 2004. DNA analysis from mixed biological materials. *Forensic Sci Int* 146 Suppl:S123.

Barbaro, A. et al. 2004. Anonymous letters? DNA and fingerprints technologies combined to solve a case. *Forensic Sci Int* 146 Suppl:S133.

Berg, L., Martin, D., and Solomon, E. 2005. *Biology*, 7th ed. Belmont: Thomson Brooks/Cole.

Bond, J. W. 2007. Value of DNA evidence in detecting crime. *J Forensic Sci* 52 (1):128.

Burger, M. F., E. Y. Song, and J. W. Schumm. 2005. Buccal DNA samples for DNA typing: New collection and processing methods. *Biotechniques* 39 (2):257.

Chiou, F. S. et al. 2001. Extraction of human DNA for PCR from chewed residues of betel quid using a novel "PVP/CTAB" method. *J Forensic Sci* 46 (5):1174.

Cina, M. S. et al. 2000. Isolation and identification of male and female DNA on a postcoital condom. *Arch Pathol Lab Med* 124 (7):1083.

Dimo-Simonin, N., F. Grange, and C. Brandt-Casadevall. 1997. PCR-based forensic testing of DNA from stained cytological smears. *J Forensic Sci* 42 (3):506.

Fridez, F., and R. Coquoz. 1996. PCR DNA typing of stamps: Evaluation of DNA extraction. *Forensic Sci Int* 78 (2):103.

Fujita, Y., and S. Kubo. 2006. Application of FTA technology to extraction of sperm DNA from mixed body fluids containing semen. *Leg Med (Tokyo)* 8 (1):43.

Grubwieser, P. et al. 2003. Systematic study on STR profiling on blood and saliva traces after visualization of fingerprint marks. *J Forensic Sci* 48 (4):733.

Gurvitz, A., L. Y. Lai, and B. A. Neilan. 1994. Exploiting biological materials in forensic science. *Australas Biotechnol* 4 (2):88.

Hellmann, A. et al. 2001. STR typing of human telogen hairs: A new approach. *Int J Legal Med* 114 (4–5):269.
Herber, B., and K. Herold. 1998. DNA typing of human dandruff. *J Forensic Sci* 43 (3):648.
Higuchi, R. et al. 1988. DNA typing from single hairs. *Nature* 332 (6164):543.
Hochmeister, M. N. 1998. PCR analysis of DNA from fresh and decomposed bodies and skeletal remains in medicolegal death investigations. *Methods Mol Biol* 98:19.
Hochmeister, M. N. et al. 1991. PCR-based typing of DNA extracted from cigarette butts. *Int J Legal Med* 104 (4):229.
Hochmeister, M. N., O. Rudin, and E. Ambach. 1998. PCR analysis from cigarette butts, postage stamps, envelope sealing flaps, and other saliva-stained material. *Methods Mol Biol* 98:27.
Holland, M. M. et al. 2003. Development of a quality, high throughput DNA analysis procedure for skeletal samples to assist with the identification of victims from the World Trade Center attacks. *Croat Med J* 44 (3):264.
Hopkins, B. et al. 1994. The use of minisatellite variant repeat polymerase chain reaction (MVR-PCR) to determine the source of saliva on a used postage stamp. *J Forensic Sci* 39 (2):526.
Hopwood, A. et al. 1997. Rapid quantification of DNA samples extracted from buccal scrapes prior to DNA profiling. *Biotechniques* 23 (1):18.
Hopwood, A. J., A. Mannucci, and K. M. Sullivan. 1996. DNA typing from human faeces. *Int J Legal Med* 108 (5):237.
Kline, M. C. et al. 2002. Polymerase chain reaction amplification of DNA from aged blood stains: Quantitative evaluation of the "suitability for purpose" of four filter papers as archival media. *Anal Chem* 74 (8):1863.
Li, R. C., and H. A. Harris. 2003. Using hydrophilic adhesive tape for collection of evidence for forensic DNA analysis. *J Forensic Sci* 48 (6):1318.
Lijnen, I., and G. Willems. 2001. DNA research in forensic dentistry. *Methods Find Exp Clin Pharmacol* 23 (9):511.
Lorente, M. et al. 1998. Dandruff as a potential source of DNA in forensic casework. *J Forensic Sci* 43 (4):901.
Martini, F. H., M. J. Timmons, and R. B. Tallitsch. 2006. *Human Anatomy*, 5th ed. San Francisco: Pearson Education.
Medintz, I. et al. 1994. Restriction fragment length polymorphism and polymerase chain reaction–HLA DQ alpha analysis of casework urine specimens. *J Forensic Sci* 39 (6):1372.
Office of Justice Programs, National Institute of Justice, U.S. Department of Justice. 2002. *Using DNA to Solve Cold Cases.*
Pfeiffer, H. et al. 1999. Influence of soil storage and exposure period on DNA recovery from teeth. *Int J Legal Med* 112 (2):142.
Prado, V. F. et al. 1997. Extraction of DNA from human skeletal remains: Practical applications in forensic sciences. *Genet Anal* 14 (2):41.
Roberts, K. A., and C. Calloway. 2007. Mitochondrial DNA amplification success rate as a function of hair morphology. *J Forensic Sci* 52 (1):40.
Salvador, J. M., and M. C. De Ungria. 2003. Isolation of DNA from saliva of betel quid chewers using treated cards. *J Forensic Sci* 48 (4):794.

Shaw, K. et al. 2007. Comparison of the effects of sterilisation techniques on subsequent DNA profiling. *Int J Legal Med.*

Springer, E. E., T. L. Laber, and B. B. Randall. 1988. Examination of vaginally inserted plastic tampon applicators for genetic markers and evidence of prior sexual intercourse. *J Forensic Sci* 33 (5):1139.

Tanaka, M. et al. 2000. Usefulness of a toothbrush as a source of evidential DNA for typing. *J Forensic Sci* 45 (3):674.

Tsuchimochi, T. et al. 2002. Chelating resin-based extraction of DNA from dental pulp and sex determination from incinerated teeth with Y-chromosomal alphoid repeat and short tandem repeats. *Am J Forensic Med Pathol* 23 (3):268.

van Oorschot, R. A. et al. 2005. Beware of the possibility of fingerprinting techniques transferring DNA. *J Forensic Sci* 50 (6):1417.

von Wurmb-Schwark, N. et al. 2006. Fast and simple DNA extraction from saliva and sperm cells obtained from the skin or isolated from swabs. *Leg Med (Tokyo)* 8 (3):177.

Webb, L. G., S. E. Egan, and G. R. Turbett. 2001. Recovery of DNA for forensic analysis from lip cosmetics. *J Forensic Sci* 46 (6):1474.

Wiegand, P., T. Bajanowski, and B. Brinkmann. 1993. DNA typing of debris from fingernails. *Int J Legal Med* 106 (2):81.

Yasuda, T. et al. 2003. A simple method of DNA extraction and STR typing from urine samples using a commercially available DNA/RNA extraction kit. *J Forensic Sci* 48 (1):108.

## Study Questions

1. A decomposed body remains unidentified. What types of samples should be collected for possible DNA testing?
   (a) Solid tissue samples such as tooth, bone, hair, and deep muscle
   (b) Biological fluids such as blood, urine, stomach contents, and vitreous humor

2. Which of the following can be isolated from naturally shed hair?
   (a) Nuclear DNA
   (b) Mitochondrial DNA
   (c) Both of the above

3. Mitochondrial DNA can usually be found in:
   (a) Shed hair
   (b) Pulled hair
   (c) Animal hair
   (d) All of the above

4. Which of the following can be isolated from hair roots or follicles?
   (a) Nuclear DNA
   (b) Mitochondrial DNA
   (c) Both of the above

5. Which of the following is true?
   (a) Nuclear DNA can be isolated from white blood cells
   (b) Nuclear DNA can be isolated from red blood cells

6. Bite mark evidence discovered at a crime scene:
   (a) Contains biological materials for forensic DNA analysis
   (b) Should not be collected for forensic DNA analysis since it is usually contaminated
   (c) Should not be collected for forensic DNA analysis since it is usually degraded

# 3 Crime Scene Investigation and Laboratory Analysis of Biological Evidence

A forensic investigation involving biological evidence requires crime scene investigation and laboratory analysis. The crime scene investigation process includes preservation of scene security, preparation of documentation, and collection and preservation of physical evidence. Laboratory analysis utilizes scientific techniques for evidence examination, identification of biological fluids, and comparisons of individual characteristics of biological evidence.

## 3.1 Crime Scene Investigation

### 3.1.1 Protection of Crime Scene

A forensic investigation begins with the initial response to a crime scene. Securing and protecting the scene are critical steps of an investigation, and the first responding officer is responsible for protecting and securing the scene. Suspects, witnesses, and living victims should be evacuated from the scene. Only authorized personnel should be admitted to the scene and their entries should be documented on a log sheet. If a victim is wounded, medical attention should be sought. All these activities should be carefully carried out with the intent to prevent contamination of evidence. Any disturbance of evidence should be noted.

Proper supplies and devices should be used both to prevent the contamination of evidence by investigators and also to ensure that biosafety procedures are followed for the protection of personnel from infectious blood-borne pathogens and other biohazardous materials (Figure 3.1). The following devices should be used:

- Disposable coverall body suit
- Face mask or shield
- Safety eye glasses
- Disposable gloves
- Disposable shoe covers
- Disposable hair net

**Figure 3.1** Crime scene reconstruction; note disposable coverall body suit.

### 3.1.2 Documentation

Both the individual items of evidence and the overall scene must be documented in order to provide vital information for investigators and for the courts. The most common documentation methods are drawing sketches and taking photographs and videographs. An investigator first makes a rough sketch that is later turned into a finished sketch showing positions of persons and objects with a scale (Figure 3.2). Photographs should also include scales to accurately depict the sizes of items such as bloodstains or bite marks. This can be achieved simply by placing a ruler adjacent to the evidence when it is photographed. See Figure 3.3, Figure 3.4, and Figure 3.5. A log sheet can be used to record the chronological order of photographs taken at a crime scene and note additional relevant information such as filming conditions (Figure 3.6). Similar documentation should be prepared for videographs when appropriate.

### 3.1.3 Chain of Custody

Custody information should be recorded at every point when evidence is handled or transferred from one person to another. Usually, a custody form listing a specific evidence item is used to document the chain. Each individual who acquires custody of the evidence must sign a chain of custody document. An incomplete chain of custody may make the evidence inadmissible in court because possible tampering or contamination could be inferred.

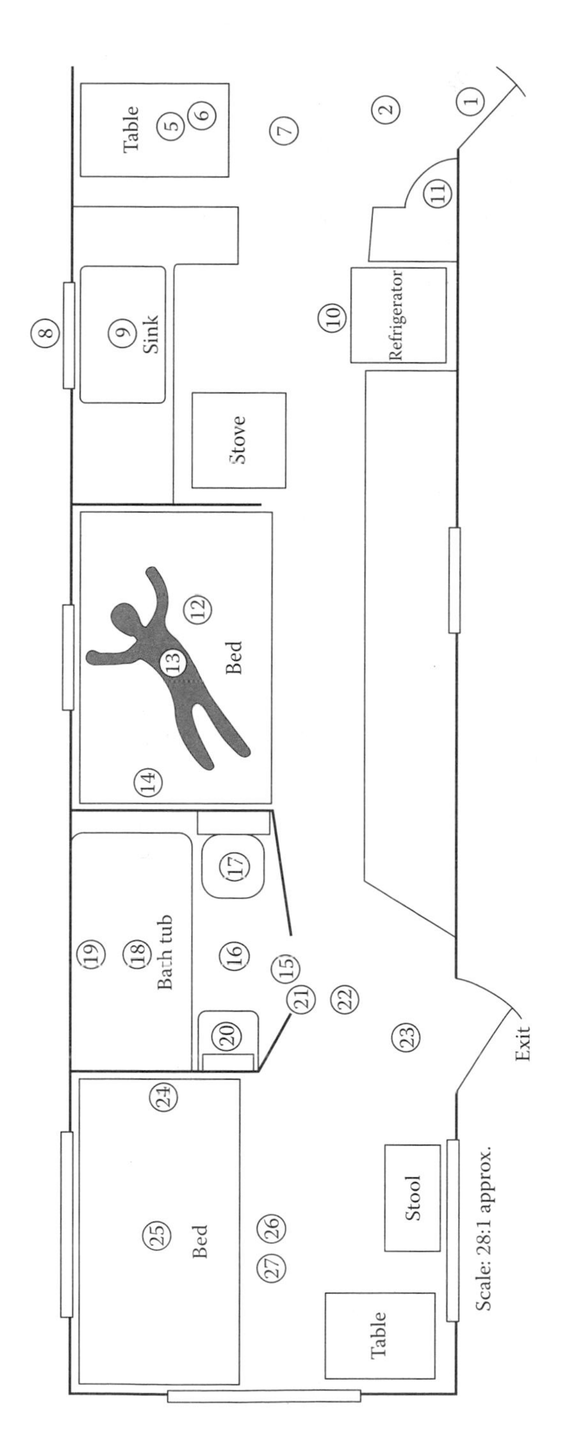


**Figure 3.2** Sketch documentation.

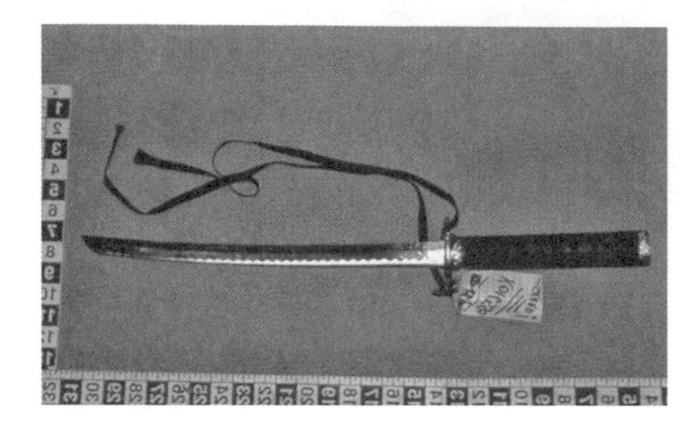

**Figure 3.3** Photographic documentation of a knife.

**Figure 3.4** Photographic documentation of blood stain patterns.

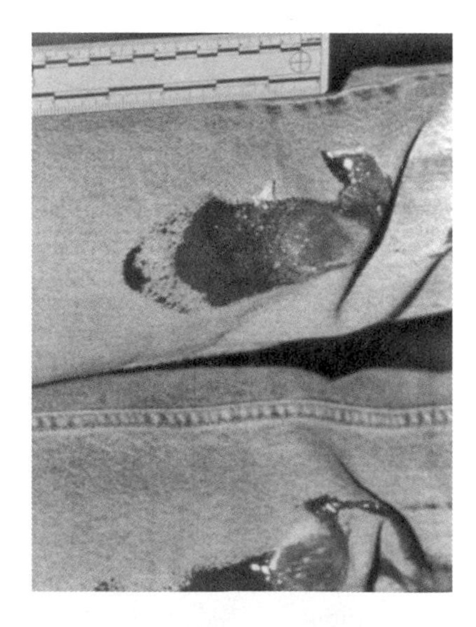

**Figure 3.5** Photographic documentation of contact blood stain patterns.

Crime Scene Photography Log

Date: 09 / 18 / 97 Case Number: FB Competency Analyst
Roll # 1 Film Type & ASA: 36 E100S Ektachrome DX

| EXP | Lens | F-stop | Category | Description |
|---|---|---|---|---|
| 1 | | | | |
| 2 | | | | |
| 3 | | | | |
| 4 | | | | |
| 5 | | | | |
| 6 | | | | |
| 7 | | | | |

**Figure 3.6** Photographic log.

### 3.1.4 Recognition of Biological Evidence

Priority of the potential evidence at crime scenes should be assessed based on each item's relevance and significance. Higher priority should be assigned to evidence with more probative value to the case. For example, the evidence related to a ***corpus delicti***, which demonstrates that a crime has occurred, is considered to be of the highest priority. A victim's blood found at a crime scene is obviously critical evidence in a criminal case.

Higher priority should also be attached to evidence that can establish connections such as ***victim-to-perpetrator linkage***. For example, items found in a suspect's possession may be linked to a victim. This also applies to transfer evidence based on the principles of transfer theory, also known as the ***Locard exchange principle***, which theorizes that cross-transfer of evidence occurs when a perpetrator has any physical contact with an object or another person. Thus, trace evidence may be transferred from a perpetrator to a victim or vice versa. This explains why it is important to make sure that suspects and their belongings are thoroughly searched for trace evidence. Likewise, victims and their belongings should be examined for the same reason.

***Victim-to-scene*** and ***perpetrator-to-scene linkages*** can also be established. Blood belonging to a perpetrator or victim found at a crime scene can establish such a linkage. Additionally, reciprocal transfers of trace evidence from crime scenes can be used to link a suspect or victim to a crime scene. A perpetrator may present a unique ***modus operandi*** **(MO)** — a pattern of characteristics that may be present as a result of committing a crime. Evidence may reveal an MO that establishes a ***case-to-case linkage*** for serial offender cases. Evidence that provides information on the MO is also vital to an investigation.

### 3.1.5 Searches

Some investigations require a search for specific items of evidence such as biological stains and perpetrators' weapons. A search usually has a specific purpose. Searching for biological stains usually requires devices such as alternate light sources (ALSs); see Figure 3.7. The use of search patterns can also be helpful, especially in cases involving large outdoor crime scenes (Figure 3.8). Search patterns may include the spiral, grid, strip, quadrant or zone (Figure 3.9). The method that is ultimately used depends on the type and size of the scene. Additionally, the points of entry and exit and the paths followed by a perpetrator should also be searched.

### 3.1.6 Collection of Biological Evidence

The collection of evidence can be initiated only after the crime scene documentation (photographs, sketches, and notes) is completed. Small or portable

**Figure 3.7** Portable alternate light source device.

**Figure 3.8** Search for outdoor scene.

**Figure 3.9** Grid search pattern for outdoor scene.

items, such as bloodstained knives can be collected and submitted to a crime laboratory. Large or unmovable items of evidence must be properly collected and may be submitted in sections. Table 3.1 and Figure 3.10 through Figure 3.12 summarize and illustrate representative collection techniques. Specific care is required for the collection of biological evidence in the following situations:

- **Bloodstain pattern evidence** — It is especially important to thoroughly document the bloodstain pattern evidence at a crime scene prior to collection. Bloodstain patterns can be especially useful in crime scene reconstruction.
- **Multiple analysis of evidence** — If multiple analyses are needed for a single item of evidence, non-destructive analyses should be carried out first. For example, a bloody fingerprint or shoe print should be collected for ridge detail analysis prior to collecting blood for DNA analysis.
- **Trace evidence** — Trace evidence can be present in bloodstained evidence and should be identified and properly collected.
- **Control samples** — Control (known or blank) samples should be collected from a control area (e.g., unstained area near a collected stain).
- **Size of stain** — Polymerase chain reaction (PCR)-based forensic DNA techniques are highly sensitive and allow for successful analysis of very small bloodstains. All bloodstains, even if barely visible, should be collected at a crime scene.
- **Wet evidence** — Wet evidence should be air dried (without heat) prior to packaging for submission to prevent degradation.

**Table 3.1 Methods of Collection of Biological Evidence**

| Type of Evidence | Condition | Method of Collection | Procedure |
|---|---|---|---|
| Blood | Dry | Swab | Best on non-absorbent surfaces; lightly moisten sterile swab with distilled or sterile water, rub over stain while rotating; allow to air dry; combination of first moistened swabbing followed by second dry swabbing (both swabs submitted) recommended |
| | | Cutting | Cut stain from item |
| | | Scraping | Scrape bloodstain into clean piece of paper using clean blade; wrap sample using druggist's fold |
| | | Lifting | Works for non-absorbent surface; use fingerprint lifting tape that does not interfere with DNA testing to lift stain; lifted stain should be covered with a piece of lifter's cover |
| | | Collect entire item | Collect if item contains blood stain pattern; difficult to swab; requires multiple exams |
| | Wet | Swab | Absorb blood sample onto sterile cotton swabs; stain should be concentrated on tip and allowed to air dry |
| | | FTA paper | Use sterile disposable pipet to collect liquid blood; spot on FTA paper; allow to air dry |
| | Reference liquid blood sample | Venous blood Collection | Collect blood in purple-topped Vacutainer tube containing EDTA anticoagulant; refrigerate but never freeze |
| Semen | Dry or wet | See "Blood" above | See "Blood" above |
| | Condom | Collect entire item | Secure condom with tie and place in container in refrigerator; submit to laboratory as soon as possible |
| | Various conditions | Victim sexual assault kit | Standardized kit to collect biological evidence from body of victim includes swabs, microscope slides, and envelopes |
| | Various conditions | Suspect standard kit | Standardized kit to collect biological evidence from body of suspect includes swabs, microscope slides, and envelopes |
| Victim Vaginal Fluid | Dry or wet | See "Blood" above | Often collected from suspect's pubic area or fingers |

**Table 3.1 Methods of Collection of Biological Evidence** (continued)

| Type of Evidence | Condition | Method of Collection | Procedure |
|---|---|---|---|
| Saliva | Dry or wet | See "Blood" above | Often collected from bite mark or victim's pubic area |
| | Reference saliva (buccal) samples | Swab | Swab inside of cheek using two swabs, rotating them during collection; allow swabs to air dry |
| | | Filter paper | Place donor saliva sample on marked area of filter paper; allow to air dry |
| Hairs | Head and pubic hairs | Lifting | Refer to dry bloodstain lifting method |
| | | Transfer | Use forceps to transfer hair into piece of paper that can be folded |
| | | Vacuum | Vacuum can be used if necessary; generally not recommended |
| Fingernails and scrapings | Various conditions | Clipping | Use clean clipper to clip nails into clean paper; wrap samples using druggist's fold |
| | | Scraping | Scrape undersides of nails onto clean paper; wrap samples using druggist's fold |
| Bones | Various conditions | Freeze in container | Collect if blood not available; collect bone with marrow if available; rib bone and vertebrae are preferred; place specimen in container and freeze if wet |
| Teeth | Various conditions | Container | Collect teeth with dental pulp if possible into container |
| Tissues and organs | Wet | Freeze in container | Collect if blood not available; place specimen in container and freeze |
| DNA database samples | Various conditions | Various | Follow jurisdiction protocol for collecting samples from arrested individuals and/or convicted offenders |

Samples requiring refrigeration or freezing are noted; other samples may be stored at room temperature. Dry evidence should be packed in porous material such as paper (envelope, bag, box) as described in text.

*Source:* Adapted from Fisher, B. 2004. *Techniques of Crime Scene Investigation*, 7th ed. Boca Raton: CRC Press; Lee, H. C. et al. 2001. *Henry Lee's Crime Scene Handbook*. San Diego: Academic Press; Office of Justice Programs, National Institute of Justice. U.S. Department of Justice. 2002. *Using DNA to Solve Cold Cases*.

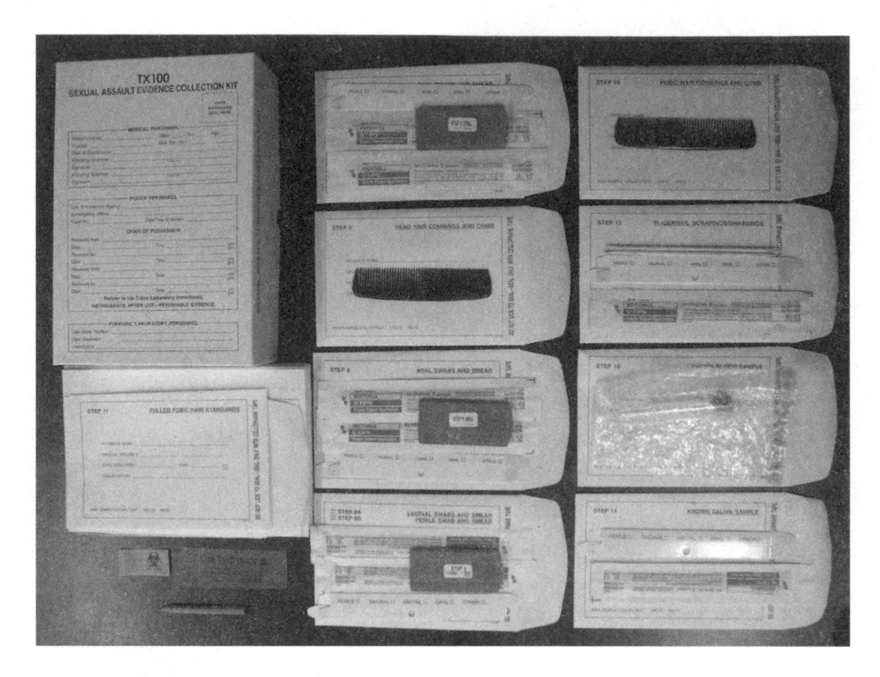


**Figure 3.10** Sexual assault evidence collection kit.

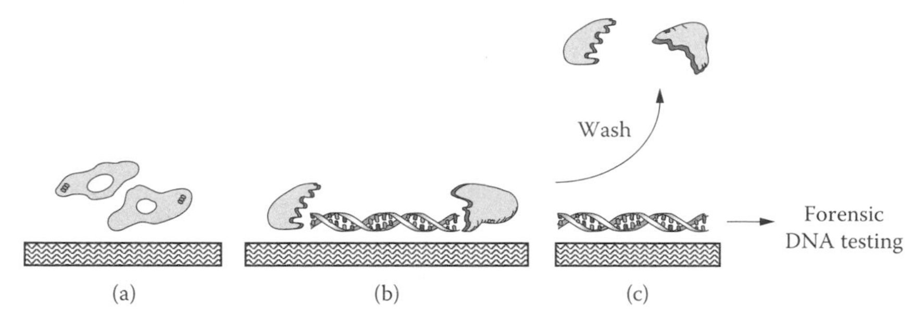


**Figure 3.11** Application of FTA paper for collection of biological evidence. (a) Biological fluid with cells is applied to FTA paper. (b) Cells are lysed and DNA is immobilized on FTA paper. (c) Cellular materials are washed away and DNA remaining on FTA paper may be used for forensic testing.

### 3.1.7 Marking Evidence

The marking of evidence is necessary for identification purposes so that it can be quickly recognized even years later. An investigator's initials, the item number, and case number are usually included in marking. Information can be marked on a tag, a label attached to the item, or directly on garment evidence. The marking of evidence should not be proximal to bullet holes or biological stains to prevent the mark from interfering with analyses.

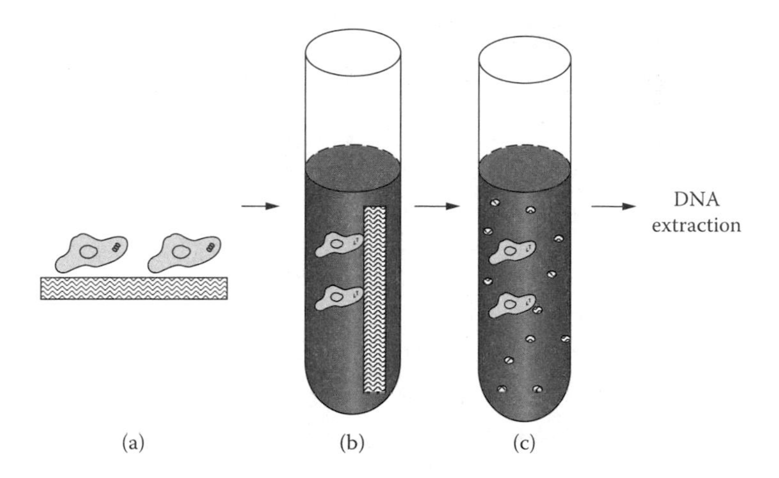


**Figure 3.12** Application of hydrophilic adhesive tape (HAT) for collection of biological evidence. (a) Cells are collected and attached on HAT. (b) Cell-containing HAT is applied to aqueous solution. (c) HAT is dissolved and DNA can be extracted from cells.

### 3.1.8 Packaging and Transportation

Packaging is intended to protect and preserve evidence. All evidence should be secured and protected from possible contamination. Fragile items should not be damaged during transportation. Exposure to heat and humidity should be avoided to protect biological evidence during transport. Packaging methods vary, depending on the type of evidence handled. The following are general considerations related to packaging of evidence.

- ***Evidence from different sources*** — Items of evidence should not be grouped in a single package because this could result in accidental transfer of evidence from different sources. However, evidence may be packed in a single container if the items were found together.
- ***Folding of evidence*** — Folding of clothing, especially items with wet bloodstains, can transfer evidence from one part of a garment to another. If a large dry garment must be folded, a piece of clean paper should be placed between different parts or layers of the garment to prevent direct contact of the parts or layers.
- ***Packing materials*** — Paper items such as envelopes, bags, and boxes are materials of choice for packaging. Dry bloodstained evidence should not be sealed in plastic bags or containers.
- ***Liquid evidence*** — Tubes containing liquid such as blood should not be frozen because the volume expands in freezing temperatures and the

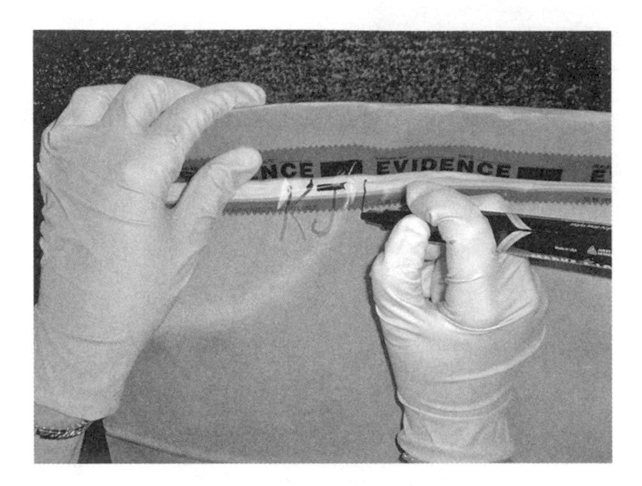


**Figure 3.13** Proper marking of sealed evidence. (*Source*: James, S. H., and J. J. Nordby. 2005. *Forensic Science: An Introduction to Scientific and Investigative Techniques,* 2nd ed. Boca Raton: CRC Press. With permission.)

expansion may lead to cracking. Tubes should be placed in plastic bags to prevent leaks in case of accidental breakage. Liquid evidence should be submitted to a laboratory as soon as possible.

- ***Trace evidence*** — All such evidence should be wrapped in paper secured with a druggist's fold and packed in paper envelope or bag.

Packaged evidence should be properly labeled and sealed with evidence tape. It is important for the person packaging the evidence to initial and date across the seal to show authenticity (Figure 3.13). A seal should not be cut when a sealed evidence bag is opened. Instead, an opening can be created by cutting at an area distal from the existing seal. After analysis is complete, the evidence packaging should be resealed. Table 3.1 summarizes additional steps for packaging evidence.

## 3.2 Laboratory Analysis of Biological Evidence

The analytical activities (Figure 3.14) of forensic scientists can be categorized as (1) the identification of biological evidence and (2) the comparison of individual characteristics of biological evidence. They may also perform crime scene reconstructions.

**Figure 3.14** Analyst working on processing biological evidence.

### 3.2.1 Identification of Biological Evidence

Identification of biological evidence is the first step before further analyses are carried out. This includes the identification of biological fluids such as blood, saliva, and semen; the process is discussed in more detail in subsequent chapters. The identification is based on the comparison of ***class characteristics*** — a set of characteristics that allows a sample to be placed in a category with similar materials. By comparing the class characteristics of a sample with known standards of its class, biological samples can be identified.

### 3.2.2 Comparison of Individual Characteristics of Biological Evidence

***Individual characteristics*** refer to the unique characteristics of both evidence and a reference sample that share a common origin to a high degree of certainty. Examples of evidence possessing individual characteristics are DNA polymorphisms, fingerprints, toolmarks, and markings on bullets.

In the case of biological evidence, current forensic DNA profiling can compare individual characteristics of DNA evidence to a known reference sample. It is possible to determine that a biological stain came from a particular individual, which is useful for human identification. Examination of individual characteristics of evidence can also exclude the possibility of a common origin. The specific methods employed for the individualization of evidence are also discussed in subsequent chapters.

### 3.2.3 Crime Scene Reconstruction

Reconstruction is usually performed in cases involving blood spatter patterns (Figure 3.15) and is intended to reveal events prior to, during, and after a crime, and thus provide important information.

### 3.2.4 Reporting Results and Expert Testimony

After analysis of evidence is completed, a report is prepared based on the results of the analysis, which may include sections discussing specific evidence analyzed, the method of analysis used, the results obtained, and conclusions drawn. In the case of DNA evidence, the strength of the conclusion is usually evaluated via statistical computations. A forensic scientist often serves as an expert witness whose testimony provides professional opinions about the evidence analyzed. Based on the federal rules of evidence, an expert witness can be qualified based on knowledge, skill, experience, training, or education, and may give an opinion to the court which is relevant to the analyses conducted. However, forensic scientists must also communicate their findings to attorneys, judges, and members of a jury. This requires the translation of technical information into lay terms.

**Figure 3.15** Reconstruction of blood stain patterns.

## Bibliography

Fisher, B. 2004. *Techniques of Crime Scene Investigation*. 7th ed. Boca Raton: CRC Press.

Lee, H. C. et al. 1991. Guidelines for collection and preservation of DNA evidence. *J Forens Ident* 41 (5):13.

Lee, H. C., and C. Ladd. 2001. Preservation and collection of biological evidence. *Croat Med J* 42 (3):225.

Lee, H. C., T. Palmbach, and M. T. Miller. 2001. *Henry Lee's Crime Scene Handbook*. San Diego: Academic Press.

Lindahl, T. 1993. Instability and decay of the primary structure of DNA. *Nature* 362 (6422):709.

Phipps, M., and S. Petricevic. 2007. The tendency of individuals to transfer DNA to handled items. *Forensic Sci Int*. 168 (2,3):162.

Sutton, T. P. 1999. *Scientific and Legal Applications of Bloodstain Pattern Interpretation*. James, S. H., Ed., Boca Raton: CRC Press.

## Study Questions

1 Which of the following is not a commonly encountered pathogen?
   (a) HIV
   (b) Hepatitis virus
   (c) Tuberculosis
   (d) Diabetes mellitus

2. Dried bloodstains may be collected by:
   (a) Cutting out a stain
   (b) Scraping a stain
   (c) Tape lifting a stain
   (d) All of the above

3. Dried blood evidence should be packed in a:
   (a) Paper bag
   (b) Plastic bag
   (c) Glass jar
   (d) Tin can

4. A wet bloody shirt should be:
   (a) Packed as soon as possible
   (b) Air dried prior to packing
   (c) Air dried and not packed
   (d) Dried with heat

5. Which of the following is true for evidence having class characteristics?
    (a) It is useful for excluding a suspect.
    (b) It is useful for including a suspect with a high degree of certainty.
    (c) It has little evidentiary value
    (d) None of the above

6. DNA evidence can be considered a tool for:
    (a) Identification
    (b) Comparison
    (c) Class characterization
    (d) Individualization

7. A forensic examination process including examination of biological evidence follows a basic sequence of steps. Which of the following lists the correct sequence?
    (a) Recognition, identification, individualization, and reconstruction
    (b) Recognition, individualization, identification, and reconstruction
    (c) Identification, recognition, individualization, and reconstruction
    (d) Individualization, recognition, identification, and reconstruction

8. Which statement is true?
    (a) Crime scene search patterns are only used for outdoor scenes.
    (b) Crime scene search patterns are only used for indoor scenes.
    (c) Systematic search patterns are only used for a search of biological evidence.
    (d) None of the above

9. A photo log should include:
    (a) Date of incident
    (b) Date and time photo taken
    (c) Roll number of film
    (d) All of the above

10. In the FTA method:
    (a) The DNA stays on the paper
    (b) The DNA is eluted off the paper

# Essential Serology II

# Serology Concepts 4

## 4.1 Serological Reagents

### 4.1.1 Antigens

An antigen is a foreign substance capable of reacting with an antibody. A foreign substance capable of eliciting antibody formation when introduced into a host is called an ***immunogen***. Natural immunogens are usually macromolecules such as proteins and polysaccharides. Other molecular structures can also act as immunogens, for example, glycolipids (such as A, B, and O blood group antigens) and glycoproteins (such as Rh and Lewis antigens). However, they must be foreign to their hosts. Another class of immunogens is called ***haptens***. Although haptens are too small to elicit antibody production, a hapten-conjugated carrier can elicit the formation of an antibody specific for the hapten. Certain controlled substances such as cocaine and amphetamines are haptens and can be detected through corresponding antibodies for forensic toxicological analysis.

All immunogens can be considered antigens, but as noted, not all antigens can elicit antibody formation. The molecular structure of an immunogen, usually a small portion recognized by an antibody, is called an ***epitope*** or determinant site. An immunogen usually consists of multiple epitopes and is thus considered ***multivalent*** (Figure 4.1). Each epitope can elicit the production of its own corresponding antibody.

### 4.1.2 Antibodies

Antibodies, also known as ***immunoglobulins***, are capable of binding specifically to antigens and are designated with an Ig prefix. The five major classes of immunoglobulins are designated IgG, IgA, IgM, IgD, and IgE. IgG is the most abundant immunoglobulin in serum. IgD, IgE, and IgG are usually monomers. IgM can be a membrane-bound monomer or a cross-linked pentamer (secreted form). IgA can be a monomer, dimer, or trimer.

Immunoglobulins have many similarities in their molecular structures. Figure 4.2 illustrates a diagram of the IgG molecule. The structures of immunoglobulins were first revealed by Gerald Edelman and Rodney Porter who shared a Nobel Prize in 1972. Immunoglobulins are composed of four polypeptide chains: two heavy (H) chains and two light (L) chains. The

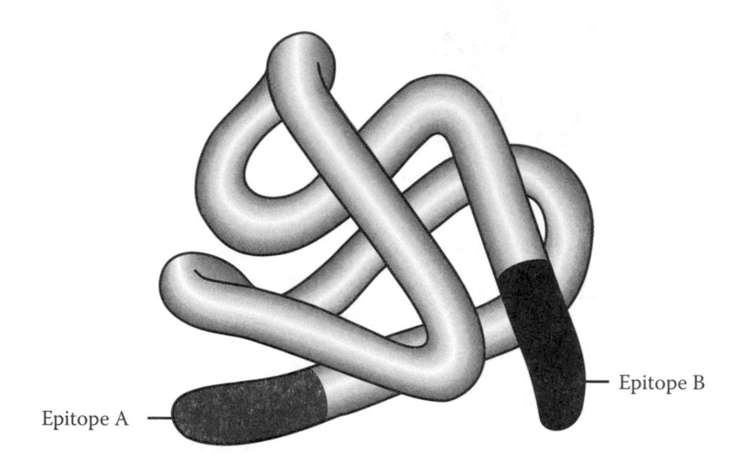


**Figure 4.1** Multivalent immunogen. A protein with two different epitopes is shown.

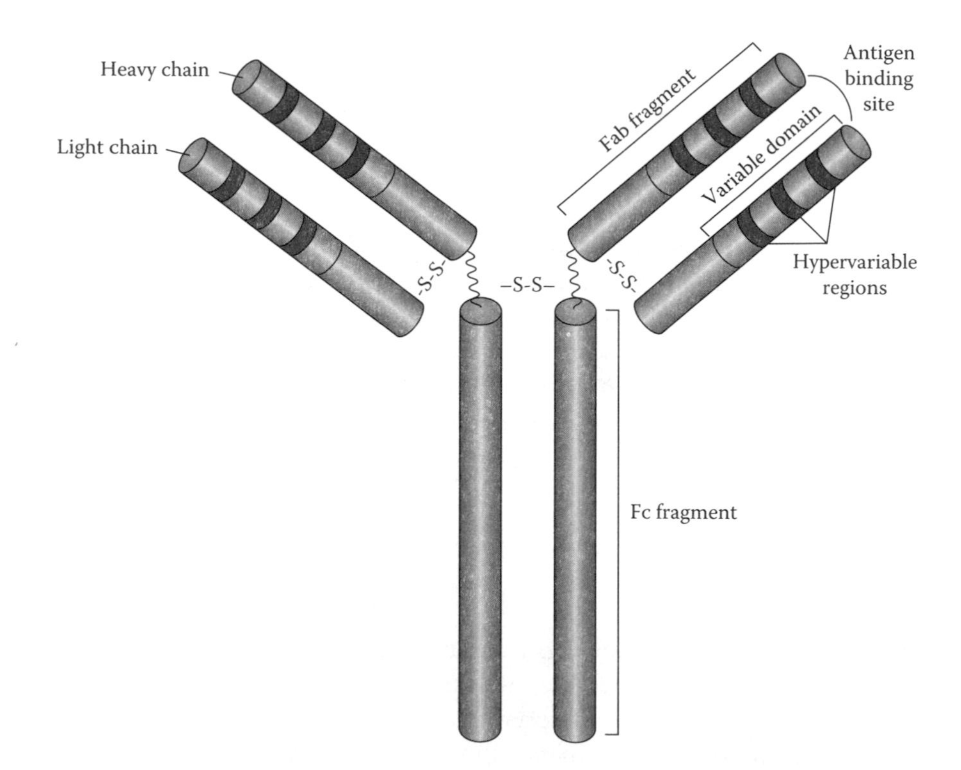


**Figure 4.2** Immunoglobulin IgG is composed of light and heavy chains that both contain Fab fragments. The remaining heavy chains form the Fc region. The Fab fragments of both light and heavy chains contain three hypervariable regions that form the antigen binding site of the immunoglobulin.

polypeptide chains are linked by disulfide bonds into a Y-shaped complex. The H chain can be divided into ***Fab*** (fragment antigen binding) and ***Fc*** (fragment crystallizable) fragments. The L chain only consists of a Fab fragment. A typical antibody has two identical antigen binding sites and is thus considered ***bivalent***. The antigen binding activity is located within the Fab fragment. In particular, the N terminal ends of the L and H chains together form antigen-binding sites. At the amino acid sequence level, both H and L chains have ***variable domains*** at their N terminal ends and three small ***hypervariable regions*** are located within the variable domain of each chain.

The diversity in the amino acid sequences of these hypervariable regions determines the specificity of antigen binding sites. The hinge regions provide flexibility to the antibody molecule and are important for the efficiency of the binding and cross-linking reactions. The basal portion of the H chain is called the ***Fc*** and consists of ***constant domains***. The binding affinity and specificity of antibodies make them useful reagents for serological testing. Two types of antibodies are commonly used.

#### *4.1.2.1 Polyclonal Antibodies*

To produce an antibody, an immunogen is usually introduced into a host animal. Depending on the type of animal used, the antibodies produced are classified as avian (B), rabbit (R), or horse (H) type. Antibodies can be circulating (in blood or other biological fluids) or tissue-bound (in cell surface antibodies). Circulating immunoglobulins are referred to as humoral antibodies. A multivalent immunogen is capable of eliciting a mixture of antibodies with diverse specificities for the immunogen. As a result, a polyclonal antibody is produced by different B lymphocyte clones in response to the different epitopes of the immunogen.

The characteristics of polyclonal antibodies may vary if they are produced from different individual host animals of the same species. Variations of reactions among different sources of antibodies should be monitored by quality control procedures. A common preparation of antibodies is called a ***polyclonal antiserum***. The blood from an immune host is removed and allowed to clot, resulting in the formation of a solid consisting largely of blood cells and a liquid portion known as ***serum*** containing antibodies (Figure 4.3).

#### *4.1.2.2 Monoclonal Antibodies*

To produce a monoclonal antibody, spleen cells are harvested from a host animal inoculated with an immunogen (Figure 4.4). Next, the plasma cells of the spleen, which produce antibodies, are fused with myeloma cells. Since only a small population of cells fuse, a selection step is needed to allow only fused cells to grow. The fused cells, called ***hybridoma*** cells, are immortal (proliferate indefinitely) in cell cultures. Pools of hybridoma cells are diluted

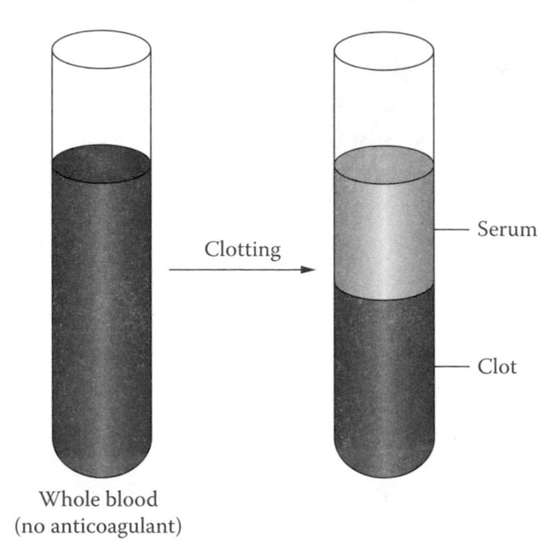


**Figure 4.3** Serum component of blood. The blood of an immunized animal is collected in the absence of anticoagulant and allowed to clot. The resulting liquid portion of the blood is serum.

into single clones and allowed to proliferate. The clones are then screened for the specific antibody of interest.

The desired hybridoma clone can be maintained indefinitely and it produces a monoclonal antibody that reacts with a single epitope. The hybridoma-derived monoclonal antibodies are specific, homogenous, and can be obtained in unlimited quantities. Monoclonal antibodies have been employed in many serological assays as discussed in later chapters. However, they have certain limitations in serology assays. For instance, monoclonal antibodies react with only a single epitope of a multivalent antigen and therefore cannot form cross-linked networks. Thus, they are not applicable for precipitation and agglutination assays.

#### *4.1.2.3 Antiglobulins*

Immunoglobulins are proteins that can function as immunogens. If a purified foreign immunoglobulin is introduced into a host, the specific antibodies produced are known as ***antiglobulins***. In addition to specific antiglobulins, it is possible to produce nonspecific antiglobulins against immunoglobulin ***isotypes***, which recognize an epitope that is common to all members of an immunoglobulin class. Antiglobulins are important reagents in many serological tests.

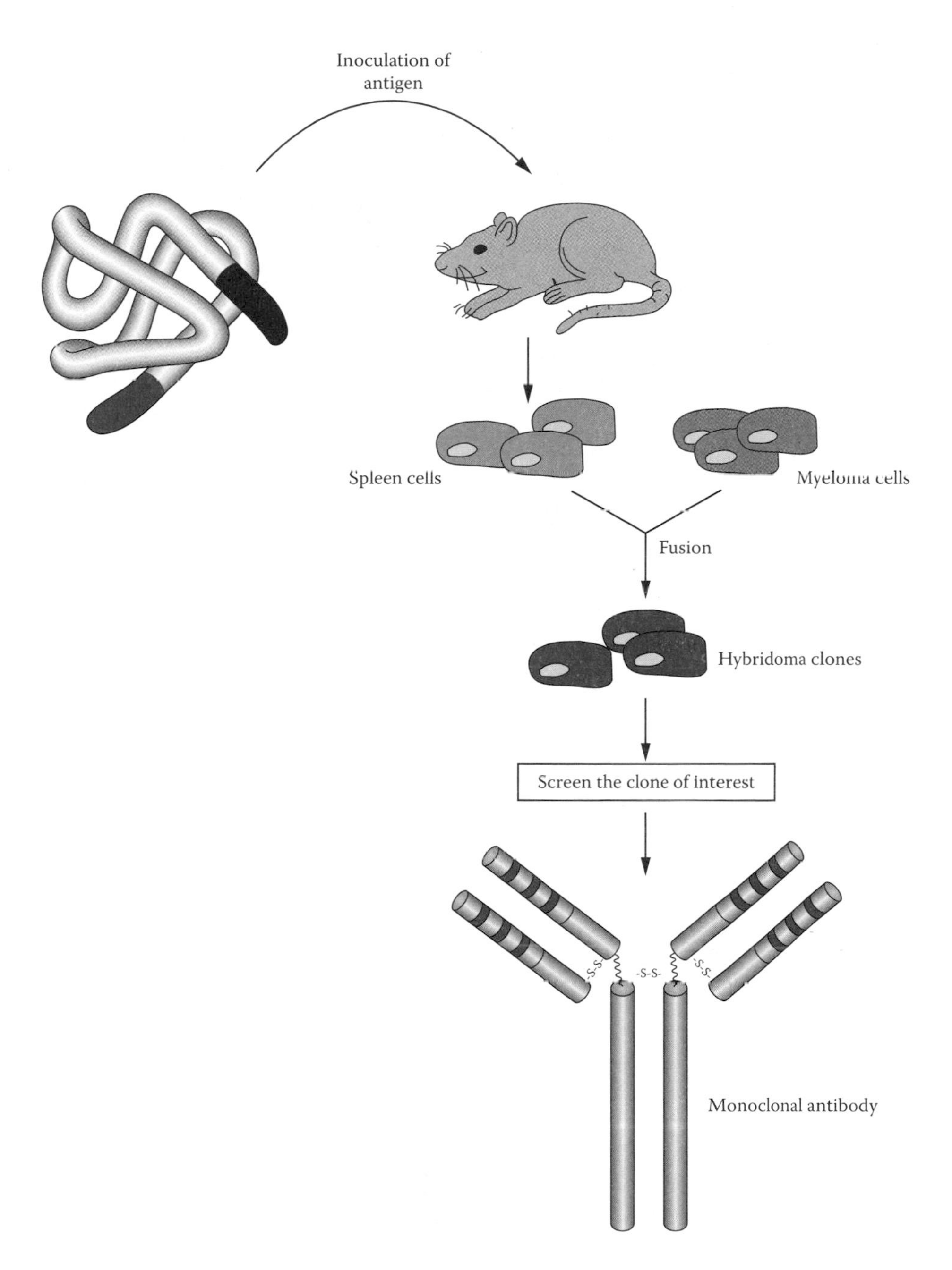


**Figure 4.4** Preparation of monoclonal antibody. A mouse is immunized with an antigen and its spleen cells are fused with myeloma cells to generate hybridoma cells. The clone that synthesizes and secretes a monoclonal antibody of interest can then be identified.

## 4.2 Strength of Antigen–Antibody Binding

The binding of antigen and its specific antibody is mediated by the interaction of the epitope of an antigen and the binding site of its antibody. Various forces act cooperatively during antigen–antibody binding. These include hydrogen bonding, hydrophobic interactions that exclude water molecules from the area of contact, and van der Waals forces arising from asymmetric distribution of the charges of electrons. As a result, noncovalent bonds can be formed during antigen–antibody binding. The binding process occurs rapidly and the formation of the antigen–antibody complex is reversible. Such binding occurs at short distances when the antigen and antibody are in close proximity. Additionally, the strongest binding occurs only if the shape of the epitope fits the binding site of the antibody. The strength of interaction between the antigen and antibody depends on two characteristics designated affinity and avidity.

***Affinity*** is the energy of the interaction of a single epitope on an antigen and a single binding site on a corresponding antibody (Figure 4.5). The strength of the interaction depends on the specificity of the antibody for the antigen. Nevertheless, antibodies can bind with lower strength to antigens that are structurally similar to the immunogen. Such a binding is known as ***cross-reaction***.

***Avidity*** is the overall strength of binding of antibody and antigen (Figure 4.6). Since an antigen is usually considered multivalent and an antibody is bivalent, the avidity reflects the sum of the binding affinities of all the binding sites of antigens and antibodies. It also reflects the overall stability of an antigen–antibody complex that is essential for many serological assays.

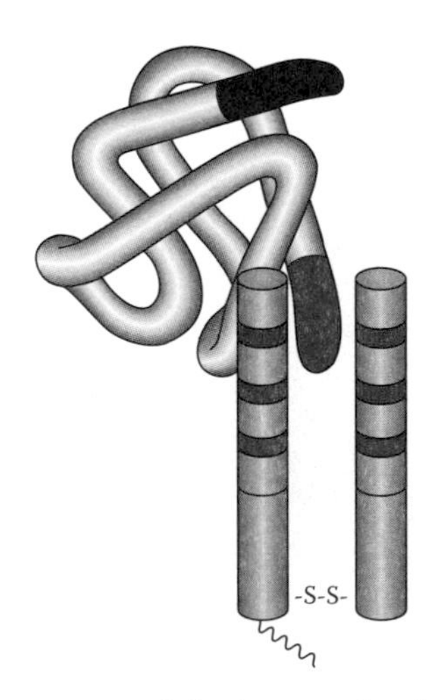


**Figure 4.5** Affinity is a measure of the interaction between a single epitope on an antigen and a single binding site on an antibody.

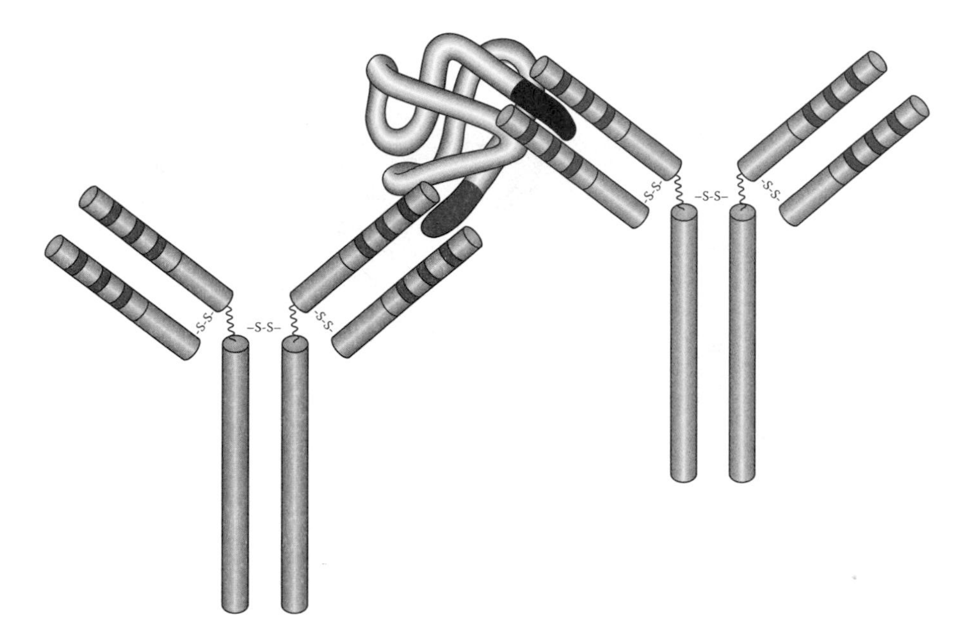


**Figure 4.6** Avidity is a measure of overall strength of the binding between antigens and antibodies.

## 4.3 Antigen–Antibody Binding Reactions

The binding of an antigen to an antibody is an equilibrium reaction consisting of three types of reactions. The primary and secondary reactions form the basis for many forensic serological assays and will be discussed below. The third type is called the tertiary reaction. It is used to measure *in vivo* immune responses such as inflammation and phagocytosis. The tertiary reaction is not commonly employed in forensic serology testing and will not be discussed here.

### 4.3.1 Primary Reactions

A primary reaction is the initial binding of a single epitope of an antigen (Ag) and a single binding site of an antibody (Ab) to form an antigen–antibody complex (Figure 4.7). This rapid and reversible binding reaction can be expressed as

$$Ag + Ab \rightleftarrows AgAb$$

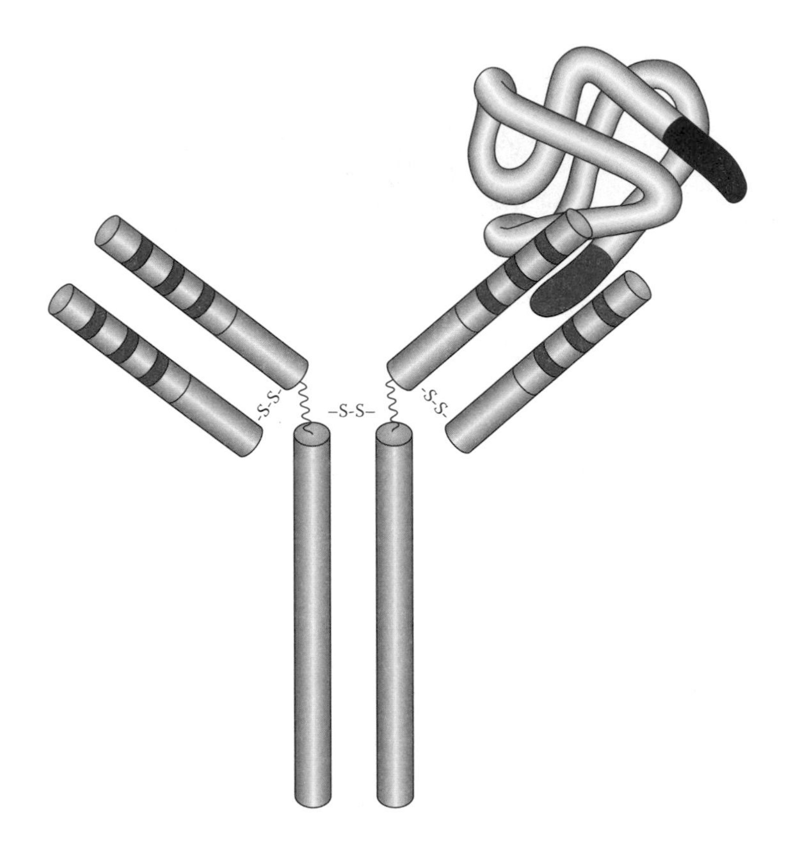


**Figure 4.7** Primary reaction. Initial binding forms an antigen–antibody complex.

At equilibrium, the strength of the interaction can be expressed as the affinity constant ($K_a$) that reflects the affinity of binding, where

$$K_a = \frac{[AgAb]}{[Ag][Ab]}$$

The square brackets indicate the concentration of each component at equilibrium. $K_a$ is the reciprocal of the concentration of free epitopes when half of the antibody-binding sites are occupied. Thus, a higher $K_a$ corresponds to a stronger binding interaction.

Techniques such as enzyme immunoassays, immunofluorescence assays, radioimmunoassays, and dye-labeled immunochromatography can measure the concentrations of antigen–antibody complexes formed by primary reactions. These techniques are the most sensitive for detecting amounts of antigen and antibody in a sample. Additionally, many forensic serology assays are based on the detection of primary reactions and will be discussed in later chapters.

### 4.3.2 Secondary Reactions

The primary reaction between antigen and antibody is often followed by a secondary reaction. The three types of secondary reactions are precipitation, agglutination, and complement fixation. Various techniques can detect secondary reactions. They are usually less sensitive than primary reaction assays but easier to perform. The precipitation and agglutination reactions form the basis for many serologic assays performed in forensic laboratories. These reactions will be discussed later in detail. The third type of reaction is called ***complement fixation***. If an antigen is located on a cell surface, the binding of the antigen and antibody may activate the classical complement pathway and lead to cell lysis also known as a complement fixation reaction. The detection of this type of reaction is not commonly used in forensic serology.

#### *4.3.2.1 Precipitation*

If a soluble antigen is mixed and incubated with its antibody, the antigen–antibody complexes can form cross-linked complexes at the optimal ratio of antigen-to-antibody concentration. The cross-linked complex is insoluble and eventually forms a precipitate that settles to the bottom of a test tube. Antibodies that produce such precipitation are also called ***precipitins***.

This precipitation reaction can be characterized by examining the effect of varying the relative ratio of antigen and antibody. If an increasing amount of soluble antigen is mixed with a constant amount of antibody, the amount of precipitate formed can be plotted. A precipitin curve (Figure 4.8) illustrates the results observed when antigens and antibodies are mixed in various concentration ratios. The curve can be divided into three zones known as the prozone, the zone of equivalence, and the postzone.

**Prozone** — At this zone, the ratio of antigen–antibody concentrations is low. In other words, the antibody is in excess. Each antigen molecule is rapidly saturated with antibody, thus preventing cross-linking (Figure 4.9a). No precipitate is formed at the prozone stage.

**Zone of Equivalence** — As the concentrations of antigen increase, the amount of precipitate increases until it reaches a maximum. The amount of precipitation depends on the relative proportions of antigens and antibodies present. The maximum precipitation occurs in what is called the zone of equivalence. In the zone of equivalence, the ratio of antibody to antigen concentration is optimal and precipitation occurs as a result of forming cross-linked networks (Figure 4.9b). The quantity of precipitation in the zone of equivalence is very important for many serology assays.

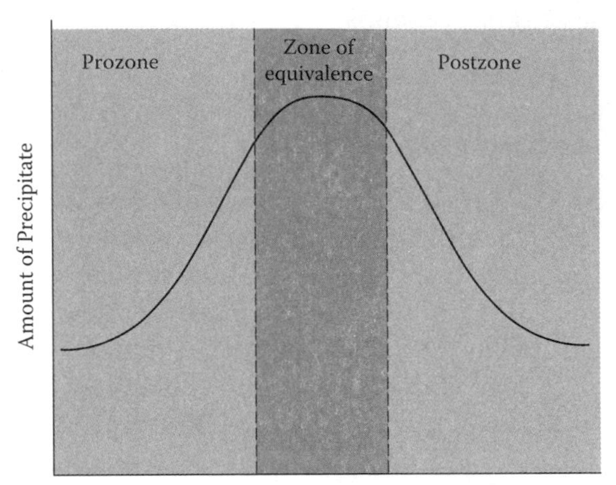


**Figure 4.8** Precipitin curve.

**Postzone** — With the addition of more antigens, the ratio of antigen–antibody concentrations is high. In other words, the antigen is in excess. The amount of precipitate decreases and eventually diminishes. Each antibody molecule is saturated with antigen molecules (Figure 4.9c). Cross-linkage cannot form and precipitation does not occur.

#### *4.3.2.2 Agglutination*

As discussed above, precipitation reactions involve soluble antigens. If the antigens are located on the surfaces of cells or carriers (carrier cells, bacteria, or latex particles), the interactions of antibodies and antigens will cause the cells or carriers to aggregate and form larger complexes. This is referred to as agglutination. If the antigen is located on an erythrocyte, the agglutination reaction is designated ***hemagglutination***. In agglutination, a visible clumping can be observed as an indicator of the reaction of antigen and antibody. The antibody will bind to the antigens on the surfaces of the cells (or carriers). Antibodies can form cross-links among cells (or carriers) and produce agglutination. Like precipitation, agglutination is a two-step process that includes initial binding and lattice formation (Figure 4.10).

**Initial Binding** — The first step of the reaction involves antigen–antibody binding at a single epitope on the cell surface (Figure 4.10b). This initial binding is rapid and reversible.

**Lattice Formation** — The second step involves the formation of a cross-linked network resulting in visible aggregates that constitute a lattice (Figure 4.10c). This involves an antibody binding to multiple epitopes because

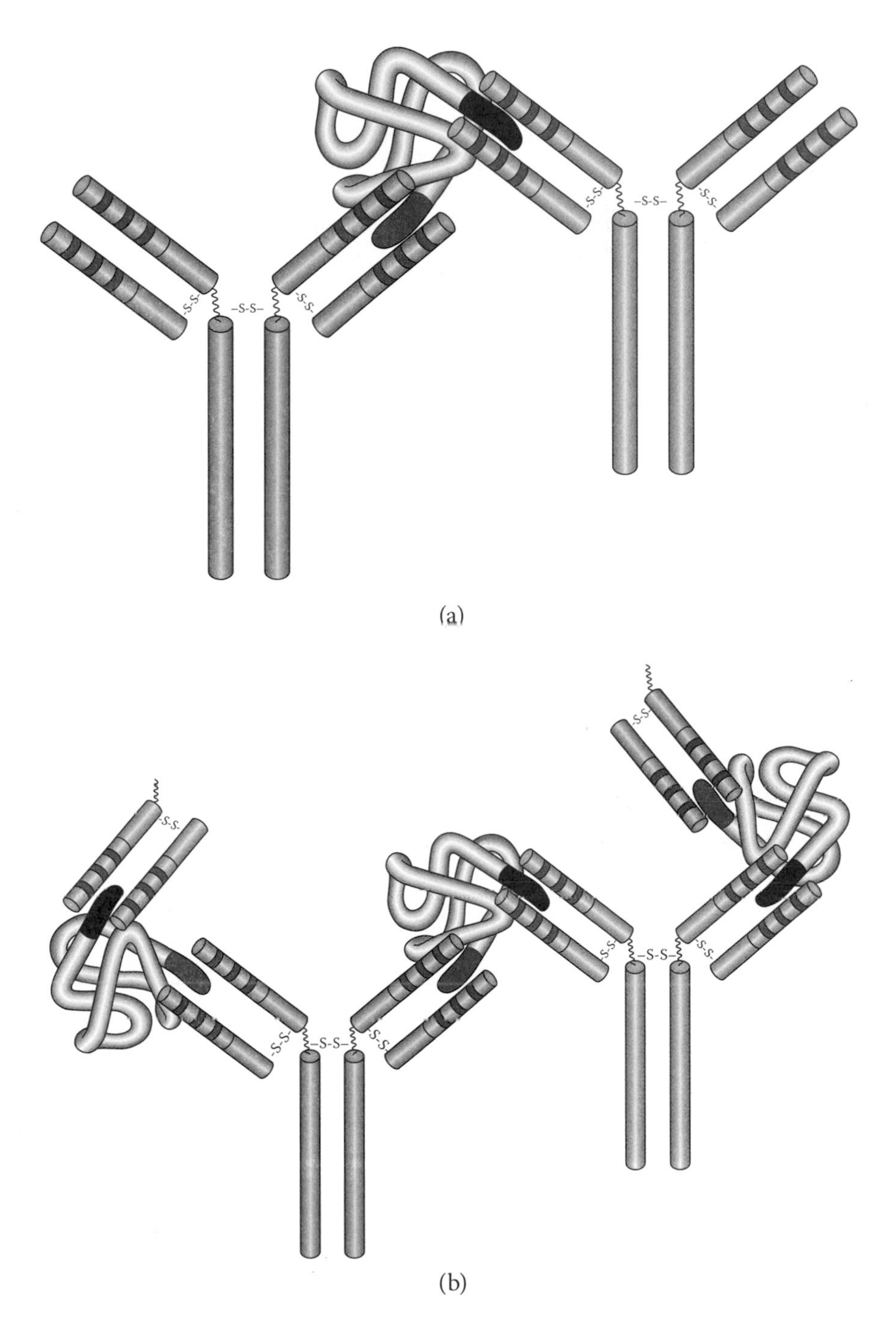


**Figure 4.9** Antigen–antibody binding in (a) prozone, (b) zone of equivalence, and (c) postzone.

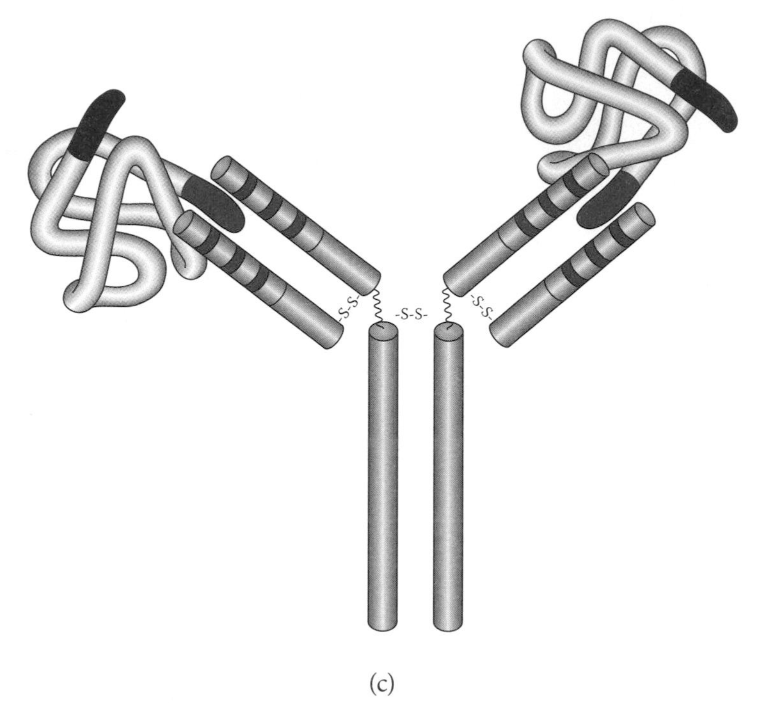


(c)

**Figure 4.9** (continued)

each antibody has two binding sites and antigens are multivalent. Lattice formation is a much slower process than the initial binding step. Cross-linking of cells requires physical contact. Additionally, an antibody must bind to epitopes on two different cells. The ability to cross-link cells depends on the nature of the antibody.

Antibodies that produce such reactions are often called ***agglutinins***. Additionally, a ***complete antibody*** is capable of carrying on both primary and secondary interactions that result in agglutination. An antibody that can carry out initial binding but fails to form agglutination is called an ***incomplete antibody***. This type of antibody is believed to have only one active antigen binding site and is thus not capable of agglutination. However, other incomplete antibodies have two active sites but cannot bridge the distance between cells, thus failing to form lattices. Certain antibodies such as IgG are small and lack flexibility at the hinge region and this prevents agglutination. In contrast, the large IgM antibodies produce agglutination much more easily than IgG.

Agglutination reactions have a wide variety of applications in the detection of antigens and antibodies. Such assays have high degrees of sensitivity and have been used for many years in forensic serology.

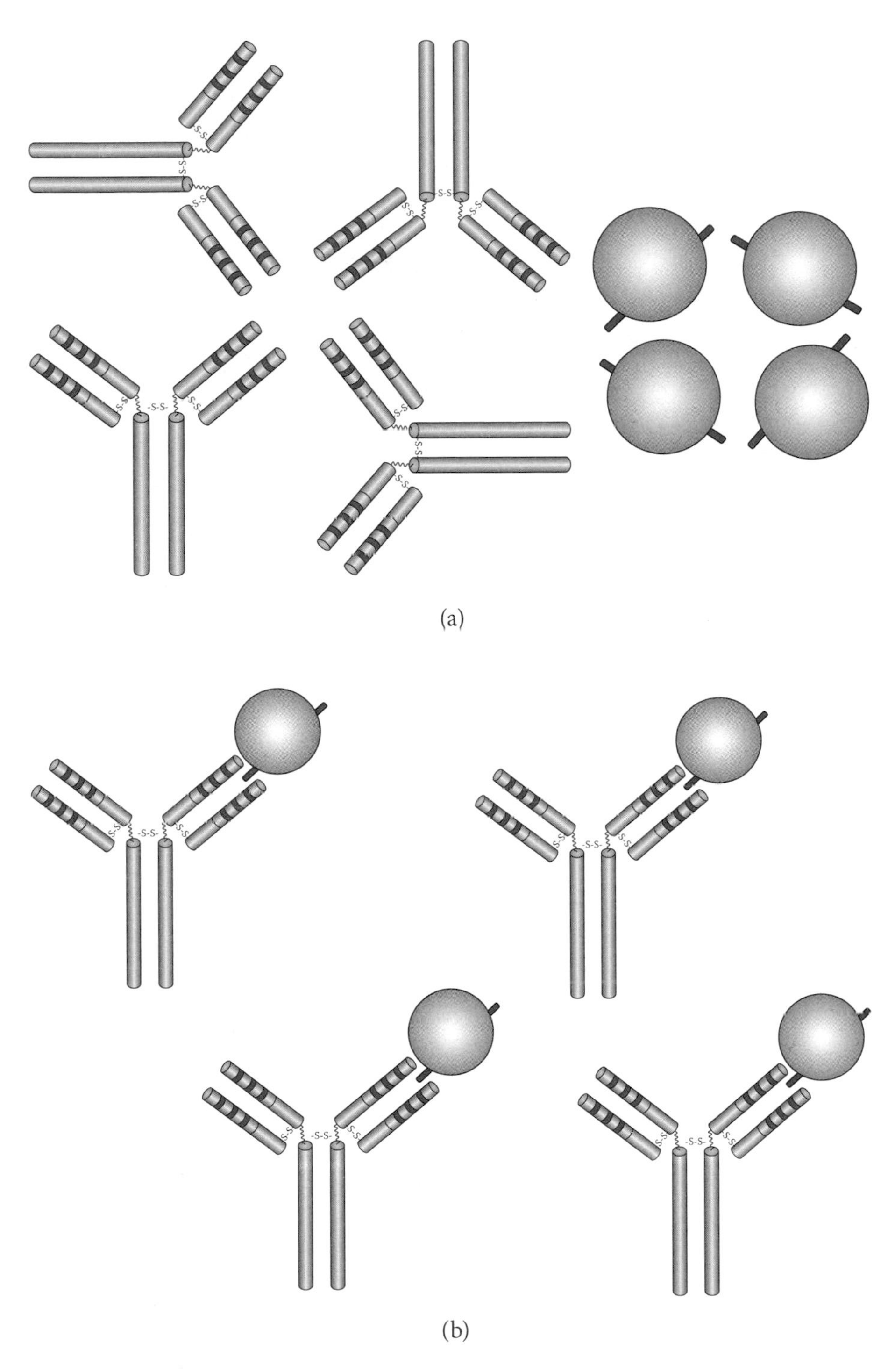


**Figure 4.10** Agglutination reaction: (a) antigens are mixed with antibodies; (b) antigen–antibody complex is formed during initial binding; (c) lattice is formed.

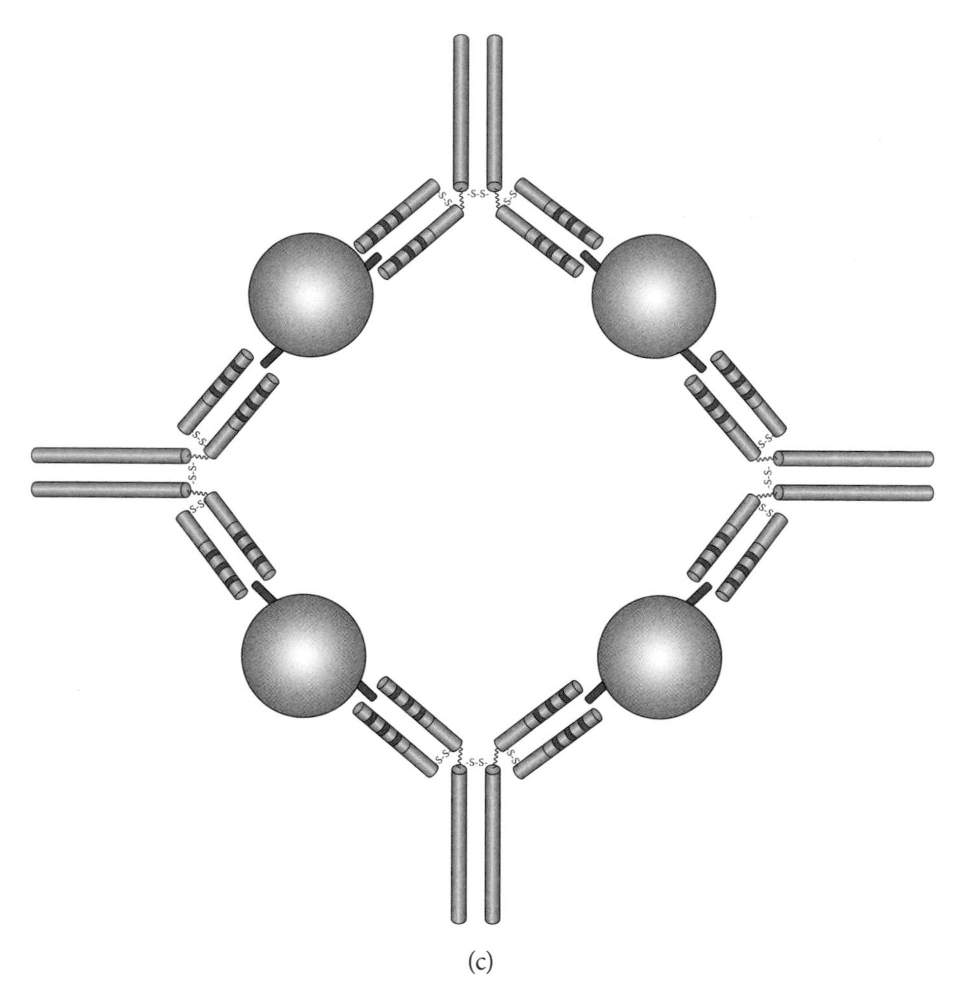

(c)

**Figure 4.10** (continued)

## Bibliography


Baxter, S. J., and B. Rees. 1974. The use of anti-human haemoglobin in forensic serology. *Med Sci Law* 14 (3):159.

Dodd, B. E. 1972. Some recent advances in forensic serology. *Med Sci Law* 12 (3):195.

Fletcher, S., and M. J. Davie. 1980. Monoclonal antibody: A major new development in immunology. *J Forensic Sci Soc* 20 (3):163.

Giusti, G. V. 1982. Leone Lattes: Italy's pioneer in forensic serology. *Am J Forensic Med Pathol* 3 (1):79.

Goldsby, R., T. Kindt, and B. Osborne. 2007. *Immunology*, 6th ed. New York: W.H. Freeman.

Lee, H. C. et al. 1991. Bits and pieces: Serology, criminalistics, anthropology and odontology. *J Forensic Sci Soc* 31 (2):293.

Patzelt, D. 2004. History of forensic serology and molecular genetics in the sphere of activity of the German Society for Forensic Medicine. *Forensic Sci Int* 144 (2–3):185.

Tietz, N. 1987. *Fundamentals of Clinical Chemistry*, Manke, D., Ed. Philadelphia: W.B. Saunders.

## Study Questions

1. What is the major difference between a polyclonal antibody and a monoclonal antibody?

2. Which of the following is true for a primary binding reaction?
   (a) Binding between a single epitope of an antigen and a single binding site of an antibody
   (b) Binding between a single epitope of an antigen and bivalent binding sites of an antibody
   (c) Binding between multiepitope of an antigen and a single binding sites of an antibody
   (d) Binding between a multiepitope of an antigen and bivalent binding sites of an antibody

3. Explain the difference between affinity and avidity.

4. What do antiglobulins recognize?

5. Which of the following involves excess antigen?
   (a) Prozone
   (b) Postzone
   (c) Zone of equivalence

# Serology Techniques 5

The detection and measurement of the antigen–antibody binding reactions serve as the bases of forensic serology. These assays fall into two categories: primary and secondary binding assays. The primary binding assays are the most sensitive. The secondary assays consist of precipitation-based and agglutination-based assays. Precipitation-based assays are usually used for species identification while agglutination-based assays are normally applied to blood group typing.

## 5.1 Primary Binding Assays

### 5.1.1 Enzyme-Linked Immunosorbent Assay (ELISA)

ELISA is an immunoenzyme assay that can be used to detect and measure an antibody or antigen in question. The most common ELISA used in forensic serology is the antibody-sandwich ELISA (Figure 5.1). It was employed to detect prostate-specific antigen (PSA) to identify seminal stains and amylase for identification of saliva.

An antibody (usually monoclonal) coating is formed by nonspecific adsorption onto a solid phase such as wells of a polystyrene microtiter plate. A sample containing the antigen to be tested is then added and binds to the solid phase antibody. Subsequently, a second antibody is added to form an antibody–antigen–antibody sandwich complex. The second antibody is usually polyclonal and binds to different epitopes of the antigen. Next, an enzyme-labled antiglobulin is added to bind the sandwich. Excess enzyme-labeled antiglobulin is then removed by washing and then the bound antiglobulin can be detected.

A number of enzymes such as alkaline phosphatase and horseradish peroxidase have been used as reporting enzymes to label the antiglobulin for ELISA. The enzyme catalyzes the substrate and produces colorimetric or fluorometric signals. The intensity of the signals can be detected spectrophotometrically and is proportional to the amount of bound antigen. The amount of antigen can be quantified by comparing the standard with known concentrations.

Alternatively, antibodies in a sample can also be detected and may be quantified by an ELISA system in which the antigen is bound to a solid phase instead of the antibody. After antibody in a sample binds to the solid phase antigen, an enzyme-labeled antiglobulin is added to bind to the bound

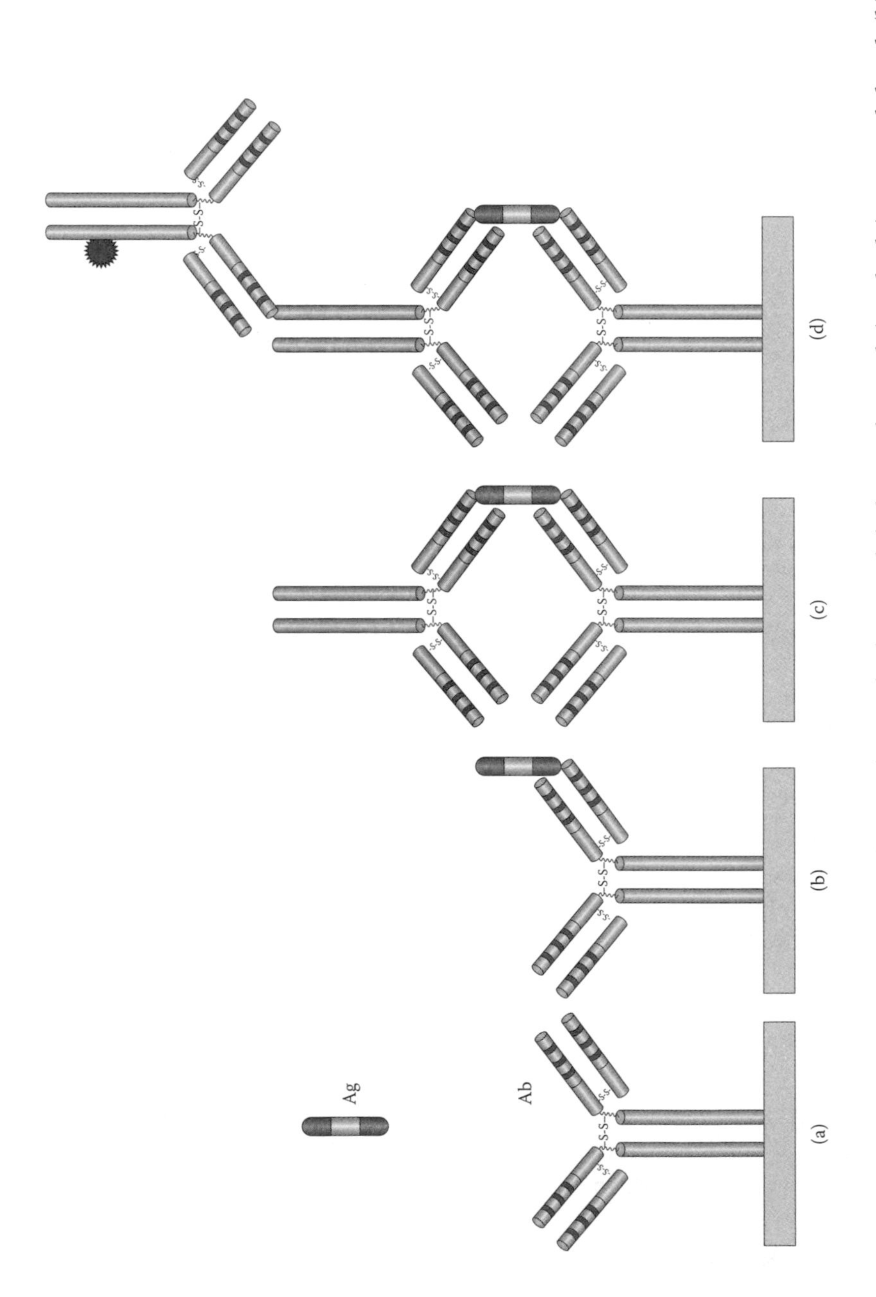


**Figure 5.1** ELISA assay. (a) Sample containing Ag (antigen) is applied to a solid phase where Ab (antibody) is immobilized. (b) Ag binds to immobilized Ab to form Ag–Ab complex. (c) Second Ab to a different epitope is added to form a Ab–Ag–Ab sandwich. (d) Labeled antiglobulin binds to the sandwich. The bound antiglobulin can be detected by various reporting schemes.

antibody. The bound antiglobulin can be detected and measured by the addition of an enzyme substrate. The enzymatic catalytic reaction is similar to the antibody sandwich ELISA procedure described above.

### 5.1.2 Immunochromatographic Assays

Figure 5.2 depicts a test using an immunochromatographic membrane device. A dye-labeled monoclonal antibody is contained in a sample well and a polyclonal antibody for the antigen (or a second monoclonal antibody to a different epitope of the antigen) is immobilized onto a test zone of a nitrocellulose membrane. An antiglobulin that recognizes the antibody is immobilized onto a control zone.

The assay is carried out by loading a sample into the sample well. The antigen in the sample binds to the dye-labeled antibody in the sample well to form an antigen–antibody complex, the complex then diffuses across the nitrocellulose membrane until it reaches the test zone. The antibody immobilized at the test zone traps the antigen–antibody complex to form an antibody–antigen–antibody sandwich. The presence of antigen in the sample results in a colored vertical line at the test zone.

The immunochromatographic device also utilizes a control zone to ensure that the device works properly and that the sample has diffused

**Table 5.1 Common Immunochromatographic Assays for Forensic Applications**

| Assay | Antigen | Labeled Antibody | Immobilized Antibody | Forensic Application |
|---|---|---|---|---|
| ABAcard® HemaTrace® (Abacus Diagnostics) | Hemoglobin (Hb) | Monoclonal antihuman Hb antibody | Polyclonal antihuman Hb antibody | Blood and species identification |
| RSID™-Blood (Independent Forensics) | Glycophorin A (GPA) | Monoclonal antihuman GPA antibody | Monoclonal antihuman GPA antibody | Blood and species identification |
| RSID™-Saliva (Independent Forensics) | Human salivary α-amylase (HAS) | Monoclonal antihuman HAS antibody | Monoclonal antihuman HAS antibody* | Saliva identification |
| One-Step ABA card PSA® (Abacus Diagnostics) | Prostate-specific antigen (PSA) | Monoclonal antihuman PSA antibody | Polyclonal antihuman PSA antibody | Semen identification |
| RSID™-Semen (Independent Forensics) | Semenogelin (Sg) | Monoclonal antihuman Sg antibody | Monoclonal antihuman Sg antibody* | Semen identification |

* The epitope recognized by the immobilized antibody is different from that of the labeled antibody.

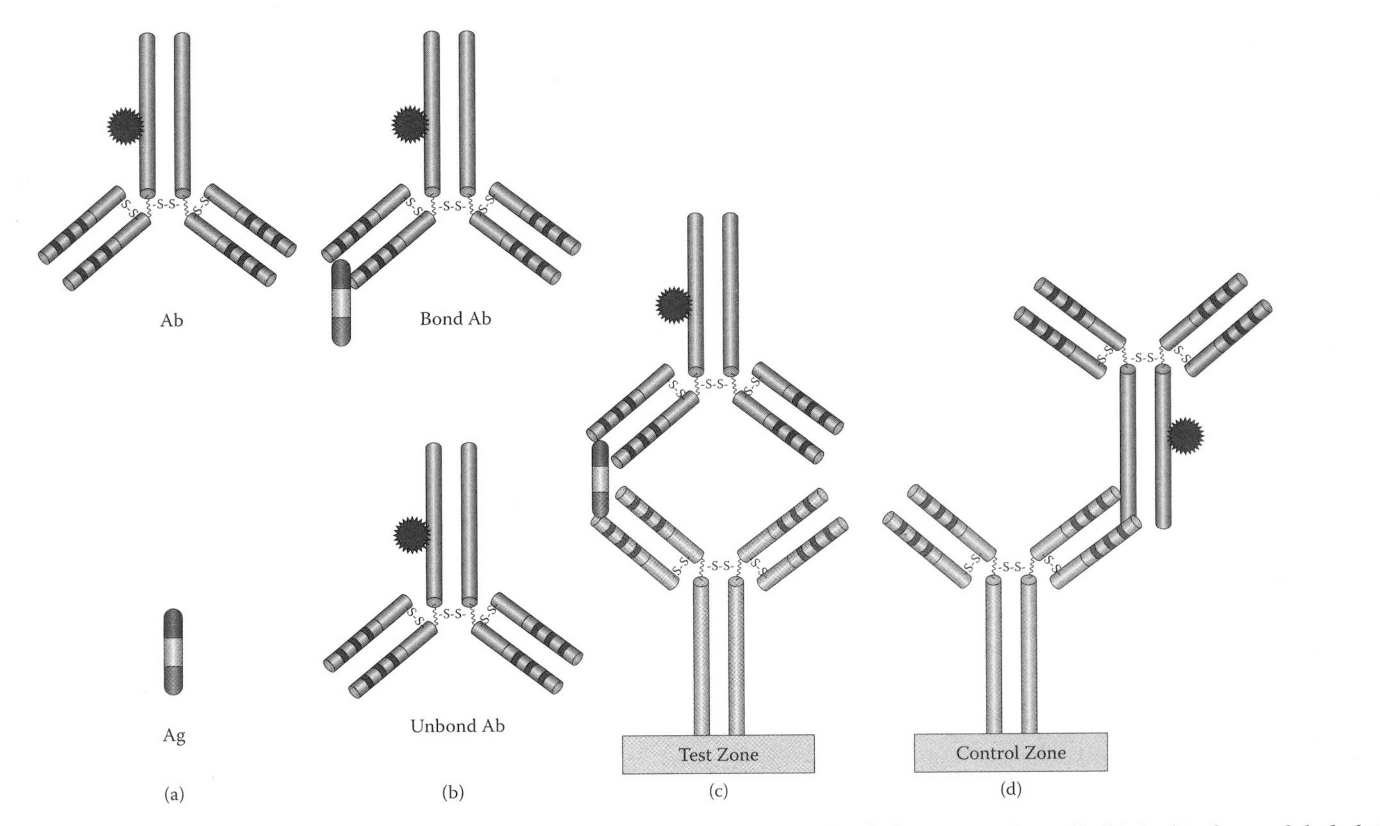


**Figure 5.2** Immunochromatographic assay. (a) Sample containing Ag (antigen) is loaded in a sample well. (b) Ag binds to a labeled Ab (antibody) to form a labeled Ab–Ag complex. (c) At the test zone, the labeled Ab–Ag complex binds to an immobilized Ab to form a labeled Ab–Ag–Ab sandwich. (d) At the control zone, labeled Ab binds to an immobilized antiglobulin and is captured at the control zone.

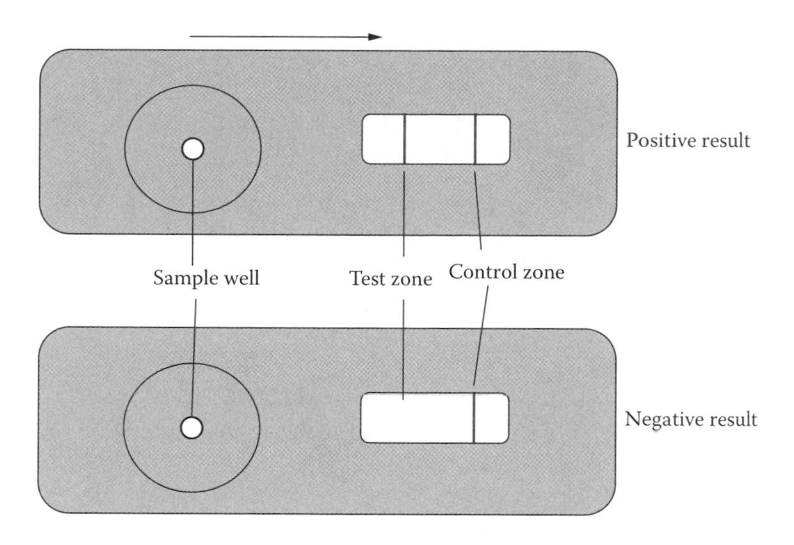


**Figure 5.3** Immunochromatographic device. Positive and negative results are shown.

completely along the test strip. Unbound monoclonal antibodies diffuse across the membrane until they reach the control zone where they are trapped by the immobilized antiglobulin. This antibody–antiglobulin complex at the control zone also results in a colored vertical line. The test is considered valid only if the line in the control zone is observed. The presence of antigen results in a line at both the test zone and control zone, while the absence of antigen results in a line in the control zone only (Figure 5.3). This method is rapid and simple and thus can be used as a screening test in laboratories and as a field test at crime scenes to identify semen, saliva, and species (Table 5.1).

However, a false negative result may be obtained if a sample contains a very high concentration of antigen (Figure 5.4). Under this condition, the dye-labeled antibody will become saturated with the antigen to form antigen–antibody complexes. The unbound antigen will diffuse along with the antigen–antibody complex toward the test zone. At the test zone, the unbound antigen will compete with the antigen–antibody complexes for the antibody immobilized at the test zone. Since the antigen is in excess, the unbound antigen binds to the immobilized antibody, preventing the formation of the antibody–antigen-antibody sandwich. The resulting reading appears as a negative result. This artifact is known as ***high-dose hook*** effect. To prevent the high-dose hook effect, a smaller volume of sample can be applied, or the sample can be diluted to reduce the amount of antigen applied.

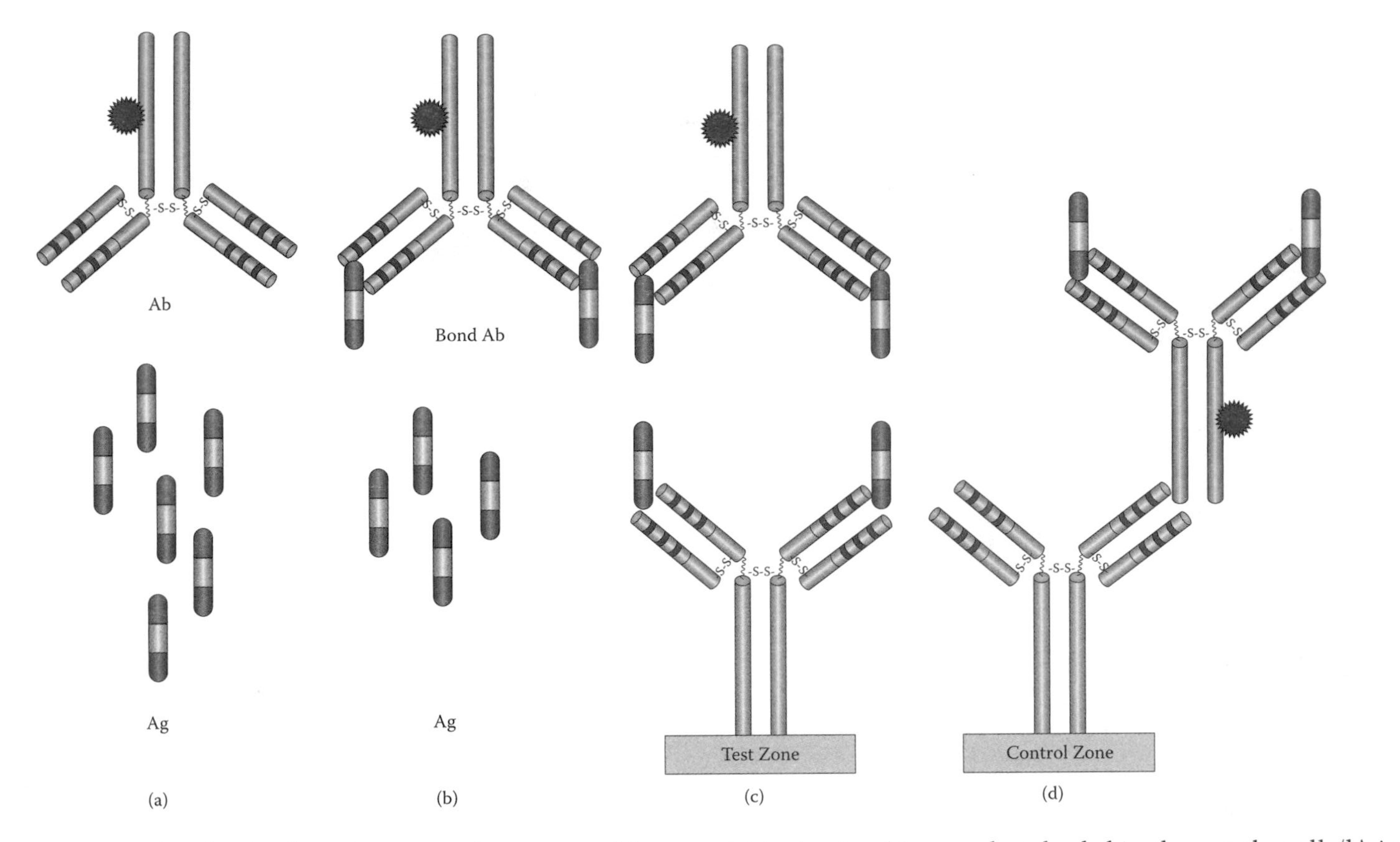


**Figure 5.4** High-dose hook effect of immunochromatographic assay. (a) Ag (antigen) in sample is loaded in the sample well. (b) Ag binds to the labeled Ab (antibody) to form labeled Ag–Ab complex. (c) At the test zone, free Ag binds to the immobilized Ab to form an unlabeled Ab–Ag complex and prevent the formation of the labeled Ab–Ag–Ab sandwich. (d) At the control zone, a labeled Ab binds to immobilized antiglobulin and is captured at the control zone.

## 5.2 Secondary Binding Assays

### 5.2.1 Precipitation-Based Assays

These techniques are based on the precipitation reaction and are used primarily for species identification in forensic laboratories.

#### *5.2.1.1 Immunodiffusion*

Immunodiffusion is a passive method in which a concentration gradient is established for an antigen and/or an antibody. The antigen and/or antibody are allowed to diffuse until precipitation occurs. The assay can be carried out in a liquid or semisolid medium such as agarose gel. The semisolid medium can stabilize the diffusion process and reduce interference such as convection. The two types are single immunodiffusion and double immunodiffusion.

**5.2.1.1.1 Single Immunodiffusion** A concentration gradient is established for either an antigen or antibody. One serology technique based on this principle is ***radial immunodiffusion*** — a single diffusion method in which a concentration gradient is established for an antigen. The antibody is uniformly distributed in the gel matrix (Figure 5.5). The antigen is loaded into a sample well and allowed to diffuse from the well into the gel until a precipitation reaction occurs. The precipitate ring around the well is observed; the

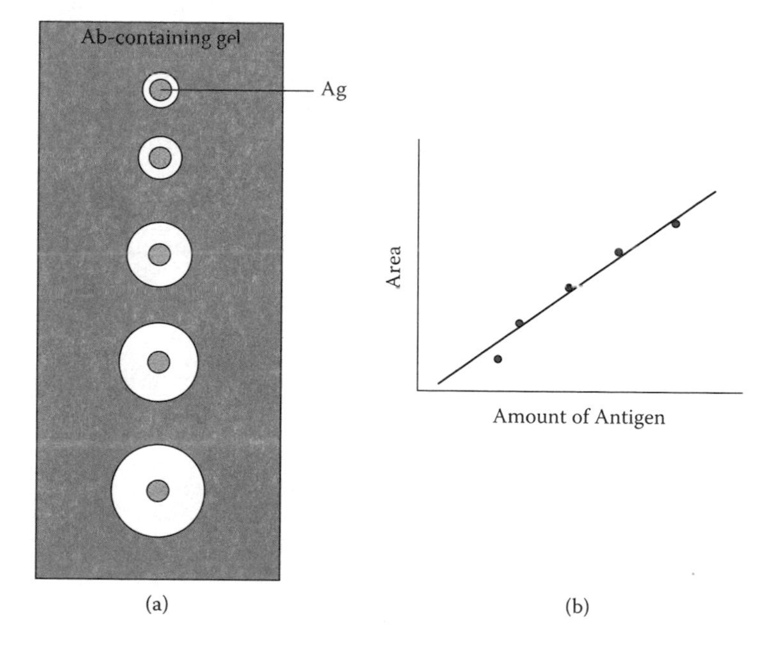


**Figure 5.5** Radial immunodiffusion assay. (a) Immunodiffusion of antigens from the well into an antibody-containing gel. (b) Standard curve based on results of standards with known amounts of antigens.

area within the ring of precipitate is proportional to the amount of antigen loaded in the well. Standards using known concentrations of antigen can be included in the same assay along with the samples and a standard curve can be plotted. The amounts of antigen in samples can then be quantified by comparing the results with the standard curve.

**5.2.1.1.2 Double Immunodiffusion** The second type of assay is double diffusion in which a concentration gradient is established for both an antigen and an antibody. The most common examples are the ring assay and the Ouchterlony assay. The ***ring assay*** can be performed in a test tube or capillary tube (Figure 5.6). An antiserum, a denser phase, is placed in a small tube. An antigen solution is carefully layered on top of an antibody solution without mixing. Both the antigen and the antibody will diffuse toward each other. In a positive reaction, a ring of precipitate can be observed at the interface of the two solutions. A negative reaction is indicated by a lack of precipitation. The assay requires positive and negative controls along with questioned samples.

The ***Ouchterlony assay*** is named after the Swedish immunologist who developed it. The assay can be performed in an agarose gel supported by a glass slide or polyester film (Figure 5.7). Wells are created by punching holes in the gel layer at desired locations. Often, a pattern with six wells surrounding a center well is used. The antibody is loaded in the central well while the questioned samples and the controls are loaded in the surrounding wells. The double diffusion of the antigen and the antibody from the wells is allowed to occur during incubation. If the reaction is positive, a precipitate line between wells can be observed at the end of incubation. The precipitate can be stained to aid the observation. A single assay can compare more than one antigen to determine whether the antigens in question react the same way or differently with the antibody (See chapter 7). This method is sometimes used to determine whether samples have come from the same or different origins.

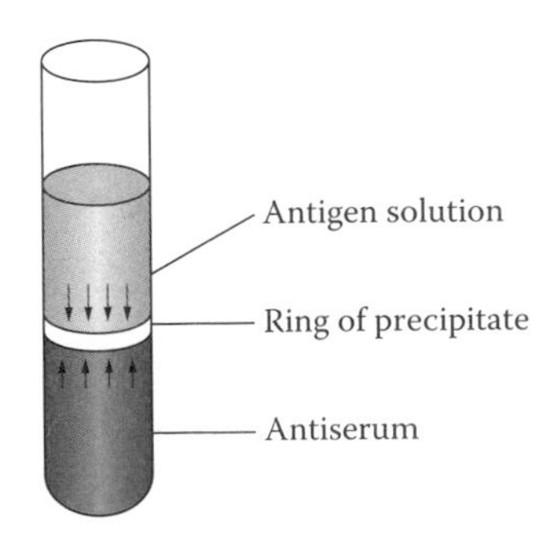

**Figure 5.6** Ring assay.

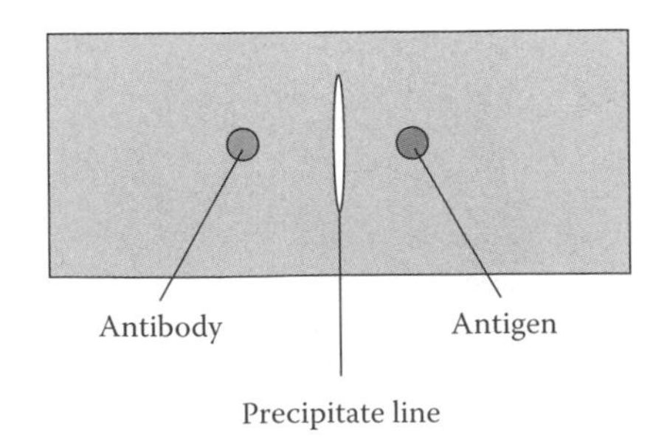


**Figure 5.7** Ouchterlony assay.

### 5.2.1.2 *Electrophoretic Methods*

Diffusion techniques can be combined with electrophoresis to enhance test results. Electrophoresis separates molecules according to differences in their electrophoretic mobility.

**5.2.1.2.1 Immunoelectrophoresis (IEP)** IEP is a two-step procedure that can analyze a wide range of antigens. This technique uses electrophoresis to separate the antigen mixture prior to immunodiffusion (Figure 5.8). In the first step, the antigens in a sample are separated using agarose gel electrophoresis. Then a trough is cut in the gel parallel to the array of the antigens separated by electrophoresis. In the second step, an antibody is loaded in the trough and the gel is incubated for double diffusion. An arc-shaped precipitate line can be observed for a positive reaction. Multiple precipitate lines can occur if more than one antigen reacts with the antibody. The shapes, intensities, and locations of the precipitate lines of a known control and questioned sample can be compared.

**5.2.1.2.2 Crossed Immunoelectrophoresis (CRIE)** CRIE, also known as two-dimensional immunoelectrophoresis, is a modification of IEP. The first step employs electrophoresis to separate antigens contained in a sample (Figure 5.9). In the second step, the gel is turned at a 90-degree angle and further separated by a second dimension electrophoresis. This drives the antigens into an agarose gel that contains uniformly distributed antibodies. Following the second dimension electrophoresis, an arc-shaped precipitate line is

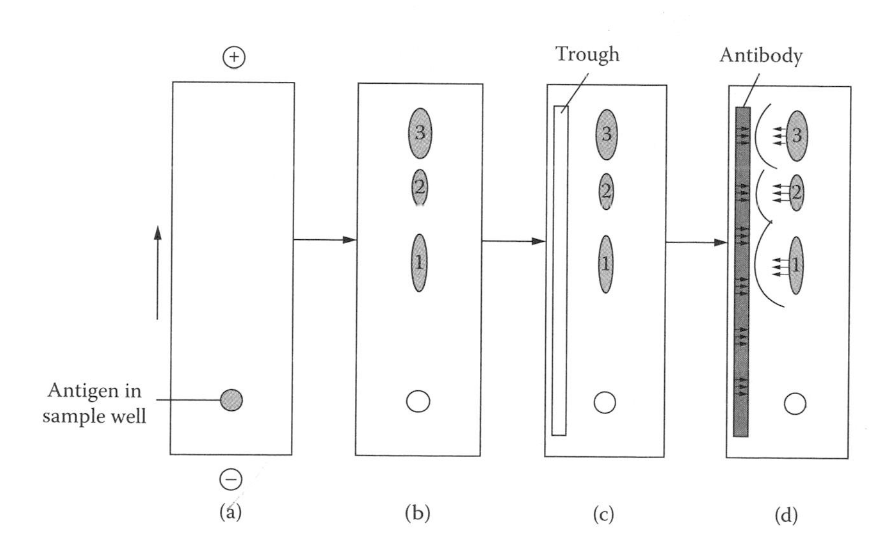


**Figure 5.8** IEP assay. (a) Electrophoresis of antigens is carried out. (b) Various antigens are separated after electrophoresis. (c) Trough is cut. (d) Antibody is applied to the trough, allowing diffusion to occur and forming precipitate lines. + = anode; - = cathode.

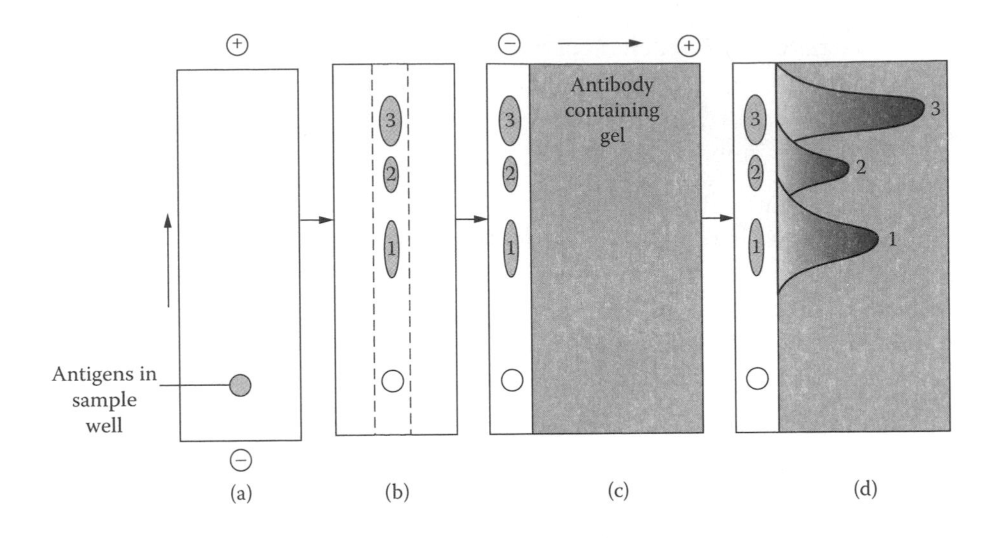


**Figure 5.9** CRIE assay. (a) Electrophoresis of antigens is carried out. (b) Various antigens are separated after electrophoresis. Strip of gel containing separated antigens is cut. (c) Second dimension electrophoresis is carried out to drive the antigens into an antibody-containing gel; (d) arc-shaped precipitate lines are formed as a result.

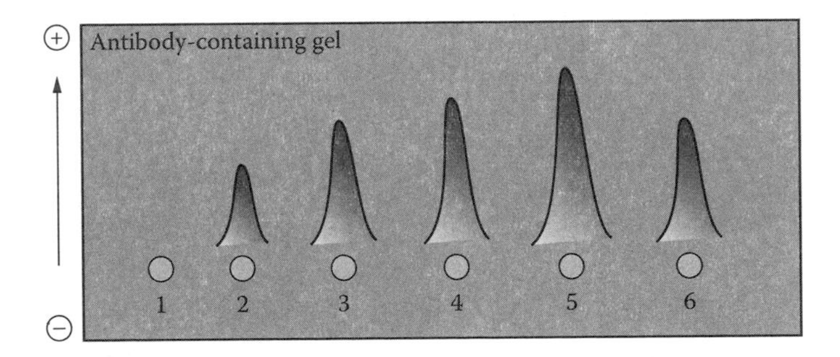


**Figure 5.10** Rocket immunoelectrophoresis assay. Antigens are driven from the wells into an antibody-containing gel forming precipitate lines.

formed. The area of the arc can be measured, and a sample can be identified by comparison to a known standard. This technique is more sensitive than IEP.

**5.2.1.2.3 Rocket Immunoelectrophoresis** An antibody-containing agarose gel is used. The antigen is then loaded into the well (Figure 5.10). Electrophoresis then drives the antigen from the well into the agarose gel. In a positive reaction, a rocket-shaped precipitate line can be observed. The height of the rocket is in proportion to the amount of antigen in the sample. Quantitation can be achieved by comparing standards and the sample in the same gel.

**5.2.1.2.4 Crossed-over Immunoelectrophoresis** This technique is also known as counterimmunoelectrophoresis (CIE). Two arrays of opposing wells are created by punching holes in the agarose gel (Figure 5.11). The antibody and samples are loaded in opposing wells arranged by pairs. Electrophoresis is used to drive the antigen and the antibody toward each other. The wells containing the antibody should be proximal to the anode and wells containing samples should be proximal to the cathode. During gel electrophoresis, the antigen, usually negatively charged, migrates toward the anode. The antibody migrates in the opposite direction as a result of electroendosmosis — a phenomenon in which the movement of molecules is caused by fluid flow. A precipitate line is formed between opposing wells if the antigen reacts with its specific antibody.

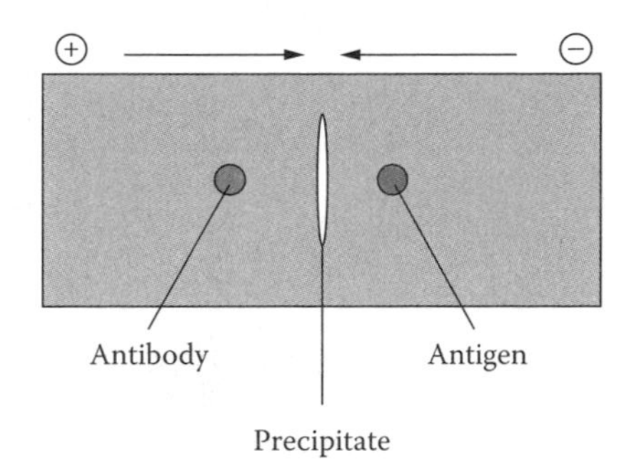


**Figure 5.11** Crossed-over immunoelectrophoresis assay.

## 5.2.2 Agglutination-Based Assays

Agglutination reactions are used as serological assays for blood group typing and species identification in forensic testing. They can identify antigens or antibodies and are more sensitive than precipitation techniques. Agglutination assays are qualitative, indicating the absence or presence of antigens or antibodies. Semiquantitative results can be obtained by titration (diluting the antigen or antibody). Many types of agglutination reactions are available; only those used in forensic serology are discussed below.

### *5.2.2.1 Direct Agglutination Assay*

Direct agglutination involves reactions in which an antibody interacts with antigens originally located on cell surfaces. This technique is different from the passive agglutination reactions in which the antigen is coated on a carrier surface as discussed below.

In a ***hemagglutination*** reaction (Figure 5.12), an antibody binds to the antigens located on erythrocytes. This method is widely used in blood banks and in forensic laboratories for identification of blood types, for example, the testing of erythrocytes for ABO blood group typing and other antigenic components. The assay can be carried out on a glass slide and the agglutination

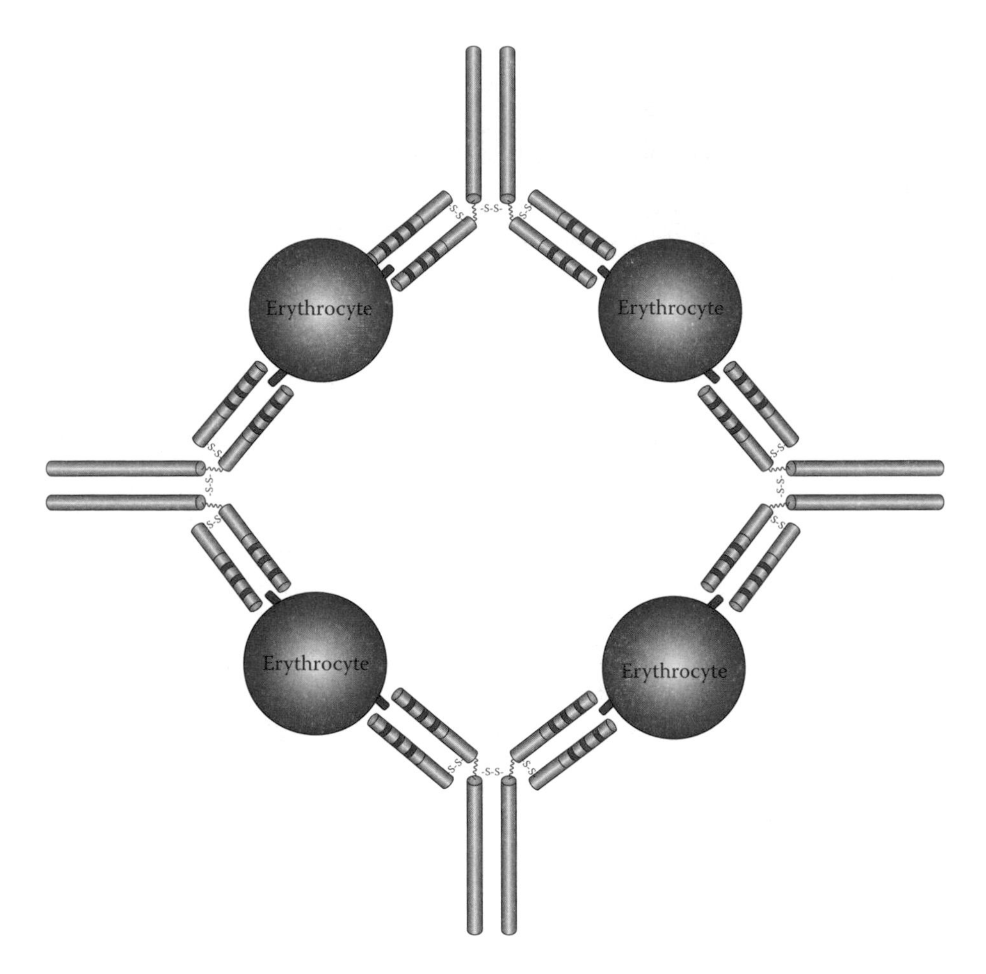


**Figure 5.12** Agglutination-based assay. Hemagglutination reaction is shown; antigens are located on erythrocytes.

can be observed under a microscope. The assay can also be carried out in a test tube. In a positive reaction, agglutinated cells are spread on the bottom of the test tube. Test tubes also can be centrifuged and then swirled to determine whether the cell clumps can be resuspended. Agglutinated cells cannot be resuspended.

### *5.2.2.2 Passive Agglutination Assay*

A coating of antigens can be formed on the surfaces of carriers such as tannic acid-treated erythrocytes. Such cells can then be agglutinated using an antibody specific to the absorbed antigens. The assays that employ coated erythrocytes are also called ***passive hemagglutination tests***. This technique is used for species identification. The erythrocyte carriers are incubated with human blood stain samples. The human antigen-coated erythrocytes can then be agglutinated by antihuman antiglobulins. Conversely, no agglutination occurs if antianimal antiglobulins are applied.

#### 5.2.2.3 Agglutination Inhibition Assay

If a known antigen is added to a mixture consisting of antigen-containing cells and antibody, the added antigen will compete with the antigen located on cell surfaces for antibody binding and inhibit the agglutination reaction. This is an indirect way to identify an antigen in question. A variation of this technique can be used for species identification.

## Bibliography

Allen, S. M. 1995. An enzyme-linked immunosorbent assay (ELISA) for detection of seminal fluid using a monoclonal antibody to prostatic acid phosphatase. *J Immunoassay* 16 (3):297.

Anderson, J. R. 1952. The agglutination of sensitized red cells by antibody to serum, with special reference to non-specific reactions. *Br J Exp Pathol* 71:96.

Berti, A. et al. 2005. Expression of seminal vesicle-specific antigen in serum of lung tumor patients. *J Forensic Sci* 50 (5):1114.

Culliford, B. J. 1964. Precipitin reactions in forensic problems: A new method for precipitin reactions on forensic blood, semen and saliva stains. *Nature* 201:1092.

Hochmeister, M. N. et al. 1999. Evaluation of prostate-specific antigen (PSA) membrane test assays for the forensic identification of seminal fluid. *J Forensic Sci* 44 (5):1057.

Hochmeister, M. N. et al. 1999. Validation studies of an immunochromatographic one-step test for the forensic identification of human blood. *J Forensic Sci* 44 (3):597.

Johnston, S., J. Newman, and R. Frappier. 2003. Validation study of Abacus Diagnostics' ABAcard HemaTrace membrane test for the forensic identification of human blood. *Can Soc Forensic Sci J* 36 (3):173.

Lincoln, P. J., and B. E. Dodd. 1978. The use of low ionic strength solution (LISS) in elution experiments and in combination with papain-treated cells for the titration of various antibodies, including eluted antibody. *Vox Sang* 34 (4):221.

Oakley, C. L., and A. J. Fulthorpe. 1953. Antigenic analysis by diffusion. *J Pathol Bacteriol* 65 (1):49.

Ouchterlony, O. 1949. Antigen–antibody reactions in gels. *Acta Pathol Microbiol Scand* 26:507.

Pang, B. C., and B. K. Cheung. 2007. Identification of human semenogelin in membrane strip test as an alternative method for the detection of semen. *Forensic Sci Int* 169 (1):27.

Petrie, G. F. 1932. A specific precipitin reaction associated with growth on agar plates on Meningococcus, Pneumococcus, and *B. dysenteriae* (shiga). *Br J Exp Pathol* 13:380.

Quarino, L. et al. An ELISA method for the identification of salivary amylase. *J Forensic Sci* 50 (4):873.

Sato, I. et al. 2004. Rapid detection of semenogelin by one-step immunochromatographic assay for semen identification. *J Immunol Methods* 287 (1-2):137.
Sweet, G. H., and J. W. Elvins. 1976. Studies by crossed electroimmunodiffusion on the individuality and sexual orgin of bloodstains. *J Forensic Sci* 21 (3):498.
Wiener, A. S., M. A. Hyman, and L. Handman. 1949. A new serological test (inhibition test) for human serum globulin. *Proc Soc Exp Biol Med* 71:96.
Yoshida, K. et al. 2003. Quantification of seminal plasma motility inhibitor/semenogelin in human seminal plasma. *J Androl* 24 (6):878.

## Study Questions

1. What is the difference between the high-dose hook effect and the prozone effect?

2. What is the difference between precipitation- and agglutination-based assays?

3. The antibody–antigen–antibody sandwich can be formed using:
   (a) Polyclonal antibodies
   (b) A single monoclonal antibody
   (c) Two monoclonal antibodies that recognize different epitopes
   (d) Both polyclonal and monoclonal antibodies

4. Which of the following is more sensitive?
   (a) Primary binding assays
   (b) Secondary binding assays

5. In immunodiffusion assays:
   (a) An antigen-containing gel is usually employed.
   (b) An antibody-containing gel is usually employed.

# Forensic Serology III

# Identification of Blood

# 6

Blood evidence, for example, a dried bloodstain, is commonly associated with criminal investigations. Preliminary and confirmatory assays are performed to ensure that a sample is in fact blood. A positive reaction of a ***preliminary assay*** indicates that the presence of blood is a possibility, but the presumptive assays are not specific only for blood and therefore should not be considered to be conclusive. In contrast, a negative assay suggests that blood is absent. Preliminary assays are quick, sensitive, and easy to perform. Thus, they may be used for screening and preliminary examination of stains prior to utilization of other types of analysis such as forensic DNA testing. Additionally, these assays can be used as search methods to locate stains of possible blood origin at a crime scene. Additional confirmatory assays should be conducted later if necessary.

***Confirmatory assays*** were performed in the past when a sample had to be proven to be blood. Such assays can produce better certainty than that provided by preliminary assays but they are not routinely performed. Blood evidence identified by preliminary assays is usually further analyzed by forensic DNA testing. Additionally, the determination of human or animal origin of blood evidence may be performed if necessary and is discussed in Chapter 7.

## 6.1 Biological Properties

Normal blood volume constitutes about 8% of body weight. ***Plasma*** is the fluid portion of the blood. The cellular portion of the blood consists of ***red blood cells***, ***white blood cells***, and ***platelets***, all of which are suspended in the plasma (Figure 6.1).

**Red Blood Cells** — These cells are also called ***erythrocytes***. Their life span in humans is approximately 3 months. Additionally, mature human erythrocytes do not have nuclei, and therefore lack nuclear DNA. Erythrocytes consist of hemoglobins — proteins responsible for the transportation of oxygen. Most adult human hemoglobin consists of four polypeptide chains, two $\alpha$ chains and two $\beta$ chains. Thus, adult hemoglobin is designated as $\alpha_2\beta_2$. Other forms of hemoglobin will be discussed in Chapter 10. Under normal physiological conditions, each hemoglobin subunit contains a heme moiety that binds to oxygen (Figure 6.2). A heme molecule consists of an organic

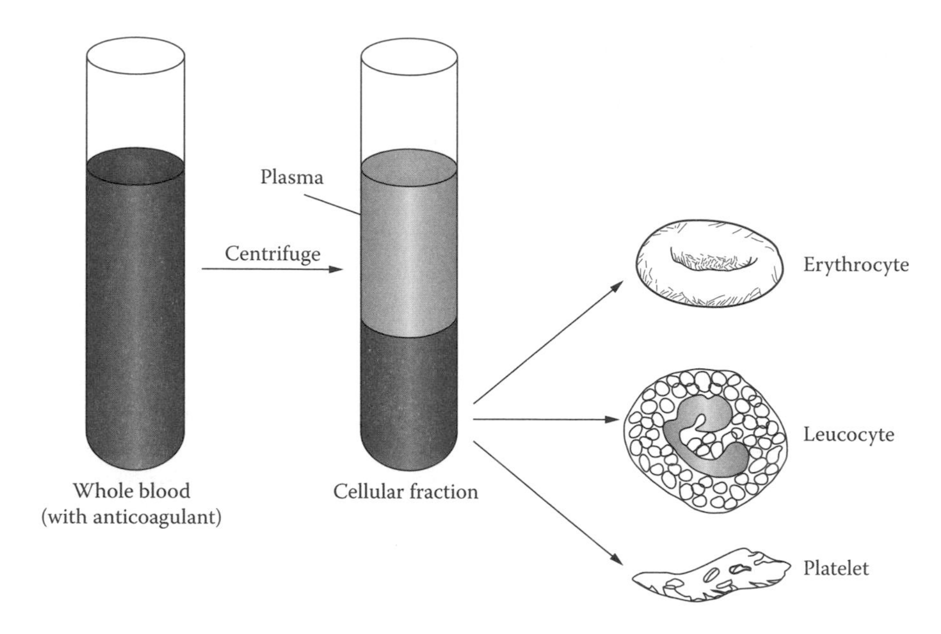


**Figure 6.1** Basic composition of blood. Blood can be separated into two phases in the presence of anticoagulant. The liquid portion called plasma accounts for approximately 55% of blood volume. The cellular elements include erythrocytes, leucocytes, and platelets.

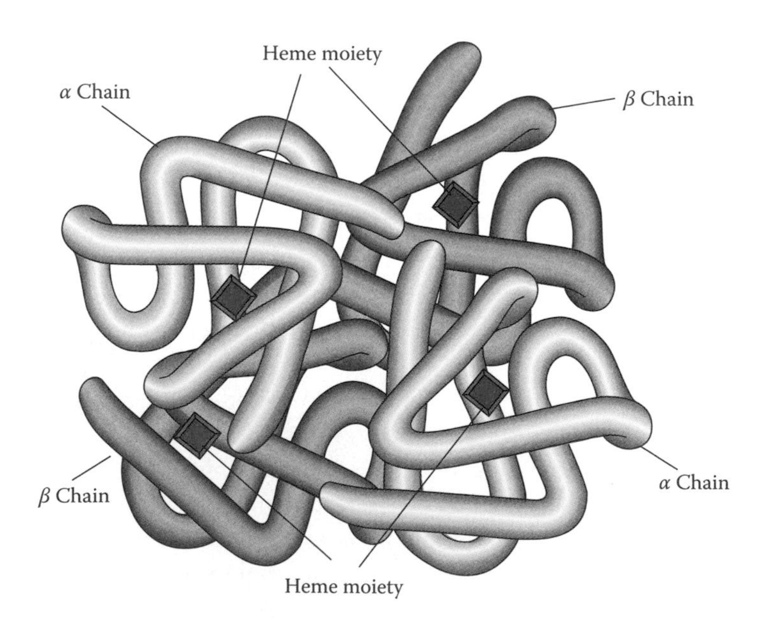


**Figure 6.2** Human adult hemoglobin. Four subunits, two α and two β chains, of human adult hemoglobin are shown. Each hemoglobin subunit contains a heme moiety.

component known as protoporphyrin IX and a ***ferrous*** ($Fe^{2+}$) iron atom (Figure 6.3). A heme molecule is also known as ***ferroprotoporphyrin***. The ferrous ion of heme forms four bonds with the nitrogens of protoporphyrin IX, along with a fifth bond with a hemoglobin chain and a sixth bond with a molecule of $O_2$. Other chemicals such as carbon monoxide and cyanide also bind to the ferrous iron of heme and can cause chemical asphyxia. Heme groups are also present in other proteins such as myoglobin in muscles and neuroglobin in the brain.

**White Blood Cells** — Also called leucocytes, white blood cells are subdivided into three types: granulocytes, lymphocytes, and monocytes. White blood cells are involved in defending the body against infection. They have nuclei and, thus represent main sources for nuclear DNA from the blood.

**Platelets** — These cells are also known as thrombocytes and they play a role in blood clotting. They aggregate at sites of vascular and blood vessel injury. Like red blood cells, they lack nuclei.

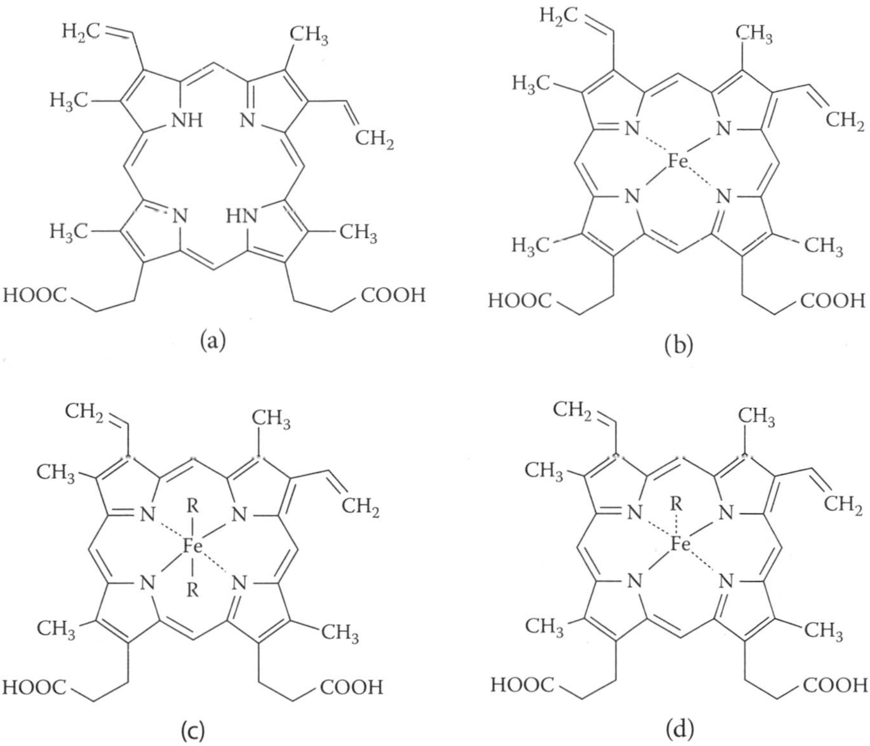

(a) (b) (c) (d)

**Figure 6.3** Chemical structures of heme and its precursor and derivatives. (a) Protoporphyrin IX. (b) Heme (ferroprotoporphyrin). (c) Hemochromagen. R = pyridine (pyridineferroprotoporphyrin. (d) Hematin hydroxide, R = OH (ferriprotoporphyrin hydroxide); hematin chloride, R = Cl (ferriprotoporphyrin chloride).

## 6.2 Presumptive Assays for Identification

### 6.2.1 Mechanisms of Presumptive Assays

Presumptive blood assays are designed to detect traces of blood. The chemistry employed in these assays is an oxidation–reduction reaction catalyzed by heme, a component of hemoglobin. The heme catalyzes various colorless substrates to undergo an oxidation reaction that results in a change of color, chemiluminescence, or fluorescence.

#### *6.2.1.1 Oxidation–Reduction Reactions*

An oxidation–reduction reaction involves changes of the oxidation state. Specifically, the ***oxidation*** of a molecule means the molecule has lost electrons, and ***reduction*** of a molecule means the molecule has gained electrons. Chemicals that can be reduced and therefore gain electrons from other molecules are called ***oxidants***. In contrast, ***reductants*** are chemicals that can be oxidized and therefore lose electrons to other molecules. In biochemical reactions, oxidation often coincides with a loss of hydrogen. Figure 6.4 depicts an example of an oxidation–reduction reaction for blood identification. In the presumptive assays, hydrogen peroxide is usually employed as an oxidant. Heme serves as the catalyst for the oxidation–reduction reaction. In the presence of heme, a colorless substrate is oxidized to a product with color, chemiluminescence, or fluorescence.

### 6.2.2 Colorimetric Assays

Many procedures are available for detecting heme in blood through color reactions. The most common agents are phenolphthalin, leucomalachite green, and benzidine derivatives. The color reactions produced by these assays can be observed immediately with the naked eye. The assays are very sensitive and can detect blood in samples with $10^{-5}$-fold dilutions. A positive reaction indicates the possible presence of blood.

#### *6.2.2.1 Phenolphthalin Assay*

The phenolphthalin assay is also known as the Kastle-Meyer test. Kastle published a study in 1901 presenting the results of a reaction that phenolphthalin,

$$\underset{\text{(Colorless)}}{AH_2} + H_2O_2 \xrightarrow{\text{Heme}} \underset{\text{(Color)}}{A} + 2H_2O$$

**Figure 6.4** Oxidation–reduction reaction as basis for presumptive assays for blood identification. $AH_2$ = substrate. A = oxidized substrate.

a colorless compound, is catalyzed by heme with hydrogen peroxide as the oxidant (Figure 6.5). The oxidized derivative is phenolphthalein which appears pink under alkaline conditions. In addition to its use in detecting blood, phenolphthalein, a member of a class of indicators and dyes, is also used in titrations of mineral and organic acids as well as most alkalies.

### *6.2.2.2 Leucomalachite Green (LMG) Assay*

Malachite green is a triphenylmethane dye. The leuco base form of malachite green is colorless and can be oxidized by the catalysis of heme to produce a green color. The reaction is carried out under acid conditions with hydrogen peroxide as the oxidant (Figure 6.6).

### *6.2.2.3 Benzidine and Derivatives*

Certain members of this group are proven carcinogens and thus not suitable for forensic testing. Tetramethylbenzidine continues to be used as noted below.

**Benzidine** — Historically, benzidine was used as an intermediate in dye manufacturing (Figure 6.7). Subsequently, it was used as a presumptive assay for the presence of blood after the discovery that the oxidation of benzidine

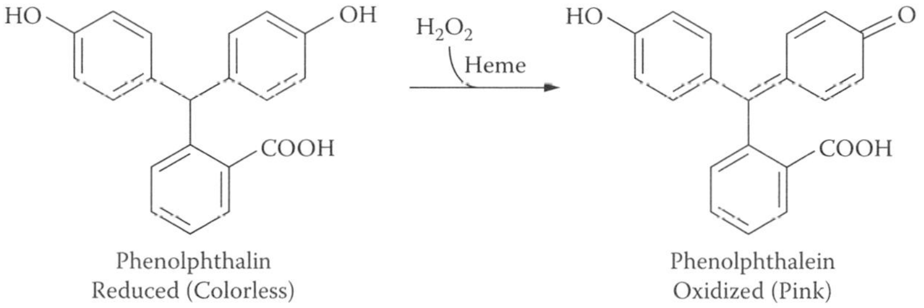

HO OH $H_2O_2$ Heme HO O
COOH COOH
Phenolphthalin
Reduced (Colorless)
Phenolphthalein
Oxidized (Pink)

**Figure 6.5** Chemical reaction of phenolphthalin assay.

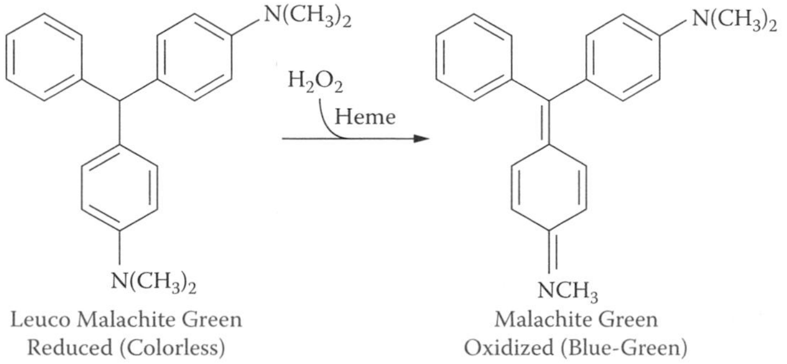

$N(CH_3)_2$ $H_2O_2$ Heme $N(CH_3)_2$
$N(CH_3)_2$ $NCH_3$
Leuco Malachite Green
Reduced (Colorless)
Malachite Green
Oxidized (Blue-Green)

**Figure 6.6** Chemical reaction of leucomalachite green assay.

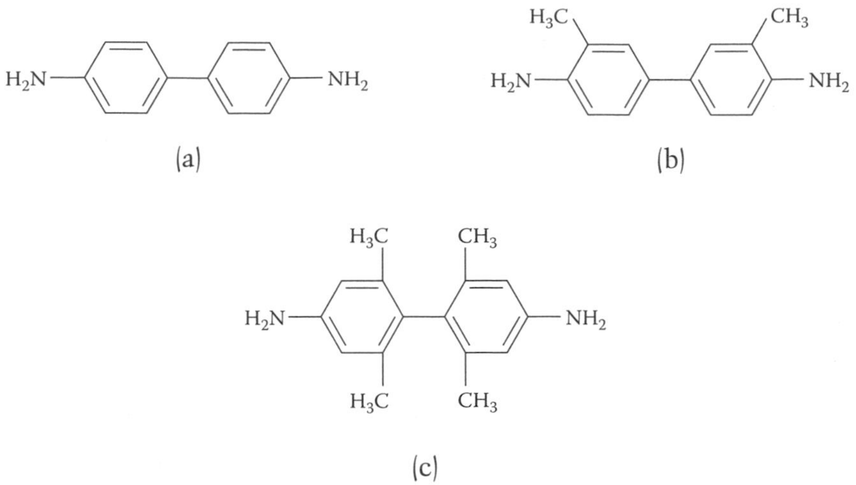

(a)

(b)

(c)

**Figure 6.7** Chemical structures of benzidine and derivatives. (a) Benzidine. (b) Ortho-tolidine. (c) Tetramethylbenzidine.

$H_2O_2$ Heme

ortho-Tolidine Reduced (Colorless)

ortho-Tolidine oxidized (Blue)

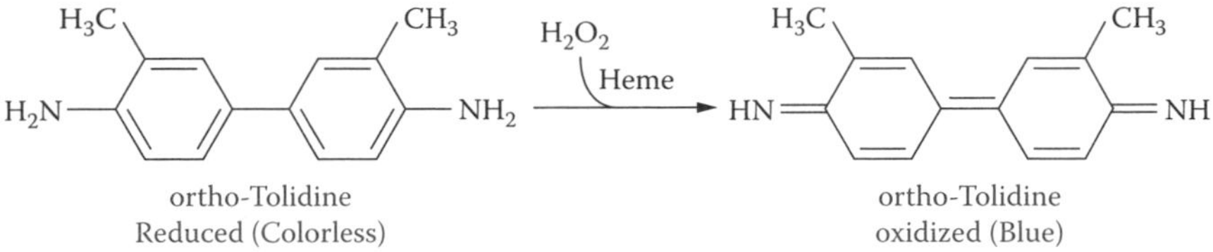

**Figure 6.8** Chemical reaction of ortho-tolidine assay.

can be catalyzed by heme to produce a blue to dark blue color (carried out in an acid solution). Since the blue may eventually turn brown, the reaction must be read immediately. Benzidine was found to be a carcinogen and is therefore no longer used for forensic testing.

**Ortho-Tolidine** — This chemical is a dimethyl derivative of benzidine. Its oxidation reaction can be catalyzed by heme to produce a blue color reaction under acidic conditions (Figure 6.8). Ortho-tolidine is also considered a potential carcinogen based on animal studies, and for this reason has been replaced by tetramethylbenzidine.

**Tetramethylbenzidine (TMB)** — This compound is a tetramethyl derivative of benzidine. The oxidation of TMB can be catalyzed by heme to produce a green to blue-green color under acidic conditions. The Hemastix® assay (Miles Laboratories) is a TMB-based assay kit that employs a strip device containing the TMB reagent. Testing is conducted by applying a moistened sample to the strip device. The immediate color change to green or blue-green indicates the presence of blood.

### 6.2.3 Chemiluminescence and Fluorescence Assays

Other organic compounds whose oxidation products have chemiluminescent or fluorescent properties are employed for testing. In the ***chemiluminescence assay***, light is emitted as a product of a chemical reaction. Luminol produces chemiluminescence and is thus included in this category. In contrast, a ***fluorescence assay*** requires exposure of an oxidized product, such as fluorescin, to a particular wavelength of light. The fluorenscence is then emitted at longer wavelengths.

One advantage of chemiluminescent and fluorescent reagents is that they can be sprayed over large areas where bloodstains are suspected. A positive reaction identifies blood and also reveals patterns such as footprints, fingerprints, and splatters. These methods are very sensitive and can pinpoint the locations of even small traces of blood. Additionally, they are useful for detecting blood at crime scenes that have been cleaned and show no visible staining. One disadvantage of chemiluminescent and fluorescent reagents is that precautions must be taken if a sample is very small or has been washed. The spraying of presumptive assay reagents may dilute a sample and thus lead to a low DNA yield for generating a DNA profile.

#### *6.2.3.1 Luminol (3-aminophthalhydrazide)*

The oxidation products of Luminol have chemiluminescent properties. Oxidation of Luminol catalyzed by heme produces an intense light (Figure 6.9). The light emitted from a positive reaction can only be observed in a dark environment and this limits the applications of Luminol.

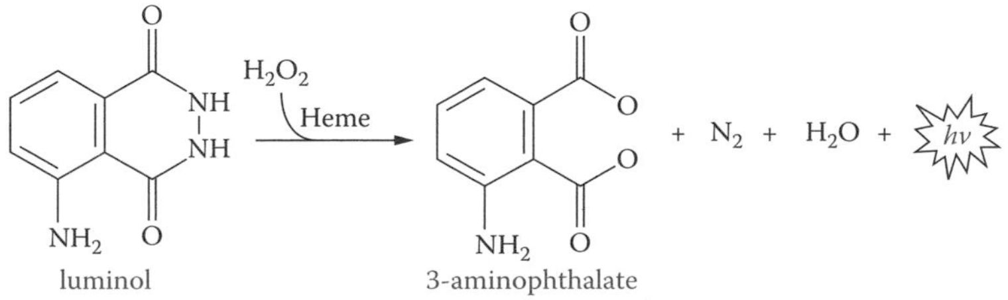

**Figure 6.9** Chemical reaction of Luminol assay.

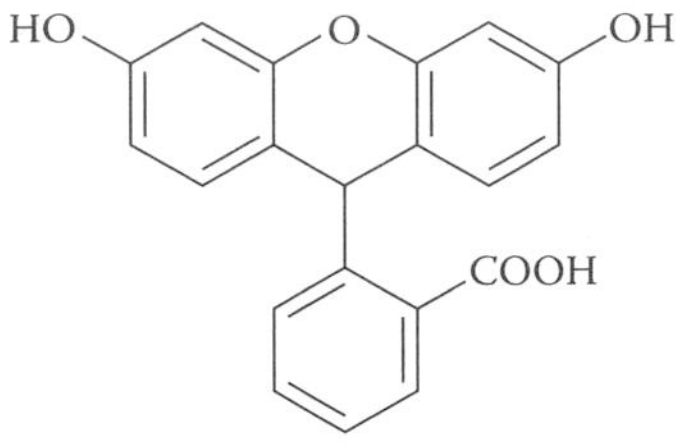

#### 6.2.3.2 *Fluorescin*

Fluorescin is another reagent used to test for the presence of bloodstains at a crime scene (Figure 6.10). When oxidized and catalyzed by heme, it demonstrates fluorescent properties. Usually, fluorescin-sprayed stains are exposed to light in the range of 425 to 485 nm using an alternate light source device. In a positive reaction, the oxidized fluorescin will emit an intense yellowish-green fluorescent light and therefore indicate the presence of a blood stain.

HO O OH COOH

**Figure 6.10** Chemical structure of fluorescein.

### 6.2.4 Factors Affecting Presumptive Assay Results

The catalytic assays discussed above are very sensitive but not specific, possibly leading to observation of false positive or false negative results (Figure 6.11).

**Oxidants** — A false positive reaction may be caused by chemicals that are strong oxidants. Such chemicals can catalyze the oxidation reaction even in the absence of heme and result in a false positive reaction. Certain metal salts, bleaches, and household cleaners work as oxidants. To address this problem, a two-step catalytic assay may be performed. The substrate is applied first to the sample in question. A color change occurring before the addition of hydrogen peroxide indicates a false positive result due to a possible oxidant in the sample. If a color change is observed after the addition of hydrogen peroxide, the result is a true positive.

**Plant Peroxidases** — Many types of plants such as horseradish contain peroxidases. Plant peroxidases may also catalyze oxidation reactions and lead to false positive results. However, plant peroxidases are usually heat sensitive and may be inactivated by high temperatures. Because the heme molecule is relatively stable at high temperatures, samples can be retested after heating. This will inactivate any plant peroxidases in a sample.

$$\underset{\text{(Colorless)}}{AH_2} + H_2O_2 \underset{\text{(3) Reductant}}{\overset{\text{(1) Oxidant} \atop \text{(2) Peroxidase}}{\rightleftharpoons}} \underset{\text{(Color)}}{A} + 2H_2O$$

**Figure 6.11** Factors affecting presumptive assay results. Strong oxidant and peroxidase may cause false positive results; reductant may cause false negative results.

**Reductants** — Although not common, a false negative result can occur when a strong reductant is present in a sample. The strong reductant may inhibit the oxidation reaction.

## 6.3 Confirmatory Assays for Identification

### 6.3.1 Microcrystal Assays

Microcrystal assays involve adding chemicals to a bloodstain to form distinctive crystals of heme derivatives. The morphologies of the resulting crystals can be visualized under a microscope. Erythrocytes contain heme and a positive crystal assay is a strong indication of the presence of blood. However, the confirmatory assays usually are not as sensitive as the presumptive assays.

#### *6.3.1.1 Hemochromagen Crystal Assay*

Hemochromagens are heme derivatives in which the ferrous iron of the heme forms two bonds with nitrogenous bases (Figure 6.3). Stockes documented the method for forming hemochromagen crystals in 1864. Since then, various modifications have been reported. The ***Takayama crystal assay*** published in 1912 is now the method preferred by many forensic laboratories. A blood stain is treated with pyridine and glucose (a reducing sugar capable of reducing ferric ion) under alkaline conditions to form crystals of pyridine ferroprotoporphyrin (Figure 6.12).

#### *6.3.1.2 Hematin Crystal Assay*

This test is also known as the ***Teichmann crystal assay.*** In 1853, Teichmann documented a method of forming crystals of blood specimens. A prismatic

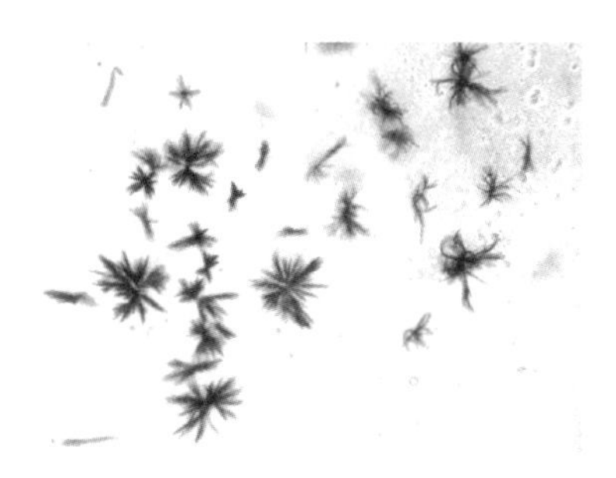

**Figure 6.12** Microcrystal assays using Takayama method. (Source: James, S. H., and J. J. Nordby. 2005. *Forensic Science: An Introduction to Scientific and Investigative Techniques,* 2nd ed. Boca Raton: CRC Press. With permission.)

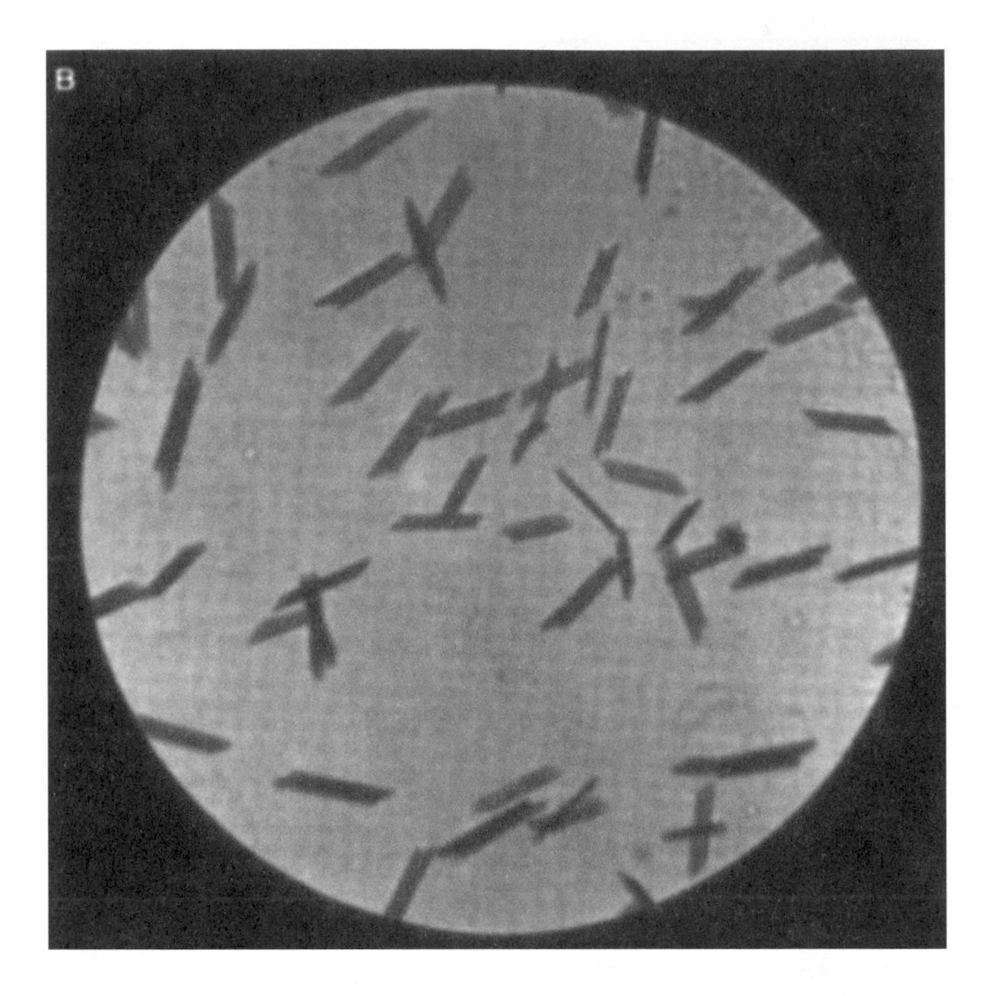


**Figure 6.13** Microcrystal assays using Teichmann method. (*Source*: James, S. H., and J. J. Nordby. 2005. *Forensic Science: An Introduction to Scientific and Investigative Techniques,* 2nd ed. Boca Raton: CRC Press. With permission.)

brown colored crystal was formed when specimens were treated with glacial acetic acid and salts and then heated (Figure 6.13). Hematin chloride (ferriprotoporphyrin chloride) crystals are formed. Hematin (Figure 6.3) is a heme derivative; its iron is in the ferric ($Fe^{+3}$) state. The hemochromagen and hematin assays have similar sensitivity and specificity. The hematin assay has the advantage of being more reliable for aged blood samples.

### 6.3.2 Other Assays

Additional techniques may be used to confirm the presence of hemoglobin. For example, ***chromatographic*** and ***electrophoretic methods*** can identify

**Table 6.1 Application of RT-PCR Assay for Blood Identification**

| Gene Symbol | Gene Product | Description | Further Reading |
|---|---|---|---|
| *HBA1* | Hemoglobin α1 | Hemoglobin α1 chain (abundant in erythrocytes) | Waye and Chui, 2001 |
| *PBGD** | Porphobilinogen deaminase | Erythrocyte-specific isoenzyme of heme biosythesis pathway | Gubin and Miller, 2001 |
| *SPTB* | β-Spectrin | Subunit of major protein of erythrocyte membrane skeleton | Amin *et al.*, 1993 |

(*Source:* Adapted from Jussola, J. and J., Ballantyne. 2005. *Forensic Sci Int* 152:1; Nussbaumer, C. E. *et al.* 2006. *Forensic Sci Int* 157, 181.)

* Also known a *HMBS* (hydroxymethylbilane synthase).

hemoglobin by its mobility characteristics. ***Spectrophotometric methods*** for identifying hemoglobin are based on measurements of the characteristic light spectra absorbed by hemoglobin and its derivatives. Finally, the ***immunological methods*** utilize antihuman hemoglobin antibodies. This antibody can be used to detect human hemoglobin and thus indicate the presence of human blood (see Chapter 7).

***RNA-based assays*** have been developed recently to identify blood. The assays are based on the fact that certain genes are specifically expressed in certain cell types. Thus, the techniques used in the identification of blood are based on the detection of specific types of messenger RNA (mRNA) expressed exclusively in erythrocytes. These assays utilize reverse transcriptase polymerase chain reaction (RT-PCR) methods to detect gene expression levels of mRNAs for blood identification. Table 6.1 summarizes the tissue-specific genes utilized for blood identification. Compared to conventional assays used for blood identification, the RNA-based assays have higher specificity and are amenable to automation. However, one limitation is that the RNA is unstable due to degradation by endogenous ribonucleases.

## Bibliography

Cheeseman, R. 1999. Direct sensitivity comparison to the fluorescein and Luminol bloodstain enhancement techniques. *J Forensic Ident* 49 (3):8.

Cheeseman, R., and L. A. DiMeo. 1995. Fluorescein as a fieldworthy latent bloodstain detection system. *J Forensic Ident* 45 (6):1.

Cheeseman, R., and R. Tomboc. 2001. Fluorescein technique performance study on bloody foot trails. *J Forensic Ident* 51 (1):12.

Cox, M. 1990. Effect of fabric washing on the presumptive identification of bloodstains. *J Forensic Sci* 35 (6):1335.

_____ . 1991. A study of the sensitivity and specificity of four presumptive tests for blood. *J Forensic Sci* 36 (5):1503.

Creamer, J. I. et al. 2003. A comprehensive experimental study of industrial, domestic and environmental interferences with the forensic Luminol test for blood. *Luminescence* 18 (4):193.

Creamer, J. I. et al. 2005. Attempted cleaning of bloodstains and its effect on the forensic Luminol test. *Luminescence* 20 (6):411.

DeForest, P., R. E. Gaensslen, and H. C. Lee. 1983. *Forensic Science: An Introduction to Criminalistics*. New York: McGraw-Hill.

Dixon, T. R. et al. 1976. A scanning electron microscope study of dried blood. *J Forensic Sci* 21 (4):797.

Dorward, D. W. 1993. Detection and quantitation of heme-containing proteins by chemiluminescence. *Anal Biochem* 209 (2):219.

Gaensslen, R. E. 1983. *Sourcebook in Forensic Serology, Immunology, and Biochemistry*. Washington, D.C.: U.S. Government Printing Office.

Garner, D. D. et al. 1976. An evaluation of tetramethylbenzidine as a presumptive test for blood. *J Forensic Sci* 21 (4):816.

Gimeno, F. E., and G. E. Rini. 1989. Fill flash luminescence to photograph Luminol blood stain patterns. *J Forensic Ident* 39 (3):1.

Grispino, R. R. J. 1990. The effect of Luminol on the serological analysis of dried human bloodstains. *Crime Lab Digest* 17 (1): 13.

Gross, A. M., K. A. Harris, and G. L. Kaldun. 1999. The effect of Luminol on presumptive tests and DNA analysis using the polymerase chain reaction. *J Forensic Sci* 44 (4):837.

Hatch, A. L. 1993. A modified reagent for the confirmation of blood. *J Forensic Sci* 38 (6):1502.

Higaki, R. S., and W. M. S. Philip. 1976. A study for the sensitivity, stability, and specificity of phenolphthalein as an indicator test for blood. *Can Soc Forensic Sci* 9 (3): 97.

Introna, F., Jr., G. DiVella, and C. P. Campobasso. 1999. Determination of postmortem interval from old skeletal remains by image analysis of Luminol test results. *J Forensic Sci* 44 (3):535.

Kent, E. J., D. A. Elliot, and G. M. Miskelly. 2003. Inhibition of bleach-induced Luminol chemiluminescence. *J Forensic Sci* 48 (1):64.

Laux, D. L. The detection of blood using Luminol, in *Principles of Bloodstain Pattern Analysis: Theory and Practice,* Ed. James, R., P. Kish, and T. Sutton, 369–389. Boca Raton, FL: CRC Press.

Laux, D. L. 1991. Effects of Luminol on the subsequent analysis of bloodstains. *J Forensic Sci* 36 (5):1512.

Lee, H. C. 1982. Identification and grouping of bloodstains, in *Forensic Science Handbook*. Saferstein, R., Ed. 267–337. Englewood Cliffs, NJ: Prentice Hall.

Lee, H. C., T. Palmbach, and M. T. Miller. 2001. *Henry Lee's Crime Scene Handbook*. San Diego: Academic Press.

Lytle, L. T., and D. G. Hedgecock. 1978. Chemiluminescence in the visualization of forensic bloodstains. *J Forensic Sci* 23 (3):550.

Ponce, A. C., and F. A. Verdu-Pascual. 1999. Critical revision of presumptive tests for bloodstains. *Forensic Sci Commun* 1 (2): 1.

Quickenden, T. I., and P. D. Cooper. 2001. Increasing the specificity of the forensic Luminol test for blood. *Luminescence* 16 (3):251.
Quickenden, T. I., and J. I. Creamer. 2001. A study of common interferences with the forensic Luminol test for blood. *Luminescence* 16 (4):295.
Quickenden, T. I., C. P. Ennis, and J. I. Creamer. 2004. The forensic use of Luminol chemiluminescence to detect traces of blood inside motor vehicles. *Luminescence* 19 (5):271.
Shaler, R. C. 2002. Modern forensic biology, in *Forensic Science Handbook*. Saferstein, R., Ed. Upper Saddle River: Person Education Inc.
Shipp, E. et al. 1993. Effects of argon laser light, alternate source light, and cyanoacrylate fuming on DNA typing of human bloodstains. *J Forensic Sci* 38 (1):184.
Spalding, R. 2005. The identification and characterization of blood and bloodstains, in *Forensic Science: An Introduction to Scientific and Investigative Techniques*. James, S. and J. J. Nordby, Eds. Boca Raton, FL: CRC Press.
Stoilovic, M. 1991. Detection of semen and blood stains using Polilight as a light source. *Forensic Sci Int* 51 (2):289.
Sutton, T. P. 1999. Presumptive blood testing, in *Scientific and Legal Applications of Bloodstain Pattern Interpretation*. James, S. H., Ed. Boca Raton, FL: CRC Press.
Tsutsumi, H. and Y. Katsumata. 1993. Forensic study on stains of blood and saliva in a chimpanzee bite case. *Forensic Sci Int* 61 (2–3):101.
Vandenberg, N., and R. A. van Oorschot. 2006. The use of Polilight in the detection of seminal fluid, saliva, and bloodstains and comparison with conventional chemical-based screening tests. *J Forensic Sci* 51 (2):361.
Webb, J. L., J. I. Creamer, and T. I. Quickenden. 2006. A comparison of the presumptive Luminol test for blood with four non-chemiluminescent forensic techniques. *Luminescence* 21 (4):214.
Zweidinger, R. A., L. T. Lytle, and C. G. Pitt. 1973. Photography of bloodstains visualized by Luminol. *J Forensic Sci* 18 (4):296.

## Effects on DNA Analysis

Andersen, J., and S. Bramble. 1997. The effects of finger mark enhancement light sources on subsequent PCR-STR DNA analysis of fresh bloodstains. *J Forensic Sci* 42 (2):303.
Barnett, P. D. et al. 1992. Discussion of "Effects of presumptive test reagents on the ability to obtain restriction fragment length polymorphism (RFLP) patterns from human blood and semen stains." *J Forensic Sci* 37 (2):369.
Budowle, B. et al. 2000. The presumptive reagent fluorescein for detection of dilute bloodstains and subsequent STR typing of recovered DNA. *J Forensic Sci* 45 (5):1090.
Caldwell, J. P., W. Henderson, and N. D. Kim. 2000. ABTS: A safe alternative to DAB for the enhancement of blood fingerprints. *J Forensic Sci* 45 (4):785.
Della Manna, A., and S. Montpetit. 2000. A novel approach to obtaining reliable PCR results from Luminol-treated bloodstains. *J Forensic Sci* 45 (4):886.
Fregeau, C. J., O. Germain, and R. M. Fourney. 2000. Fingerprint enhancement revisited and the effects of blood enhancement chemicals on subsequent Profiler Plus fluorescent short tandem repeat DNA analysis of fresh and aged bloody fingerprints. *J Forensic Sci* 45 (2):354.

Gross, A. M., K. A. Harris, and G. L. Kaldun. 1999. The effect of Luminol on presumptive tests and DNA analysis using the polymerase chain reaction. *J Forensic Sci* 44 (4):837.
Hochmeister, M. N., B. Budowle, and F. S. Baechtel. 1991. Effects of presumptive test reagents on the ability to obtain restriction fragment length polymorphism (RFLP) patterns from human blood and semen stains. *J Forensic Sci* 36 (3):656.
Tobe, S. S., N. Watson, and N. N. Daeid. 2007. Evaluation of six presumptive tests for blood, their specificity, sensitivity, and effect on high molecular-weight DNA. *J Forensic Sci* 52 (1):102.

### RNA-Based Assays

Amin, K. M. et al. 1993. The exon–intron organization of the human erythroid beta-spectrin gene. *Genomics* 18 (1):118.
Chu, Z. L. et al. 1994. Erythroid-specific processing of human beta spectrin I pre-mRNA. *Blood* 84 (6):1992.
Gubin, A. N., and J. L. Miller. 2001. Human erythroid porphobilinogen deaminase exists in two splice variants. *Blood* 97 (3):815.
Juusola, J., and J. Ballantyne. 2003. Messenger RNA profiling: a prototype method to supplant conventional methods for body fluid identification. *Forensic Sci Int* 135 (2):85.
____. 2005. Multiplex mRNA profiling for the identification of body fluids. *Forensic Sci Int* 152 (1):1.
Nussbaumer, C., E. Gharehbaghi-Schnell, and I. Korschineck. 2006. Messenger RNA profiling: A novel method for body fluid identification by real-time PCR. *Forensic Sci Int* 157 (2-3):181.

## Study Questions

1. The presumptive test for blood:
   (a) Is human-specific
   (b) Is mammal-specific
   (c) Is primate-specific
   (d) Is none of the above

2. A household cleaning product may react with Luminol and lead to:
   (a) A false positive result
   (b) A false negative result

3. A false positive blood test may be caused by:
   (a) Household cleaning agents
   (b) Certain vegetables
   (c) Reductants
   (d) Both A and B

4. The crystal test for blood:
   (a) Is human-specific
   (b) Is mammal-specific
   (c) Is primate-specific
   (d) Is none of the above

5. A stain can presumptively be identified as blood by the:
   (a) Amelogenin test
   (b) Kastle-Meyer test
   (c) Acid phosphatase test
   (d) Amylase test

6. Which of the following is true?
   (a) Heme is a component of hemoglobin.
   (b) Hemoglobin is located in the red blood cells.
   (c) Hemoglobin is responsible for carrying oxygen.
   (d) All of the above.

7. Which of the following may be used in the dark?
   (a) Kastle-Meyer test
   (b) LMG test
   (c) Luminol test
   (d) Teichmann test

8. Which of the following applies to the identification of blood?
   (a) The Takayama assay acts on hemoglobin.
   (b) The acid phosphatase test acts on hemoglobin.
   (c) The amylase assay acts on hemoglobin.
   (d) The amelogenin test acts on hemoglobin.

9. Which of the following is true?
   (a) A positive Kastle-Meyer test produces a pink color.
   (b) A positive Kastle-Meyer test produces a green color.
   (c) A positive LMG test produces a pink color.
   (d) None of the above.

10. Presumptive blood tests are usually:
    (a) Sensitive
    (b) Not sensitive
    (c) Specific
    (d) Labor-intensive

# 7 Species Identification

Chapter 6 discusses the principles of the identification of blood. If a stain is identified as blood, the evidence can be tested to determine whether the blood is of human origin. If the blood stain is non-human, further analysis is not necessary.

Before forensic DNA techniques were implemented, species identification was largely determined by serological methods. Most forensic laboratories currently perform DNA quantitation prior to DNA profile analysis. The quantitation method specifically detects higher primate DNA. The DNA revealed by the quantitation tests concurrently identifies a sample as having human origin (since crimes involving primate blood are extremely rare). Thus, species identification is usually not performed in forensic laboratories.

Nevertheless, species identification assays can be useful for screening to exclude or eliminate nonhuman samples unrelated to an investigation. Thus, it is practical for small laboratories to eliminate unnecessary analyses due to time and budget constraints. Additionally, species identification kits such as immunochromatographic devices allow field testing by crime scene investigators. In cases involving animals, it may be necessary to identify the animal species prior to further analysis.

## 7.1 General Considerations

Most assays for species identification are based on serological techniques including primary and secondary binding assays. The most common primary binding assays are immunochromatographic. The most commonly used secondary binding assays are precipitation-based assays that rely on the binding of an antigen to an antibody, causing the formation of visible precipitation. These include ring assays, Ouchterlony assays, and crossed-over immunoelectrophoresis. These assays employ antihuman and antianimal antibodies to identify human and animal species, respectively.

### 7.1.1 Types of Antibodies

An antihuman antibody used in the identification of human samples is made by introducing human serum into a host animal that then produces specific antibodies against the human serum proteins. The blood is then removed from the host animal and the serum portion is collected. The collected serum is a polyclonal antihuman antiserum containing a mixture of antibodies against various human serum proteins. Since albumin is the most abundant protein in human serum, the antihuman antibody reacts strongly with human albumin. Additionally, antibodies produced from different species of host animals may produce variations in the characteristics of reactions. Likewise, an antibody against animal serum proteins can also be made to identify animal species of interest.

Other antibodies such as antihuman hemoglobin (Hb) antibodies can also be used to identify the human origin of a sample. Hb is an oxygen transport protein found in erythrocytes. Purified Hb can be used to generate monoclonal and polyclonal antihuman Hb antibodies. Likewise, antibodies recognizing glycophorin A (GPA), a human erythrocyte membrane antigen, can also be produced in a similar manner.

### 7.1.2 Titration of Antibodies

Recall that the ratio of antigen to antibody is critical for the success of a secondary reaction. Extreme excess antigen or antibody concentrations can inhibit secondary reactions. The prozone and postzone phenomena must be considered and the concentrations of antigen and antibody must be carefully determined for forensic serology assays. For instance, in the prozone situation, a false negative reaction may occur due to the presence of a high concentration of antibody.

Quality control procedures can be used to estimate the amount of specific antibody present, often via titration (Figure 7.1). To titrate an antiserum, a series of dilutions are made and each dilution is then tested for activity using precipitation or agglutination methods. The reciprocal of the highest dilution giving a positive reaction is known as the ***titer***. This reflects the amount of antibody in the antiserum. Additionally, the polyclonal antiserum is a mixture of antibodies; thus the reaction of the antiserum may vary from animal to animal (of the same species). Each lot or batch of antiserum must be validated by titration.

### 7.1.3 Antibody Specificity

In addition to titer, the specificity of the antihuman antibody must be tested. Most antihuman antibody usually have cross-reactivity with higher primates. This is not a great concern because crimes involving non-human primates are very rare. Nevertheless, the antihuman antibody must not cross-react with other commonly encountered animals. All lots of antisera and positive control samples must be validated for cross-reactivity. Tissue specificity must also be validated. The antiserum against human serum is usually reactive with other human biological fluids such as semen and saliva.

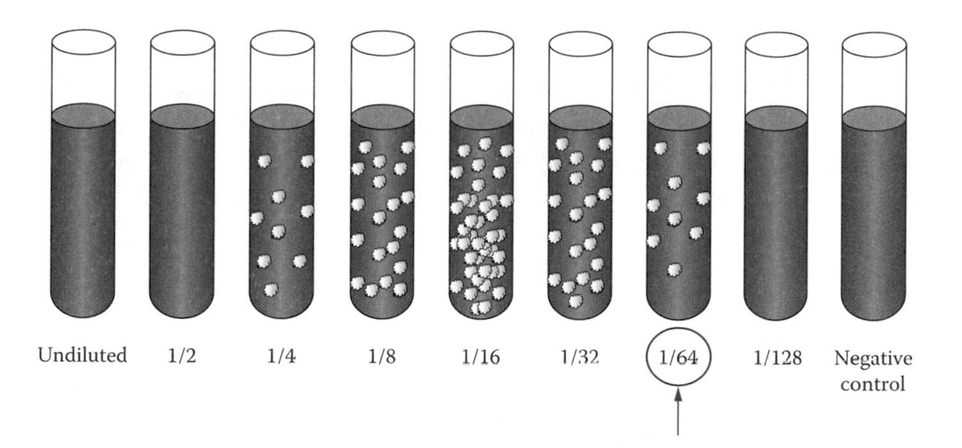


**Figure 7.1** Titration of antibodies. Serum is serially diluted and a constant amount of antigen is applied to each tube. The mixture is incubated, allowing agglutination to occur. The reciprocal of the highest dilution giving a positive agglutination reaction is 64 (the titer).

### 7.1.4 Optimal Conditions for Antigen–Antibody Binding

Antigen–antibody binding can be affected by a number of factors. For example, increasing ionic strength can inhibit the binding of antigen and antibody. Stronger inhibition is usually observed for ions with large ionic radii and small radii of hydration. It is believed that the lower degree of hydration permits interactions of ions and the antibody binding site, leading to the inhibition. A proper buffer system must be selected in serological assays to ensure reliable results. The introduction of polymers can facilitate precipitation in secondary binding reactions because the presence of a polymer in solution decreases the solubility of proteins. Linear hydrophilic polymers with high molecular weights (e.g., such as polyethylene glycol) are preferred. Additional factors such as temperature and pH can also affect antigen–antibody binding. The extent of effect of temperature and pH varies depending on the antibody used.

### 7.1.5 Sample Preparation

Samples can be prepared by cutting out a portion of a stain or scraping stains from a surface. A sample is usually extracted with a small volume of saline or buffer. The extracted sample can be tested using the assays described below. Controls should be included, e.g., using a known human serum as a positive control and an extraction blank as a negative control.

## 7.2 Assays

### 7.2.1 Immunochromatographic Assays

Immunochromatographic assays are rapid, specific, and sensitive and can be used in both laboratory and field tests for species identification. Two types of assays are discussed, including those based on the detection of human erythrocyte proteins. Chapter 5 discusses the principle of immunochromatographic assays in more detail.

#### *7.2.1.1 Identification of Human Hemoglobin Protein*

Commercially produced immunochromatographic kits such as the Hexagon OBTI (Human Gesellschaft fuer Biochemica und Diagnostica mbH) and the ABAcard HemaTrace® (Abacus Diagnostics, California) are now available. They employ the antibody–antigen–antibody sandwich method by using antibodies that recognize human Hb. The ABAcard HemaTrace assay utilizes a labeled monoclonal antihuman Hb antibody contained in a sample well and a polyclonal antihuman Hb antibody immobilized at a test zone of a nitrocellulose membrane. Additionally, an antiglobulin that recognizes the antibody is immobilized onto a control zone (Figure 7.2).

A sample can be prepared by cutting a small portion (2 mm diameter) of a stain or swab. Each sample is extracted for 5 min in 2 ml of extraction buffer. Longer extraction time may be used for older stains. The samples are loaded into the sample well and the antigen in the sample binds to the labeled antibody in the well to form an antigen–antibody complex that then diffuses across a nitrocellulose membrane. At the test zone, the solid phase antihuman Hb antibody binds to the antigen–antibody complex to form a labeled antibody-antigen-antibody sandwich.

The ABAcard HemaTrace® uses a pink dye that is visualized in a positive result as a pink vertical line at the test zone. In the control zone, unbound labeled antihuman Hb antibody binds to the solid phase antiglobulin. This antibody–antiglobulin complex at the control zone also produces a pink vertical line. The test is considered valid only if the line in the control zone is observed. The presence of human Hb results in a pink line at both the test and control zones. The absence of human Hb results in a pink line in the control zone only. A positive result can appear in less than a minute.

Validation studies have revealed that the sensitivity of the ABAcard HemaTrace® can be as low as 0.07 μg/ml of Hb (the average Hb concentration is 130 mg/ml in blood among healthy females). It is more sensitive than the Kastle-Meyer assay. Additionally, the assay was responsive to aged stains and degraded materials and was not affected by various contaminants and substrates. Specificity studies have shown that it is specific for blood of higher primates including humans. However, it is also responsive to seminal stains,

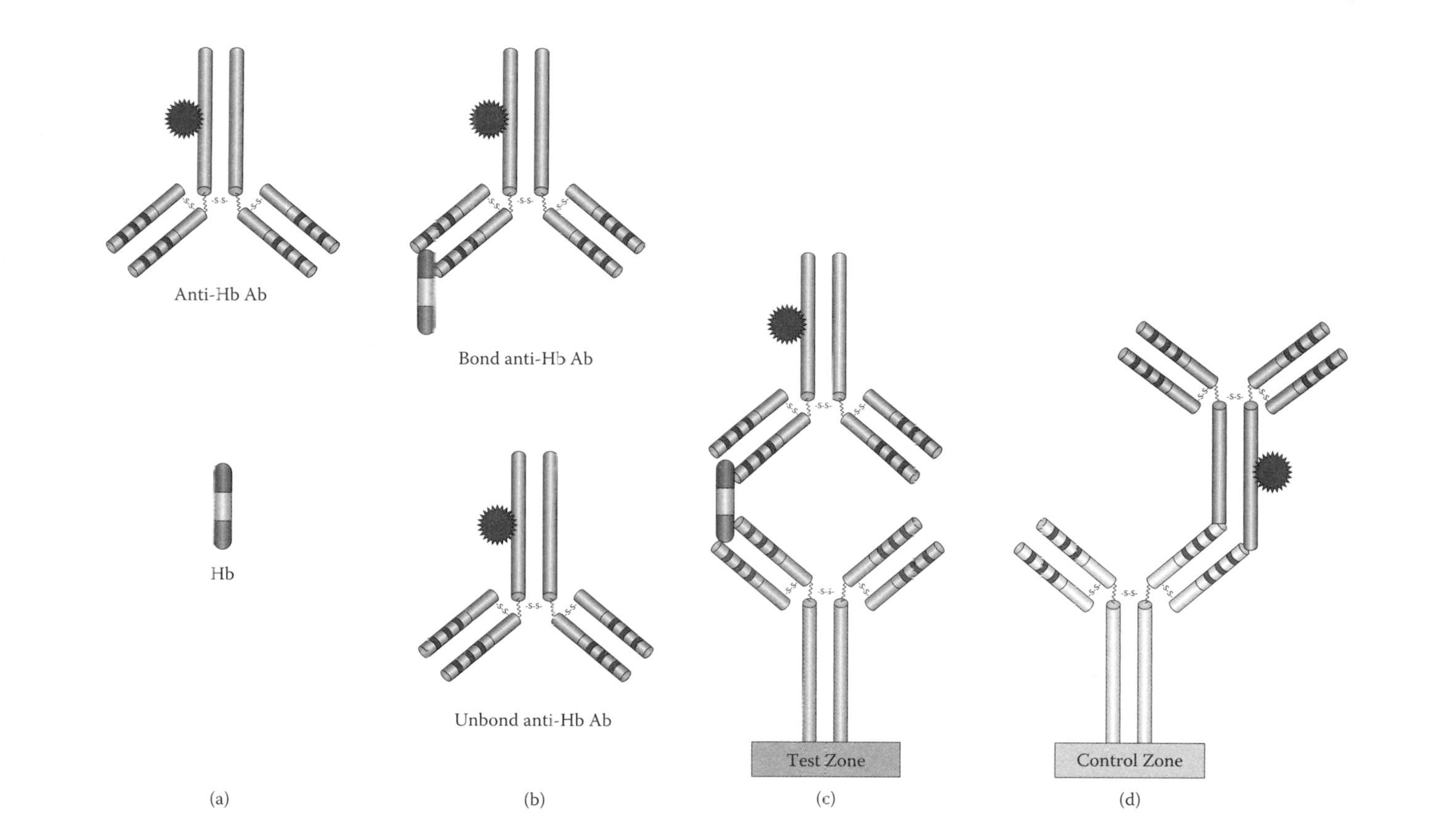


**Figure 7.2** Immunochromatographic assays for identification of Hb in human blood. (a) In a sample well, Hb in a blood sample is mixed with a labeled anti-Hb Ab. (b) The Hb binds to the labeled anti-Hb Ab to form a labeled Ab–Hb complex. (c) At the test zone, the labeled Ab–Hb complex binds to an immobilized anti-Hb Ab to form a labeled Ab–Hb–Ab sandwich. (d) At the control zone, the labeled anti-Hb Ab binds to an immobilized antiglobulin and is captured. Ab = antibody. Hb = hemoglobin.

oral, vaginal, anal, and rectal swabs. It is believed that these biological fluids contain very low amounts of Hb that can still be detected by highly sensitive assays. However, if the concentration of blood is too high, a false negative can result due to the high-dose hook effect described in an earlier chapter.

#### *7.2.2.2 Identification of Human Glycophorin A Protein*

Commercially produced immunochromatographic kits such as RSID™-Blood (Independent Forensics, Hillside, IL) use antibodies that recognize human GPA. A labeled monoclonal antihuman GPA antibody is contained in a sample well, and a second monoclonal antihuman GPA antibody, to a different epitope of GPA, is immobilized onto a test zone of the membrane. An antiglobulin that recognizes the antibody is immobilized onto a control zone (Figure 7.3).

The sample can be collected by cutting out a small portion of a stain or a swab. Each sample is then extracted overnight in an extraction buffer. The extract is removed and mixed with a running buffer. The assay is carried out by loading the extracted sample into the sample well. Again, the presence of GPA results in a pink line at both the test zone and control zone while the absence of GPA results in a pink line in the control zone only. The test is considered valid only if the line in the control zone is observed. A result can be read after 10 min.

Validation studies revealed that the sensitivity of the RSID kit can be as low as 100 nl of human blood. Species specificity studies showed no cross-reactivity with various animal species including non-human primates. Biological fluid specificity studies revealed that the kit is not responsive to other human biological fluids such as semen, saliva, urine, milk, amniotic, and vaginal fluid. No high-dose hook effects were observed in samples containing up to 5 μl of blood.

### 7.2.2 Ring Assay

This is a double immunodiffusion assay. The procedure is described in Box 1 and illustrated in Figure 7.4. Chapter 5 discusses the basic principle of ring assay in detail. In a positive reaction, a white precipitate between the two layers observed after several minutes indicates that a sample is of human origin. If the bloodstain extract is not human, no precipitation should appear.

### 7.2.3 Ouchterlony Assay

This is another double immunodiffusion assay. The procedure is described in Box 2 and discussed in further detail in Chapter 5. In a positive reaction, a line of precipitate will form between each antigen well and antibody well. This assay can also determine the similarity of antigens. During the

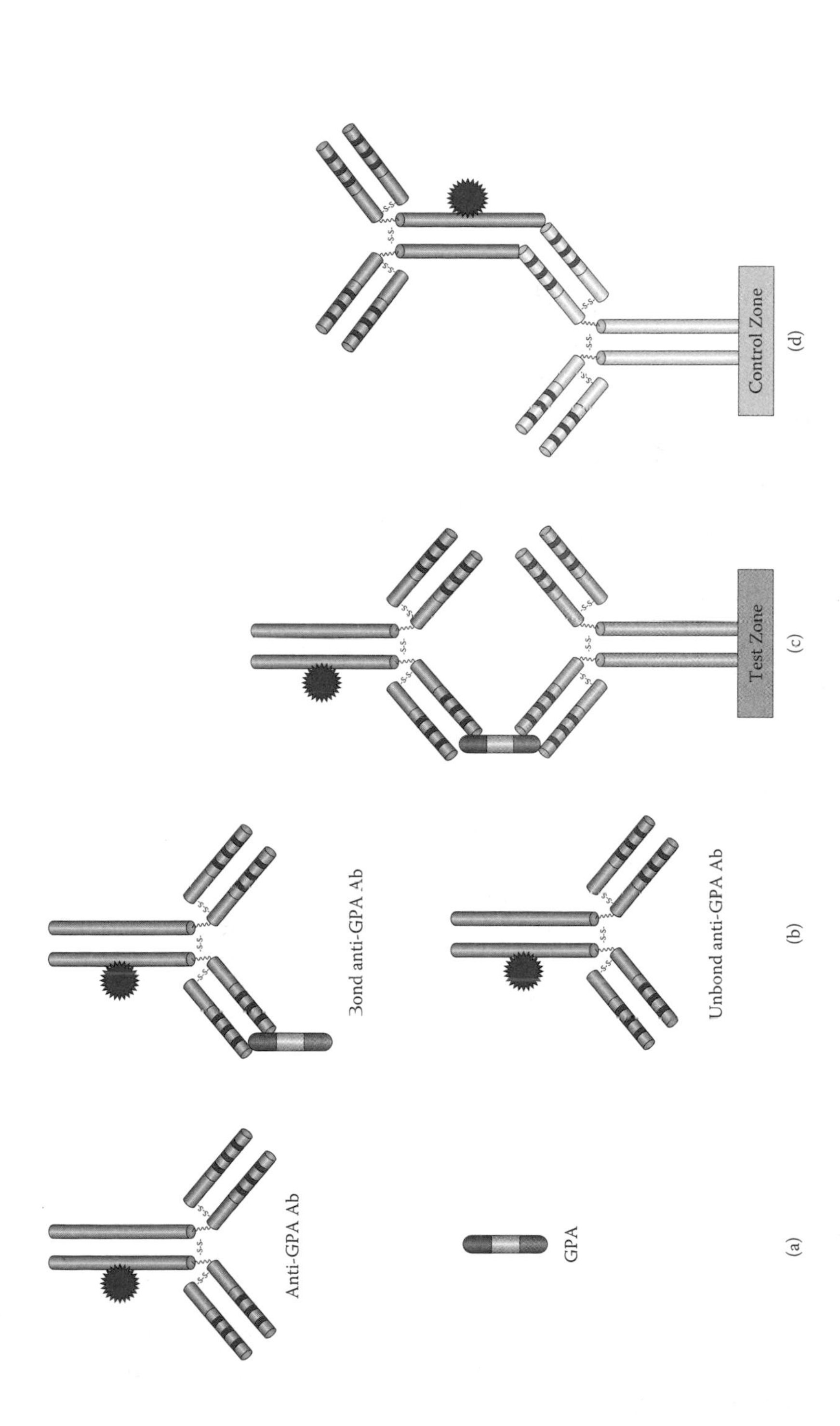


**Figure 7.3** Immunochromatographic assays for identification of GPA in human blood. (a) In a sample well, GPA in a blood sample is mixed with a labeled anti-GPA Ab. (b) The GPA binds to the labeled anti-GPA Ab to form a labeled Ab–GPA complex. (c) At the test zone, the labeled Ab–GPA complex binds to an immobilized anti-GPA Ab to form a labeled Ab–GPA–Ab sandwich. (d) At the control zone, the labeled anti-GPA Ab binds to an immobilized antiglobulin and is captured. Ab = antibody. GPA = glycophorin A.

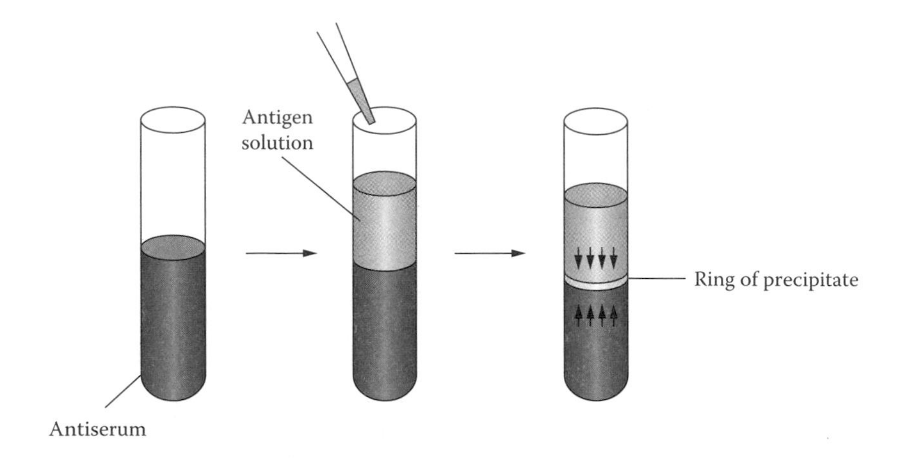


**Figure 7.4** Ring assay. The antigen solution is carefully applied over the antiserum solution. Incubation is allowed to form a ring of precipitate.

diffusion process, different antigen–antibody complexes migrate at different rates. Consequently, a separate line of precipitate will appear in the gel for each antigen–antibody complex (Figure 7.5). In an assay in which two antigens are loaded in adjacent wells and an antibody in the third well, the following results can be observed:

- If the two antigens are identical, the two lines will become fused. This phenomenon is referred to as ***identity***.
- If the two antigens are totally unrelated, the lines will cross each other but not fuse; this is known as ***non-identity***.
- If the two antigens are related (share a common epitope) but are not identical, the lines will merge with spur formation. The spurs are continuations of the line formed by the antigen due to its unique epitope. This phenomenon is known as ***partial identity***.

Thus, in a species test, a positive result is noted when the precipitate lines for the positive controls and the samples fuse. No spur formation should be observed.

### 7.2.4 Crossed-Over Electrophoresis

This method is a combination of immunodiffusion and electrophoresis. The procedure for the assay is described in Box 3 and the basic principles of this method are discussed in Chapter 5. With this technique, a sharp precipitate band is visualized in a positive reaction (Figure 7.6). However, false negative results can occur due to the postzone phenomenon in which excess antigen

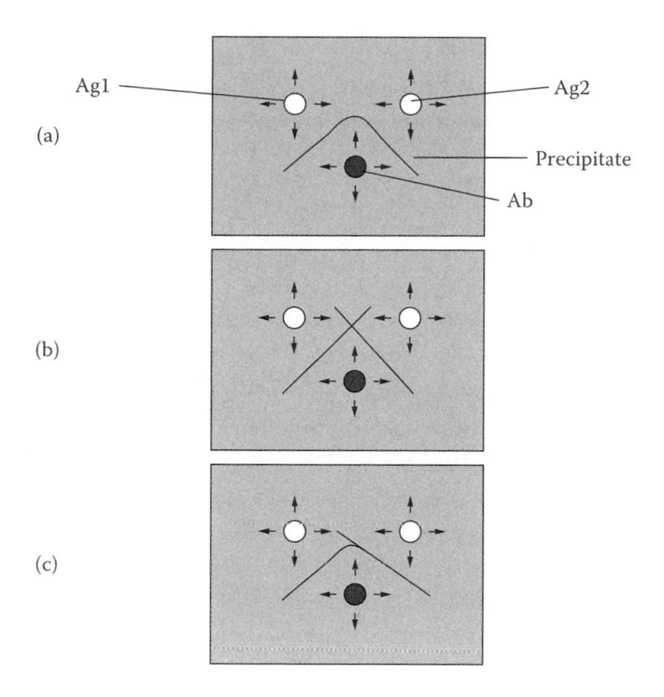


**Figure 7.5** Results of Ouchterlony assay. Ab = antibody; Agl and Ag2 = antigen samples in question. (a) Identity. The two antigens are identical (fused line). (b) Non-identity. The two antigens are unrelated (spur). (c) Partial identity. The two antigens are related but not identical (fused line with spur).

may inhibit precipitation. In this situation, the sample can be diluted and the assay can be repeated. False negative results can also occur due to simple mistakes made during electrophoresis:

- Electrophoresis was carried out in the opposite direction, which resulted in samples running off the gel.
- Electrophoresis was carried out using an incorrect buffer system.
- The amount of current applied during the electrophoresis was too strong and generated heat and denatured proteins.

**Box 1: Ring Assay Procedure**

**Sample Preparation and Extraction**

Extract a portion of a stain with saline at 4°C overnight. Spin the antihuman antibody in a microfuge and transfer the supernatant into test tubes or capillary tubes (depending on the volume of the stain and antiserum extracted).

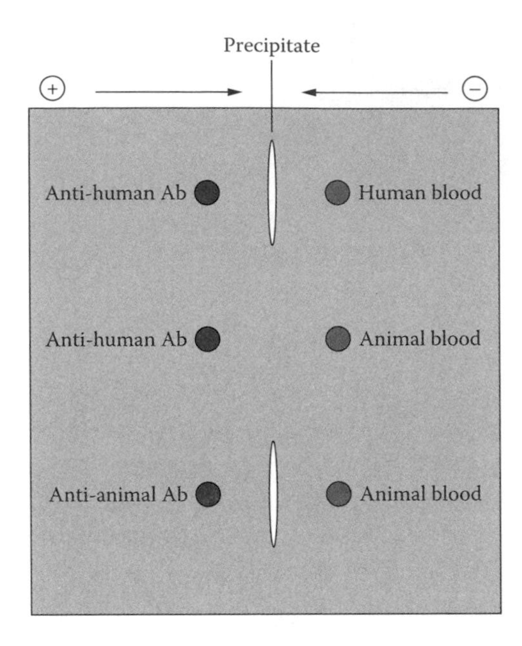


**Figure 7.6** Results of crossed-over electrophoresis. A precipitate line is formed between a human blood sample and an antihuman antibody. No precipitate line is formed when the antihuman antibody is tested for an animal blood sample. A precipitate line is formed between the animal blood sample and antianimal antibody.

### Controls

Include a positive control (known human serum sample) and a negative control (extraction blank).

### Loading Samples

Place the sample carefully over the top of the antiserum solution (which is usually denser than the sample).

### Immunodiffusion Reaction

Carry out the reaction at room temperature. In a positive reaction, white precipitate between the two layers can be observed after several minutes. This indicates that the sample is of human origin. No reaction will occur if a blood stain extract is not human.

### Box 2: Ouchterlony Assay Procedure

### Sample Preparation and Extraction

Cut out a small portion (approximately 5 × 5 mm) of a stain or a portion of a swab and extract in 100 μl water for 30 min at room temperature. The extract can be diluted if necessary. Alternatively, a very small piece of stain or swab can be inserted directly into the well.

**Controls**

Positive (known serum); negative (extraction blank); substrate controls (extraction of substrate from unstained area) if applicable.

**Agarose Gel Preparation**

Heat a suspension of agarose (4%) until liquefied. Cool the solution in a water bath at 55°C. Pour the agarose onto a piece of glass slide and let the gel solidify to a thickness of about 2 to 3 mm. Alternatively, a polyester support film such as GelBond (Cambrex, New Zealand) can be used as a gel support. The agarose should be poured onto the hydrophilic side of a piece of GelBond film (6 × 9 cm). Punch wells consisting of a central well surrounded by four wells using a template.

**Loading Samples**

Apply antihuman antibody to the central well. Apply the positive control to one of the surrounding wells. Apply the sample(s) so that a stain extract is always next to a positive control. Apply negative and substrate controls to the remaining wells; only one negative control is needed per gel.

**Immunodiffusion Reaction**

Incubate the plate overnight in a moisture chamber at 37°C.

**Staining**

Soak the gel overnight in saline solution, then soak in deionized water for 10 min. Repeat once. Dry the gel between paper towels with a weight on top for 30 min. Dry in an oven for 30 min. Stain the gel with Coomassie Blue.

**Box 3: Crossed-Over Electrophoresis Procedure**

**Sample Preparation and Extraction**

Cut out a small portion (approximately 5 × 5 mm) of a stain or a portion of a swab and extract at room temperature in 100 μl water for 30 min. The extract can be diluted if necessary. Alternatively, a very small piece of stain or swab can be inserted directly into the well.

**Controls**

Positive (known serum); negative (extraction blank); substrate (extraction of substrate from unstained area) if applicable.

**Agarose Gel Preparation**

Heat a suspension of agarose (4%) until liquefied. Cool the solution in a water bath at 55°C. Pour the agarose onto a piece of glass slide and let it solidify. Alternatively, a polyester support film such as GelBond can be used as a gel support. The agarose should be poured onto the hydrophilic side of a piece of GelBond (6 × 9 cm). Punch small wells (about 1 to 2 mm) in rows using a template.

**Loading Samples**

Apply antihuman antibody in one row of wells. Apply samples in the other row of wells. Apply the positive, negative, and substrate controls.

**Electrophoresis**

Submerge the agarose gel in an electrophoresis tank in proper orientation. The wells containing antihuman antibody should be closest to the anode (positive electrode) and wells containing samples should be closest to the cathode (negative electrode). During electrophoresis, the antibody in the antiserum should migrate toward the cathode while the antigen migrates toward the anode. Electrophoresis is carried out at 10 V/cm for 20 min.

**Staining**

Soak the gel overnight in saline solution, then soak in deionized water for 10 min. Repeat once. Dry the gel between paper towels with a weight on top for 30 min. Dry in an oven for 30 min. Stain the gel with Coomassie Blue.

## Bibliography

Allison, A. C., and J. A. Morton. 1953. Species specificity in the inhibition of antiglobulin sera: A technique for the identification of human and animal bloods. *J Clin Pathol* 6 (4):314.

Bhatia, R. Y. 1974. Specificity of some plant lectins in the differentiation of animal blood. *Forensic Sci* 4 (1):47.

DeForest, P., R. E. Gaensslen, and H .C. Lee. 1983. *Forensic Science: An Introduction to Criminalistics*. New York: McGraw-Hill.

Dorrill, M., and P. H. Whitehead. 1979. The species identification of very old human blood stains. *Forensic Sci Int* 13 (2):111.

Grobbelaar, B. G., D. Skinner, and H. N. van de Gertenbach. 1970. The antihuman globulin inhibition test in the identification of human blood stains. *J Forensic Med* 17 (3):103.

Hochmeister, M. N. et al. 1999. Validation studies of an immunochromatographic one-step test for the forensic identification of human blood. *J Forensic Sci* 44 (3):597.

James, S., and J. J. Nordby. 2005. *Forensic Science: An Introduction to Scientific and Investigative Techniques*. Boca Raton, FL: CRC Press.

Johnston, S., J. Newman, and R. Frappier. 2003. Validation study of the Abacus Diagnostics ABAcard HemaTrace membrane test for the forensic identification of human blood. *Can J Forensic Sci*.

Juusola, J., and J. Ballantyne. 2003. Messenger RNA profiling: a prototype method to supplant conventional methods for body fluid identification. *Forensic Sci Int* 135 (2):85.

Lawton, M.E., and J. G. Sutton. 1982. Species identification of deer blood by isoelectric focusing. *J Forensic Sci Soc* 22 (4):361.

Lee, H. C. 1982. Identification and grouping of bloodstains, in *Forensic Science Handbook*. Saferstein, R., Ed. Englewood Cliffs, NJ: Prentice Hall.

Lee, H. C., and P. R. De Forest. 1976. A precipitin-inhibition test on denatured bloodstains for the determination of human origin. *J Forensic Sci* 21 (4):804.

Saferstein, R. 2004. *Criminalistics: An Introduction to Forensic Science*. Englewood Cliffs, NJ: Prentice Hall.

Shaler, R. C. 2002. Modern forensic biology, in *Forensic Science Handbook*. Saferstein, R., Ed. Upper Saddle River, NJ: Person Education.

Sivaram, S. et al. 1975. Differentiation between stains of human blood and blood of monkey. *Forensic Sci* 6 (3):145.

Spear, T. F. and S. A. Binkley. 1994. The HemeSelect test: A simple and sensitive forensic species test. *J Forensic Sci Soc* 34 (1):41.

Tabata, N., and M. Morita. 1997. Immunohistochemical demonstration of bleeding in decomposed bodies by using anti-glycophorin A monoclonal antibody. *Forensic Sci Int* 87 (1):1.

Tanaka, Y. et al. 2003. Enzyme-linked immunospot assay for detecting cells secreting antibodies against human blood group A epitopes. *Transplant Proc* 35 (1):555.

Tsutsumi, H., and Y. Katsumata. 1993. Forensic study on stains of blood and saliva in a chimpanzee bite case. *Forensic Sci Int* 61 (2–3):101.

Whitehead, P. H., and A. Bech. 1974. A microtechnique involving species identification and ABO grouping on the same fragment of blood. *J Forensic Sci Soc* 14 (2):109.

## Study Questions

1. When is it necessary to perform a species identification test?

2. Which of the following can be used for species identification?
   (a) Antihuman albumin antiserum
   (b) Antihuman hemoglobin antibody
   (c) Antihuman glycophorin A antibody
   (d) All of the above

3. The titration of antibodies is intended to prevent the:
   (a) Prozone effect
   (b) Postzone effect
   (c) High-dose hook effect

4. Most antihuman sera usually react with:
   (a) Higher primates
   (b) Lower primates
   (c) Non-primates

5. Which of the following is the most sensitive method for identification of species origin of a blood sample?
   (a) Immunochromatographic assay
   (b) Ring assay
   (c) Ouchterlony assay
   (d) Crossed-over electrophoresis

# Identification of Semen

# 8

## 8.1 Biological Characteristics

A typical ejaculation releases 2 to 5 ml of semen that contain seminal fluid and sperm cells (***spermatozoa***). A normal sperm count ranges from $10^7$ to $10^8$ spermatozoa per milliliter of semen. Each testis is composed of numerous coiled seminiferous tubules in which the spermatozoa are formed from spermatogonia. This process of generating spermatozoa is referred to as spermatogenesis. The spermatozoa are then transported and stored at the tubular network of the epididymis where they undergo functional maturation (spermatogenesis and maturation take approximately 3 months). The epididymis joins the ductus deferens that transports matured sperm from the epididymis to the ejaculatory duct. From there, spermatozoa follow the ejaculatory ducts into the prostatic urethra where they are joined with secretions from the prostate. Figure 8.1 illustrates the anatomy of the male reproductive system.

Seminal fluid is a complex mixture of glandular secretions. A typical sample of seminal fluid contains the combined secretions of several accessory glands. Seminal vesicle fluid accounts for approximately 60% of the ejaculate. Various proteins secreted by the seminal vesicles play a role in the coagulation of the ejaculate. Additionally, seminal vesicle fluid contains flavin that causes semen to fluoresce under ultraviolet light, often utilized when searching for semen stain evidence.

Prostatic fluid secretions account for approximately 30% of the ejaculate. The components of this fluid are complex as well. This portion of semen contains high concentrations of ***acid phosphatase*** (AP) and ***prostate-specific antigen*** (PSA). Both are useful markers for the identification of semen in forensic laboratories.

The epididymis and the bulbourethral secretions each account for approximately 5% of the ejaculate.

A ***vasectomy*** is the surgical removal of a bilateral segment of the ductus deferens. The surgery prevents spermatozoa from reaching the distal portions of the male reproductive tract. However, a vasectomized male can still produce ejaculate that contains only seminal vesicle fluid and prostatic fluid. The condition by which males have abnormally low sperm counts is known as ***oligospermia***. ***Aspermia*** is a condition that cause males to produce no

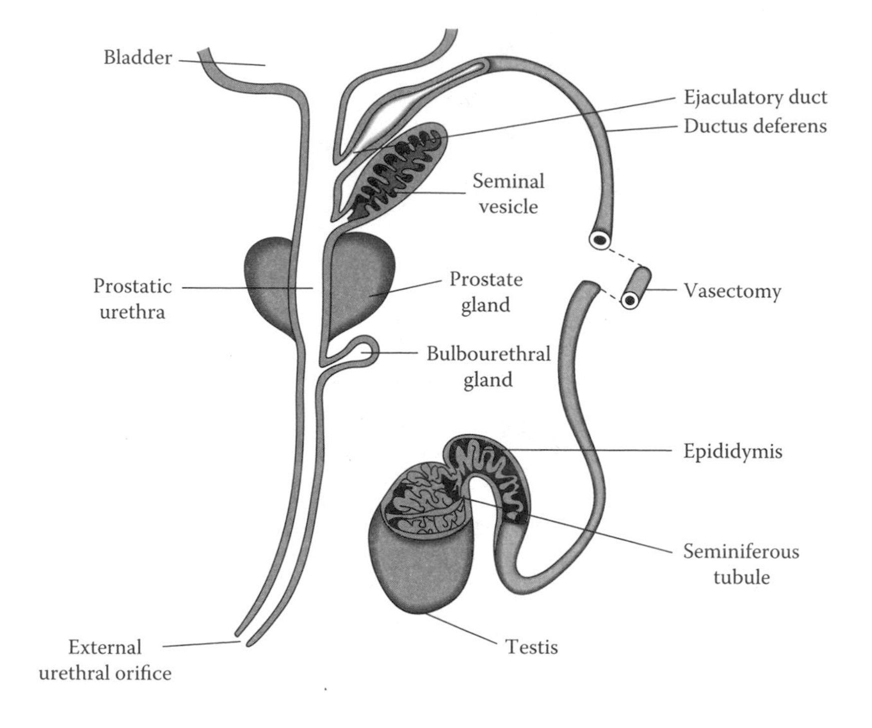


**Figure 8.1** Male reproductive system and accessory glands (unilateral view).

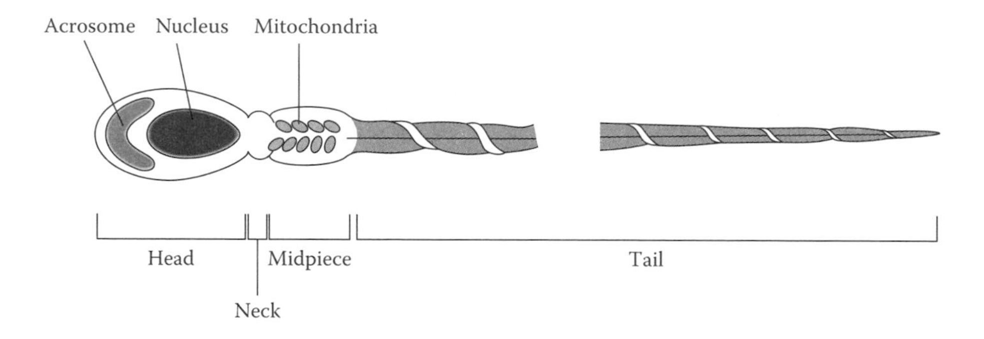


**Figure 8.2** Structure of spermatozoon.

spermatozoa. However, the secretion of seminal fluid is not affected in males who have these conditions.

### 8.1.1 Spermatozoa

A human spermatozoon has three distinct regions: the head, the middle piece, and the tail (Figure 8.2). The head is shaped like a flattened ellipse and contains a nucleus with densely packed chromosomes. At the tip of the head is the ***acrosomal cap***, a membranous compartment containing enzymes essential for fertilization. The head is attached to the ***middle piece*** through a short neck where the mitochondria that provide the energy for moving the tail are located. The tail or flagellum is responsible for spermatozoon motility. In contrast to other cell types, a mature spermatozoon lacks various intracellular organelles such as an endoplasmic reticulum, Golgi apparatus, lysosomes, and peroxisomes. In a normal male, at least 60% of spermatozoa have normal morphology, so morphological abnormalities can often be observed.

### 8.1.2 Acid Phosphatase (AP)

AP consists of a group of phosphatases with optimal activity in an acidic pH environment. AP is commonly present in lysosomes, but extralysosomal acid phosphatases are also found in other cell types. The greatest forensic importance of AP is that the prostate is the most abundant source of AP and contributes most of the AP activity present in semen. AP levels in semen are not affected by vasectomies.

The half life of AP activity at 37°C is 6 months. However, the half life is decreased if a sample is stored in a wet environment. AP activity can be detected from dry seminal stains stored at –20°C up to one year. Low levels of prostatic AP are present in the sera of healthy males. Elevated levels of prostatic AP found in serum are useful in diagnosing and monitoring prostate carcinoma. Many AP tests utilized in clinical testing may be used to identify semen for forensic applications.

### 8.1.3 Prostate-Specific Antigen (PSA)

PSA is a major protein present in seminal fluid at concentrations of 0.5 to 2.0 mg/ml. PSA is produced in the prostate epithelium and secreted into the semen. PSA can also be found in the paraurethral glands, perianal glands, apocrine sweat glands, and mammary glands. Thus small quantities can be detected in urine, fecal material, sweat, and milk. PSA can also be found at much lower levels in the bloodstream. An elevated plasma PSA is present in prostate cancer patients and it is widely used as a screening test for this disease. PSA is also elevated in benign prostatic hyperplasia and prostatitis. The synthesis of PSA is stimulated by androgen, a steroid hormone.

PSA is a protein that has a molecular weight of 30 kDa and is thus also known as ***P30***. It is responsible for hydrolyzing ***semenogelin*** (Sg) which mediates gel formation in semen. PSA is a member of the tissue ***kallikrein*** (serine protease) family and is encoded by the *KLK3* locus located on chromosome 19. In addition to PSA, other tissue kallikreins encoded by *KLK2* and *KLK4* loci are expressed in the prostate as well. The half life for PSA in a dried semen stain is about 3 years at room temperature. The half life is greatly reduced when a sample is stored in wet conditions.

### 8.1.4 Seminal Vesicle-Specific Antigen (SVSA)

SVSA includes two major semenogelins and constitutes the major seminal vesicle-secreted protein in semen. Upon ejaculation, it forms a coagulum that is liquefied after a few minutes due to the degradation of SVSA. In human semen, two major SVSAs are semenogelin I (SgI) and semenogelin II (SgII). Both are present in a number of tissues of the male reproductive system including the seminal vesicles, ductus deferens, prostate, and epididymis. They are also present in several other tissues such as skeletal muscle, kidney, colon, and trachea. They have also been found in the sera of lung cancer patients.

Use of Sg as a marker for semen identification instead of PSA presents certain advantages. The concentration of Sg in seminal fluid is much higher than that of PSA and this is beneficial for the sensitivity of detection. Sg is present in seminal fluid and absent in urine, milk, and sweat where PSA can be found. Although Sg compounds are present in skeletal muscle, kidney, and colon, this is not a great concern because tissue samples are not routinely collected for semen detection in sexual assault cases.

## 8.2 Analytical Techniques for Identifying Semen

Visual examinations, particularly with alternate light sources (ALS), facilitate searches for semen stains. The presumptive identification of semen is largely based on the presence of prostatic AP activity in a sample. However, most presumptive assays cannot completely distinguish prostatic AP from non-prostatic AP. The confirmatory identification of semen can be performed via microscopic examination of spermatozoa. Other confirmatory assays are useful for identifying semen such as the identification of PSA and SVSA and the RNA-based assay.

### 8.2.1 Presumptive Assays

#### *8.2.1.1 Lighting Techniques for Visual Examination of Semen Stains*

Lighting techniques are often used to search for semen stains. Dried semen stains fluoresce when irradiated with ultraviolet light, argon lasers, or ALSs. ALSs are most commonly used for visual examination (Figure 8.3); 450- to 495-nm excitation wavelengths can be used with colored goggles or filters (530 nm) that allow for the visualization of fluorescence. However, this approach

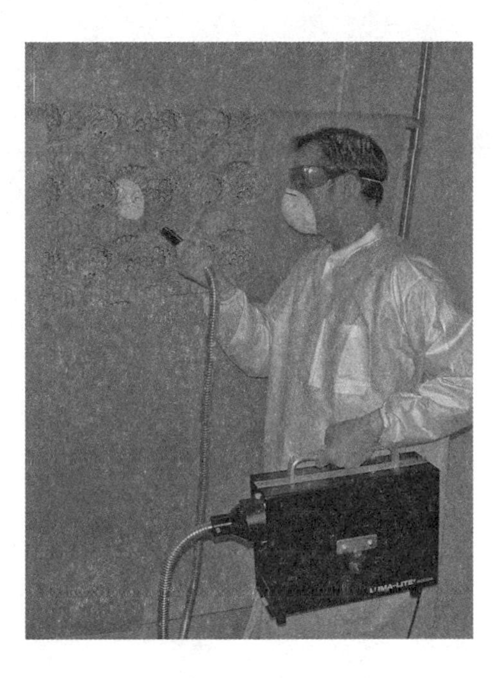

**Figure 8.3** Detection of semen stain using alternate light sources. (Source: James, S., and J. J. Nordby. 2005. *Forensic Science: An Introduction to Scientific and Investigative Techniques,* 2nd ed. Boca Raton, FL: CRC Press. With permission.)

is not specific for semen. Additionally, the intensity of the fluorescence can be affected by different types of substrates.

### *8.2.1.2 Acid Phosphatase Techniques*

**8.2.1.2.1 Colorimetric Assays** Colorimetric assays can be used for the presumptive identification of semen. The AP contained in semen can hydrolyze a variety of phosphate esters. It catalyzes the removal of the phosphate group from a substrate. Subsequent coupling of a stabilized diazonium salt results in the formation of an insoluble colored precipitate at sites of acid phospatase activity. However, interference during a test by non-prostatic AP isoenzymes (multiple forms of AP), such as contamination by AP commonly present in vaginal secretions, can create problems in specimens collected from victims.

Thus, it is desirable to be able to increase the specificity of the assay for prostatic AP. One solution is the application of substrates that are hydrolyzed rapidly by the prostatic enzyme and at a slower rate by the other forms of AP isoenzymes. For example, α-naphthyl phosphate and thymolphthalein monophosphate are more specific to prostatic AP than phenyl phosphate and 4-nitrophenyl phosphate (Figure 8.4). Additionally, the prostatic enzyme is strongly inhibited by dextrorotatory tartrate ions. Thus, these inhibitors,

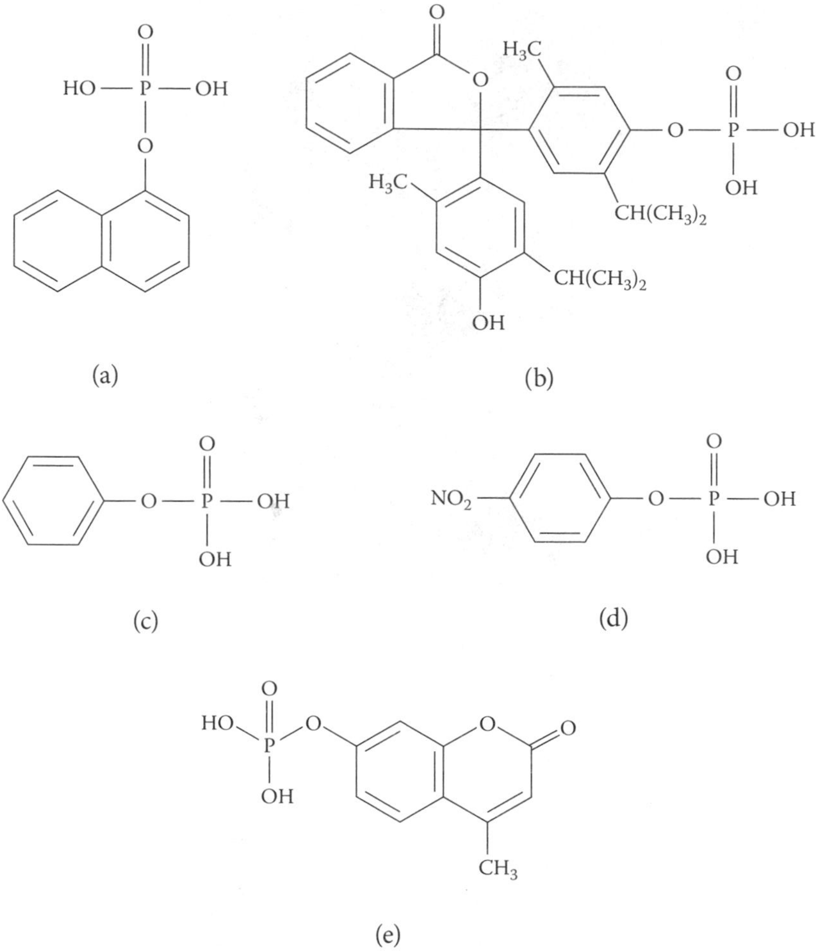

**Figure 8.4** Chemical structures of acid phosphatase substrates. (a) α-Naphthyl phosphate. (b) Thymolphthalein monophosphate. (c) Phenyl phosphate. (d) 4-Nitrophenyl phosphate. (e) MUP.

particularly tartrate, allow a distinction to be made between prostatic AP and other AP isoenzymes.

The most common method for forensic applications is the use of α-naphthyl phosphate as a substrate coupled with Brentamine Fast Blue B (Figure 8.5). Prostatic AP is water-soluble. Thus, a moistened cotton swab or piece of filter paper can be used to transfer a small amount of sample from a stain by brief pressing onto the questioned stain area. The substrate reagent is added to the swab or filter paper followed by the addition of Brentamine Fast Blue B. If a purple coloration develops within 1 min, the test is considered a

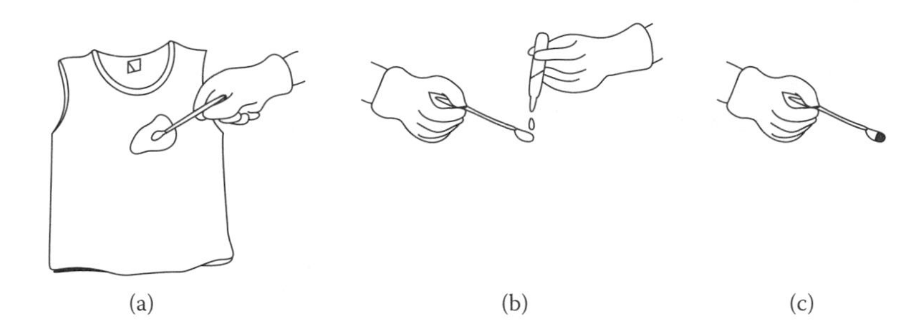


**Figure 8.5** AP colorimetric assay. (a) Transfer small amount of sample from stain using a moistened cotton swab. (b) Apply substrate reagent followed by Brentamine Fast Blue B. (c) Purple coloration indicates positive reaction.

positive indication for semen. Color that develops after more than 1 min may arise from the activity of non-prostatic AP.

**8.2.1.2.2 Fluorometric Assays** Fluorometric methods are more sensitive than the colorimetric detection of AP and are used for semen stain mapping. AP catalyzes the removal of the phosphate residue on a 4-methylumbelliferone phosphate (MUP) substrate, a reaction that generates fluorescence under ultraviolet light. A piece of moistened filter paper, marked for proper orientation and identification, is used for transferring the prostatic AP. The evidence to be tested, for example, a garment, is covered by the filter paper. Gloved hands are used to press the filter paper onto the stained area, ensuring that the evidence is in close contact with the paper. The filter paper is lifted from the evidence and examined in a dark room with long-wave ultraviolet light to detect any background fluorescence which is then marked on the paper. The paper can then be sprayed with MUP reagent in a fume hood. The AP reaction on the paper can be visualized within seconds with a long-wave ultraviolet lamp in a dark room. Semen stains will present as fluorescent areas (Figure 8.6).

### 8.2.2 Confirmatory Assays

#### *8.2.2.1 Microscopic Examination of Spermatozoa*

The cells from a questioned stain on an absorbent material can be transferred to a microscope slide by extracting a small portion of a stain with water, followed by gentle vortexing. The suspension is then transferred to a slide and evaporated at room temperature or fixed with low heat. Alternatively, it can be transferred by dampening the stain with water and rubbing or rolling it onto a microscope slide.

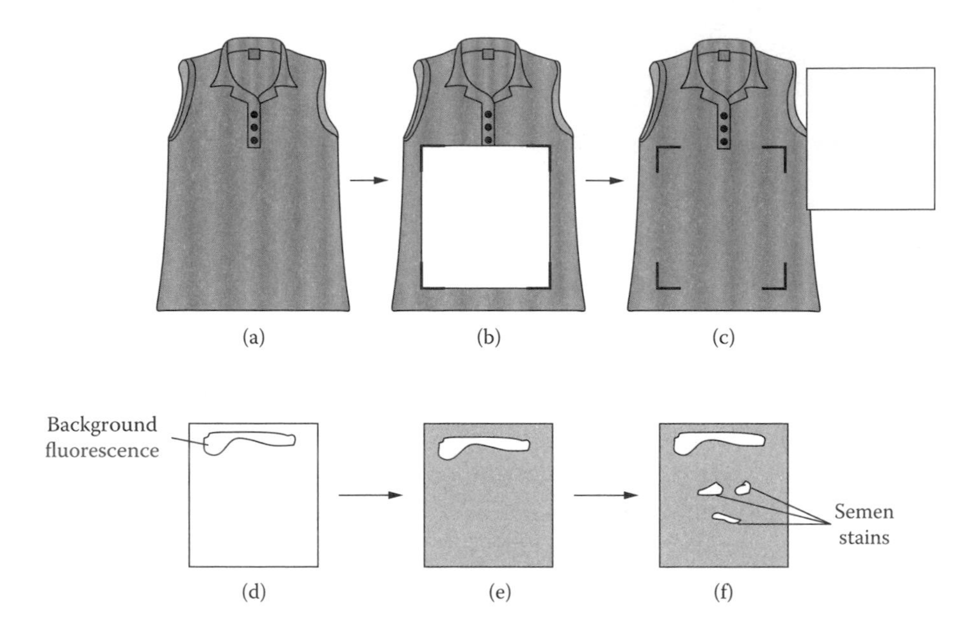


**Figure 8.6** Fluorometric assay of AP. Evidence item (a) is closely covered by a piece of moistened filter paper (b) to allow transfer of small amount of stain. The orientation of the paper is marked (c). The paper is lifted. The background fluorescence is marked under UV light (d). The filter paper is treated with MUP (e). The presence of fluorescence under ultraviolet light indicates semen stains (f).

Microscopic identification of spermatozoa provides the proof of a seminal stain. Histological staining can facilitate microscopic examination. The most common staining technique is the ***Christmas tree stain*** (Figure 8.7). The red component known as Nuclear Fast Red (NFR) is a dye used for staining the nuclei of spermatozoa in the presence of aluminum ions. The green component, picroindigocarmine (PIC), stains the neck and tail portions of the sperm. The acrosomal cap turns pink, the nucleus is red, the sperm tails stain green, and the midpiece stains blue. Any epithelial cells present in the sample will appear blue-green and have red nuclei.

***Laser capture microdissection*** (LCM) was shown to be an effective technique for separating the spermatozoa from non-sperm cells (i.e. epithelial cells from the victim) on a glass slide. This technique involves using a thin layer of a thermo-sensitive polymer which is placed on the surface of an LCM cap. Once spermatozoa are identified on the slide under a microscope, a polymer-containing LCM cap is placed over the spermatozoa on the slide. An infrared laser melts the polymer and causes it to adhere only to the targeted spermatozoa. The spermatozoa are then lifted off the slide. This allows

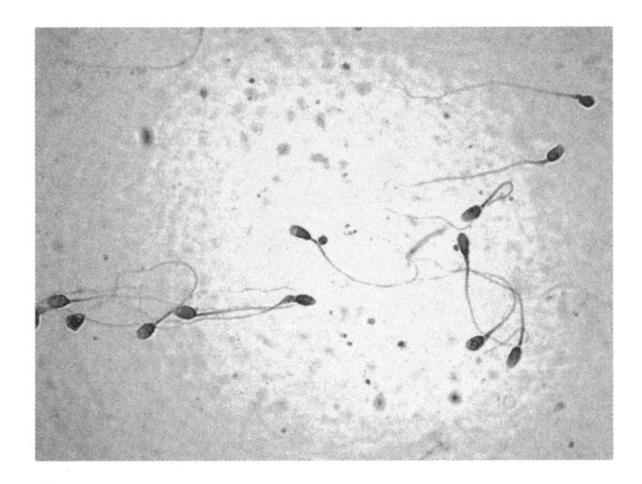

**Figure 8.7** Human spermatozoa stained with Christmas tree stain (Source: James S., and J. J. Nordby. 2005. *Forensic Science: An Introduction to Scientific and Investigative Techniques*, 2nd ed. Boca Raton, FL: CRC Press. With permission.)

spermatozoa to be separated and placed into snap-cap tubes for forensic DNA analysis.

### *8.2.2.2 Identification of Prostate-Specific Antigen*

Over the years, a number of methods have been employed to detect PSA: radial immunodiffusion, rocket immunoelectrophoresis, CIE, ELISA, and immunochromatographic assays. ELISA and immunochromatographic assays were found to be the most sensitive methods.

**8.2.2.2.1 Immunochromatographic Assays** Commercially produced immunochromatographic kits such as the PSA-check-1 (VED-LAB, Alencon); Seratec® PSA Semiquant (Seratec Diagnostica, Göttingen); and One Step ABAcard PSA® (Abacus Diagnostics, California) are available. These devices employ antihuman PSA antibodies. In the ABAcard PSA® assay, a labeled monoclonal antihuman PSA antibody is contained in a sample well, a polyclonal antihuman PSA antibody is immobilized on a test zone of a nitrocellulose membrane, and an antiglobulin that recognizes the antibody is immobilized on a control zone (Figure 8.8).

The assay is carried out by loading an extracted sample into the sample well. The antigen in the sample binds to the labeled antibody in the sample well to form an antigen–antibody complex. The complex then diffuses across the nitrocellulose membrane. At the test zone, the immobilized antihuman PSA antibody binds with the antigen–antibody complex to form an antibody–antigen–antibody sandwich. The ABAcard PSA® uses a pink dye that

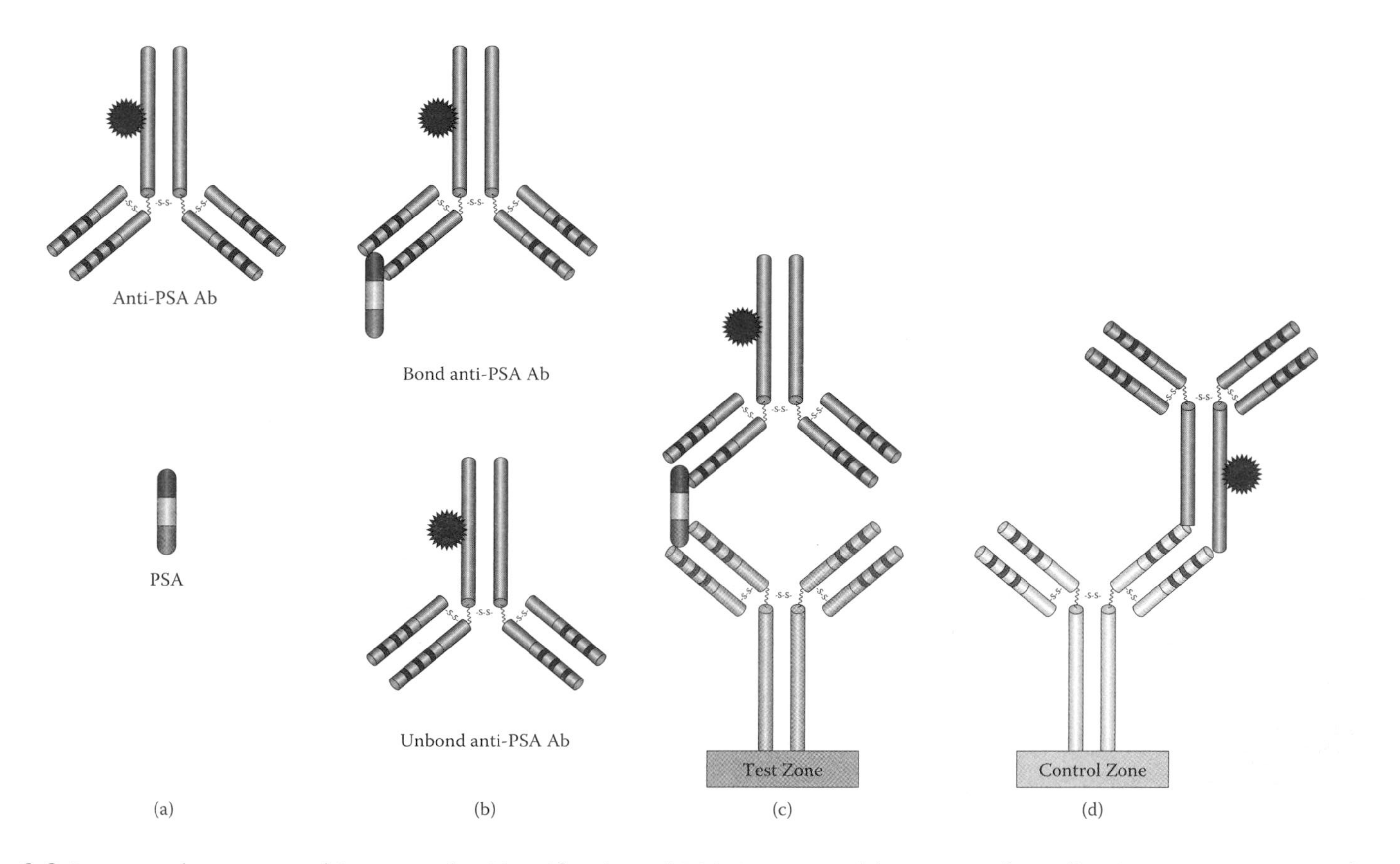


**Figure 8.8** Immunochromatographic assays for identification of PSA in semen. (a) In a sample well, PSA in a semen sample is mixed with labeled anti-PSA Ab. (b) The PSA binds to the labeled anti-PSA Ab to form a labeled Ab–PSA complex. (c) At the test zone, the labeled complex binds to an immobilized anti-PSA Ab to form a labeled Ab–PSA–Ab sandwich. (d) At the control zone, the labeled anti-PSA Ab binds to an immobilized antiglobulin and is captured at the control zone. Ab = antibody. PSA = prostate-specific antigen.

allows for the visualization of a positive test with a pink vertical line at the test zone. In the control zone, unbound labeled antihuman PSA antibody binds to the immobilized antiglobulin. This antibody–antiglobulin complex at the control zone also results in a pink vertical line. The test is considered valid only if the line in the control zone is observed.

The presence of human PSA results in a pink line at both the test and control zones. The absence of human PSA produces a pink line in the control zone only. A positive result can appear within 1 min; a negative result is read after 10 min. However, the high-dose hook effect, an artifact that may cause false negative results described in chapter 5, occurs when high quantities of seminal fluid are tested.

**8.2.2.2.2 Enzyme-Linked Immunosorbent Assay (ELISA)** The ELISA method can be used to detect and measure PSA with anti-PSA antibodies. The most common method used in forensic serology is antibody sandwich ELISA, in which an antibody–antigen–antibody sandwich complex is formed (Figure 8.9). The intensity of the signal can be detected and is proportional to the amount of bound antigen. The amount of PSA can also be quantified by comparing a standard with known concentrations. Although this method is specific and highly sensitive, it is time-consuming. Chapter 5 discusses the principle of ELISA in further detail.

### *8.2.2.3 Identification of Seminal Vesicle-Specific Antigen*

**8.2.2.3.1 Immunochromatographic Assays** Commercially produced immunochromatographic kits include the RSID®-Semen test (Independent Forensics, Hillside, IL), and the Nanotrap Sg. In the RSID®-Semen assay, a labeled monoclonal anti-Sg antibody is contained in a sample well and a second monoclonal anti-Sg antibody, to a different epitope of Sg, is immobilized on the test zone of the membrane. An antiglobulin that recognizes the antibody is immobilized on a control zone (Figure 8.10).

The sample can be prepared by cutting a small portion of a stain or a swab and extracted for 1 to 2 hr in an extraction buffer (200 to 300 μl). Approximately 10% of the extract is removed and mixed with the running buffer. The assay is carried out by loading an extracted sample into the sample well. The antigen in the sample binds to the labeled anti-Sg antibody in the sample well to form a labeled antibody–antigen complex that then diffuses across the membrane. At the test zone, the solid phase anti-Sg antibody binds with the labeled complex to form a labeled antibody–antigen–antibody sandwich. The antigen in the sample produces a pink vertical line at the test zone. In the control zone, unbound labeled anti-Sg antibody binds to the solid phase antiglobulin. This labeled antibody–antiglobulin complex at the control zone also results in a pink vertical line. The presence of Sg generates a pink line at

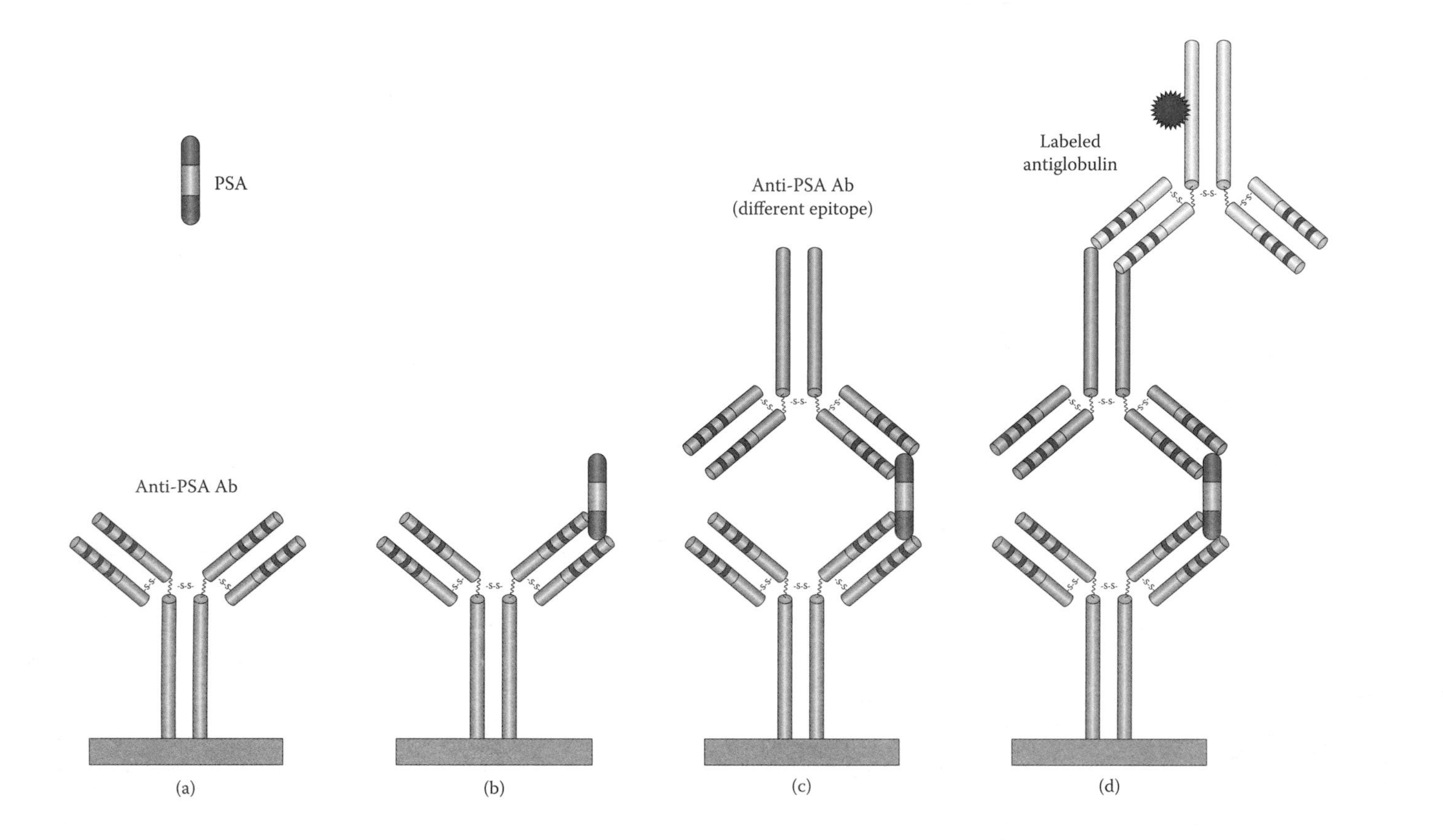


**Figure 8.9** Use of ELISA for identification of PSA in semen. (a) Sample containing PSA is applied to polystyrene tubes in which anti-PSA Ab is immobilized. (b) The PSA binds to immobilized Ab to form a PSA–Ab complex. (c) A second anti-PSA Ab , specific for a different epitope of PSA, is added to form a Ab–PSA–Ab sandwich. (d) A labeled antiglobulin then binds to the Ab–PSA–Ab sandwich. The bound antiglobulin can be detected by various reporting schemes. Ab = antibody. PSA = prostate-specific antigen.

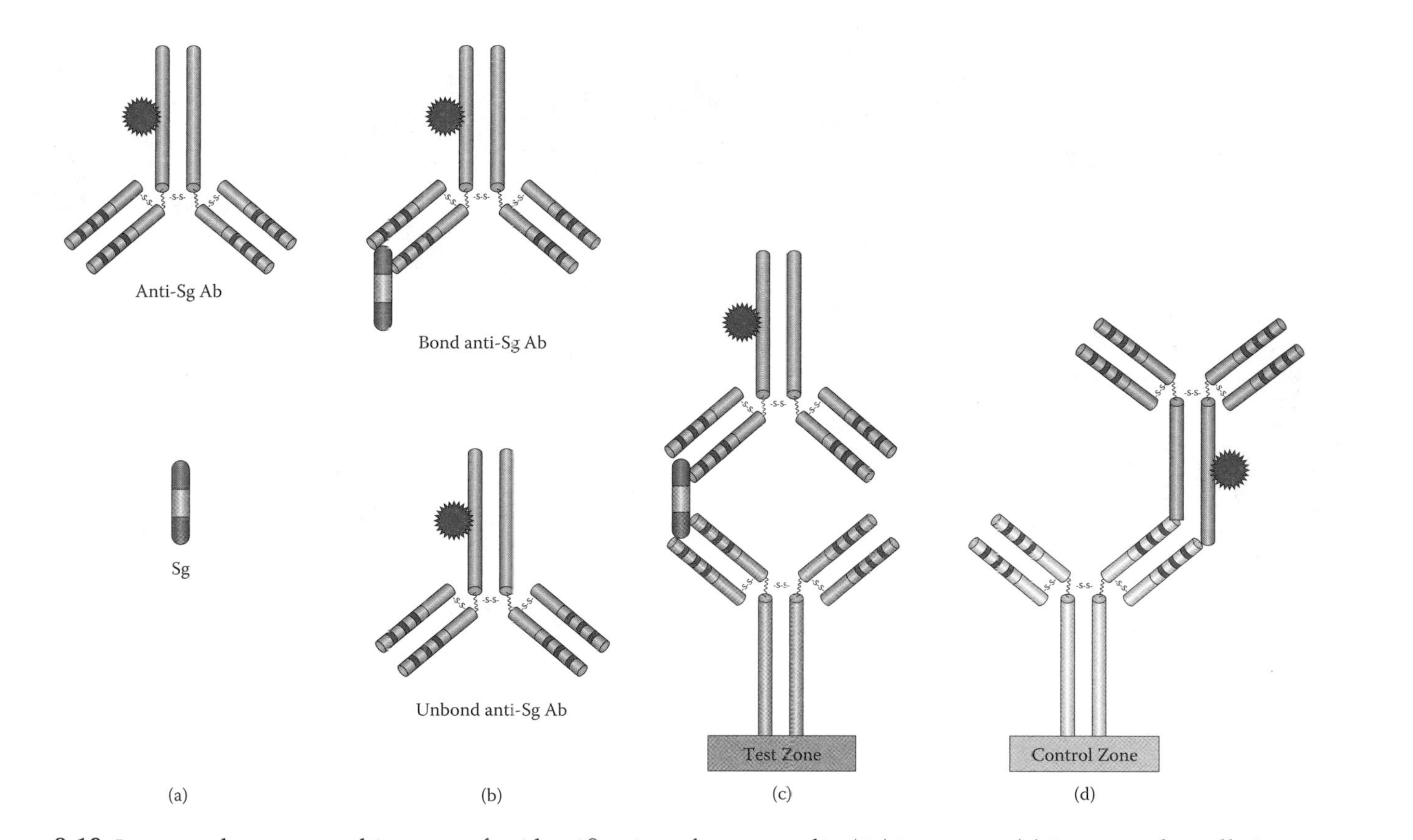


**Figure 8.10** Immunochromatographic assays for identification of semenogelin (Sg) in semen. (a) In a sample well, Sg in a semen sample is mixed with labeled anti-Sg Ab. (b) Sg binds to the labeled anti-Sg Ab to form a labeled Ab–Sg complex. (c) At the test zone, the labeled complex binds to an immobilized anti-Sg Ab to form a labeled Ab–Sg–Ab sandwich. (d) At the control zone, the labeled anti-Sg Ab binds to an immobilized antiglobulin and is captured at the control zone. Ab = antibody.

both the test and control zones. The absence of Sg results in a pink line in the control zone only. Results may be read after 10 min.

Validation studies revealed that the sensitivity of the RSID-Semen kit for detecting seminal fluid can be as low as a $5 \times 10^4$-fold dilution. Species specificity studies have shown no cross-reactivity with various animal species including ruminants and small mammals. Biological fluid specificity studies have also shown that the assay is not responsive to human blood, saliva, urine, sweat, fecal material, milk, or vaginal secretion. The assay results were not affected by condom lubricants or spermicides such as nonoxynol-9 and menfegol. However, the high-dose hook effect occurred when more than 3 μl seminal fluid was tested.

**8.2.2.3.2 ELISA** Identification of Sg for semen detection has also been carried out with ELISA. Anti-Sg antibodies are utilized. An antibody–antigen-antibody sandwich complex is formed (Figure 8.11). The intensity of the colorimetric or fluorometric signals can be detected spectrophotometrically and is proportional to the amount of bound antigen. The amount of Sg can be quantified by comparing a standard with known concentrations.

### *8.2.2.4 RNA-Based Assays*

RNA-based assays have been developed to identify semen. The assays are based on the expression of certain genes in certain cell or tissue types. Thus, the techniques used in the identification of semen are based on the detection of specific types of mRNA expressed exclusively in spermatozoa and in certain cells of male accessory glands. These assays utilize reverse transcriptase polymerase chain reaction (RT-PCR) methods to detect gene expression levels of mRNAs for semen identification. Table 8.1 lists the tissue-specific genes utilized for semen identification. Compared to conventional assays used for semen identification, the RNA-based assay has higher specificity and is amenable to automation. However, one limitation is that the RNA is unstable due to degradation by endogenous ribonucleases.

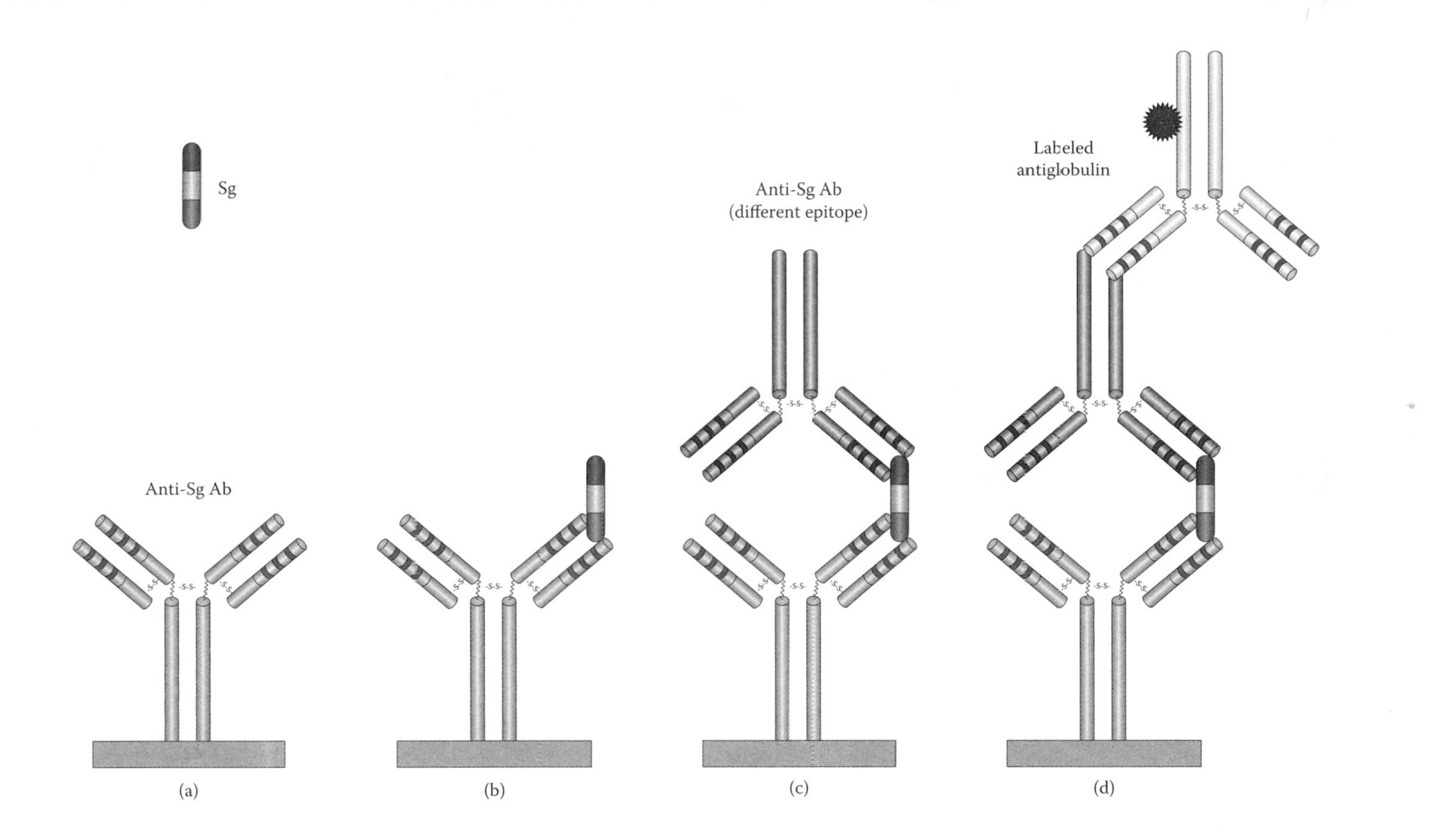


**Figure 8.11** Use of ELISA for identification of semenogelin (Sg) in semen. (a) Sample containing Sg is applied to polystyrene tubes in which anti-Sg Ab is immobilized. (b) Sg binds to immobilized Ab to form Sg–Ab complex. (c) A second anti-Sg Ab, specific for a different epitope of Sg is added to form an Ab–Sg–Ab sandwich. (d) A labeled antiglobulin then binds to the Ab-Sg-Ab sandwich. The bound antiglobulin can be detected by various reporting schemes. Ab = antibody.

**Table 8.1 Application of RT-PCR for Semen Identification**

| Gene Symbol | Gene Product | Description | Further Reading |
|---|---|---|---|
| *KLK3* | Kallikrein 3 | Also called prostate-specific antigen (PSA) | Gelmini et al., 2001 |
| *PRM1* | Protamine 1 | DNA binding proteins involved in condensation of sperm chromatin | Steger et al., 2000 |
| *PRM2* | Protamine 2 | DNA binding proteins involved in condensation of sperm chromatin | Steger et al., 2000 |

*Source:* Adapted from Jussola, J., and J. Ballantyne. 2005. *Forensic Sci Int* 152:1; Nussbaumer C. et al. 2006. *Forensic Sci Int* 157:181.

# Bibliography

Ablett, P. J. 1983. The identification of the precise conditions for seminal acid phosphatase (SAP) and vaginal acid phosphatase (VAP) separation by isoelectric focusing patterns. *J Forensic Sci Soc* 23 (3):255.

Allard, J. E. 1997. The collection of data from findings in cases of sexual assault and the significance of spermatozoa on vaginal, anal and oral swabs. *Sci Justice* 37 (2):99.

Allen, S. M. 1995. An enzyme linked immunosorbent assay (ELISA) for detection of seminal fluid using a monoclonal antibody to prostatic acid phosphatase. *J Immunoassay* 16 (3):297.

Allery, J. P. et al. 2001. Cytological detection of spermatozoa: Comparison of three staining methods. *J Forensic Sci* 46 (2):349.

Auvdel, M. J. 1988. Comparison of laser and high-intensity quartz arc tubes in the detection of body secretions. *J Forensic Sci* 33 (4):929.

Baechtel, S. 1988. The identification and individualization of semen stains, in *Forensic Science Handbook*. Saferstein, R., Ed. Englewood, NJ: Prentice Hall.

Bauer, M., and D. Patzelt. 2003. Protamine mRNA as molecular marker for spermatozoa in semen stains. *Int J Legal Med* 117 (3):175.

Baxter, S. J. 1973. Immunological identification of human semen. *Med Sci Law* 13 (3):155.

Blake, E. T., and G. F. Sensabaugh. 1976. Genetic markers in human semen: A review. *J Forensic Sci* 21 (4):785.

Chapman, R. L., N. M. Brown, and S. M. Keating. 1989. The isolation of spermatozoa from sexual assault swabs using proteinase K. *J Forensic Sci Soc* 29 (3):207.

Chen, J. T., and G. L. Hortin. 2000. Interferences with semen detection by an immunoassay for a seminal vesicle-specific antigen. *J Forensic Sci* 45 (1):234.

Clery, J. M. 2001. Stability of prostate specific antigen (PSA), and subsequent Y-STR typing, of *Lucilia (Phaenicia) sericata* (Meigen) (Diptera: Calliphoridae) maggots reared from a simulated postmortem sexual assault. *Forensic Sci Int* 120 (1-2):72.

Collins, K. A. et al. 1994. Identification of sperm and non-sperm male cells in cervicovaginal smears using fluorescence in situ hybridization: Applications in alleged sexual assault cases. *J Forensic Sci* 39 (6):1347.

Cortner, G. V., and A. J. Boudreau. 1978. Phase contrast microscopy versus differential interference contrast microscopy as applicable to the observation of spermatozoa. *J Forensic Sci* 23 (4):830.

Dahlke, M. B. et al. 1977. Identification of semen in 500 patients seen because of rape. *Am J Clin Pathol* 68 (6):740.

Davies, A. 1978. Evaluation of results from tests performed on vaginal, anal and oral swabs received in casework. *J Forensic Sci Soc* 17 (2-3):127.

Davies, A., and E. Wilson. 1974. The persistence of seminal constituents in the human vagina. *Forensic Sci* 3 (1):45.

Duenhoelter, J. H. et al. 1978. Detection of seminal fluid constituents after alleged sexual assault. *J Forensic Sci* 23 (4):824.

Elliott, K. et al. 2003. Use of laser microdissection greatly improves the recovery of DNA from sperm on microscope slides. *Forensic Sci Int* 137 (1):28.

Enos, W. F., and J. C. Beyer. 1978. Spermatozoa in the anal canal and rectum and in the oral cavity of female rape victims. *J Forensic Sci* 23 (1):231.

______. 1980. Prostatic acid phosphatase, aspermia, and alcoholism in rape cases. *J Forensic Sci* 25 (2):353.

Eungprabhanth, V. 1974. Finding of the spermatozoa in the vagina related to elapsed time of coitus. *Z Rechtsmed* 74 (4):301.

Fraysier, H. D. 1987. A rapid screening technique for the detection of spermatozoa. *J Forensic Sci* 32 (2):527.

Gaensslen, R. E. 1983. *Sourcebook in Forensic Serology, Immunology, and Biochemistry.* Washington, D.C.: U.S. Government Printing Office.

Gelmini, S. et al. 2001. Real-Time quantitative reverse transcriptase-polymerase chain reaction (RT-PCR) for measurement of prostate-specific antigen mRNA in peripheral blood of patients with prostate carcinoma using the Taqman detection system. *Clin Chem Lab Med* 39 (5):385.

Greenfield, A., and M. Sloan. 2005. Identification of biological fluids and stains, in *Forensic Science: An Introduction to Scientific and Investigative Techniques.* James, S., and J. J. Nordby, Eds. Boca Raton, FL: CRC Press.

Healy, D. A. et al. 2007. Biosensor developments: Application to prostate-specific antigen detection. *Trends Biotechnol.*

Herr, J. C. et al. 1986. Characterization of a monoclonal antibody to a conserved epitope on human seminal vesicle-specific peptides: A novel probe/marker system for semen identification. *Biol Reprod* 35 (3):773.

Herr, J. C., and M. P. Woodward. 1987. An enzyme-linked immunosorbent assay (ELISA) for human semen identification based on a biotinylated monoclonal antibody to a seminal vesicle-specific antigen. *J Forensic Sci* 32 (2):346.

Hochmeister, M. N. et al. 1997. High levels of alpha-amylase in seminal fluid may represent a simple artifact in the collection process. *J Forensic Sci* 42 (3):535.

Hochmeister, M. N. et al. 1999. Evaluation of prostate-specific antigen (PSA) membrane test assays for the forensic identification of seminal fluid. *J Forensic Sci* 44 (5):1057.

Hueske, E. E. 1977. Techniques for extraction of spermatozoa from stained clothing: A critical review. *J Forensic Sci* 22 (3):596.

Ishiyama, I. 1981. Rapid histological examination of trace evidence by means of cellophane tape. *J Forensic Sci* 26 (3):570.

Iwasaki, M. et al. 1989. A demonstration of spermatozoa on vaginal swabs after complete destruction of the vaginal cell deposits. *J Forensic Sci* 34 (3):659.

Jones, E. L. 2005. The identification of semen and other body fluids, in *Forensic Science Handbook*. Saferstein, R., Ed. Englewood Cliffs, NJ: Pearson Prentice Hall.

Jussola, J., and J. Ballantyne. 2003. Messenger RNA profiling: a prototype method to supplant conventional methods for body fluid identification. *Forensic Sci Int* 135 (2):85.

_____. 2005. Multiplex mRNA profiling for the identification of body fluids. *Forensic Sci Int* 152 (1):1.

Keil, W., J. Bachus, and H. D. Troger. 1996. Evaluation of MHS-5 in detecting seminal fluid in vaginal swabs. *Int J Legal Med* 108 (4):186.

Khaldi, N. et al. 2004. Evaluation of three rapid detection methods for the forensic identification of seminal fluid in rape cases. *J Forensic Sci* 49 (4):749.

Levine, B. et al. 2004. Use of prostate specific antigen in the identification of semen in postmortem cases. *Am J Forensic Med Pathol* 25 (4):288.

Liedtke, R. J., and J. D. Batjer. 1984. Measurement of prostate-specific antigen by radioimmunoassay. *Clin Chem* 30 (5):649.

Lilja, H. 1985. A kallikrein-like serine protease in prostatic fluid cleaves the predominant seminal vesicle protein. *J Clin Invest* 76 (5):1899.

Lin, M. F. et al. 1980. Fundamental biochemical and immunological aspects of prostatic acid phosphatase. *Prostate* 1 (4):415.

Linde, H. G., and K. E. Molnar. 1980. The simultaneous identification of seminal acid phosphatase and phosphoglucomutase by starch gel electrophoresis. *J Forensic Sci* 25 (1):113.

Maher, J. et al. 2002. Evaluation of the BioSign PSA membrane test for the identification of semen stains in forensic casework. *New Zeal Med J* 115 (1147):48.

Masibay, A. S., and N. T. Lappas. 1984. The detection of protein P30 in seminal stains by means of thin-layer immunoassay. *J Forensic Sci* 29 (4):1173.

McGee, R. S., and J. C. Herr. 1988. Human seminal vesicle-specific antigen is a substrate for prostate-specific antigen (or P-30). *Biol Reprod* 39 (2):499.

Montagna, C. P. 1996. The recovery of seminal components and DNA from the vagina of a homicide victim 34 days postmortem. *J Forensic Sci* 41 (4):700.

Nussbaumer, C., E. Gharehbaghi-Schnell, and I. Korschineck. 2006. Messenger RNA profiling: A novel method for body fluid identification by real-time PCR. *Forensic Sci Int* 157 (2–3):181.

Pang, B. C., and B. K. Cheung. 2006. Identification of human semenogelin in membrane strip test as an alternative method for the detection of semen. *Forensic Sci Int*.

Poyntz, F. M., and P. D. Martin. 1984. Comparison of p30 and acid phosphatase levels in post-coital vaginal swabs from donor and casework studies. *Forensic Sci Int* 24 (1):17.

Randall, B. 1988. Glycogenated squamous epithelial cells as a marker of foreign body penetration in sexual assault. *J Forensic Sci* 33 (2):511.

Rodrigues, R. G. et al. 2001. Semenogelins are ectopically expressed in small cell lung carcinoma. *Clin Cancer Res* 7 (4):854.

Roy, A. V., M. E. Brower, and J. E. Hayden. 1971. Sodium thymolphthalein monophosphate: A new acid phosphatase substrate with greater specificity for the prostatic enzyme in serum. *Clin Chem* 17 (11):1093.

Sato, I. et al. 2001. A dot-blot-immunoassay for semen identification using a polyclonal antibody against semenogelin, a powerful seminal marker. *Forensic Sci Int* 122 (1):27.

Sato, I. et al. 2002. Use of the "SMITEST" PSA card to identify the presence of prostate-specific antigen in semen and male urine. *Forensic Sci Int* 127 (1-2):71.

Sato, I. et al. 2004. Rapid detection of semenogelin by one-step immunochromatographic assay for semen identification. *J Immunol Methods* 287 (1-2):137.

Sato, I. et al. 2007. Urinary prostate-specific antigen is a noninvasive indicator of sexual development in male children. *J Androl* 28 (1):150.

Schiff, A. F. 1978. Reliability of the acid phosphatase test for the identification of seminal fluid. *J Forensic Sci* 23 (4):833.

Sensabaugh, G. F. 1978. Isolation and characterization of a semen-specific protein from human seminal plasma: a potential new marker for semen identification. *J Forensic Sci* 23 (1):106.

_____. 1979. The quantitative acid phosphatase test: a statistical analysis of endogenous and postcoital acid phosphatase levels. *J Forensic Sci* 24 (2):19.

Shaler, R. C., and P. Ryan. 1982. High acid phosphatase levels as a possible false indicator of the presence of seminal fluid. *Am J Forensic Med Pathol* 3 (2):161.

Simich, J. P. et al. 1999. Validation of the use of a commercially available kit for the identification of prostate specific antigen (PSA) in semen stains. *J Forensic Sci* 44 (6):1229.

Sokoll, L. J., and D. W. Chan. 1997. Prostate-specific antigen: discovery and biochemical characteristics. *Urol Clin North Am* 24 (2):253.

Standefer, J. C., and E. W. Street. 1977. Postmortem stability of prostatic acid phosphatase. *J Forensic Sci* 22 (1):165.

Steger, K. et al. 2000. Expression of protamine-1 and -2 mRNA during human spermiogenesis. *Mol Hum Reprod* 6 (3):219.

Stoilovic, M. 1991. Detection of semen and blood stains using polilight as a light source. *Forensic Sci Int* 51 (2):289.

Stowell, L. I., L. E. Sharman, and K. Hamel. 1991. An enzyme-linked immunosorbent assay (ELISA) for prostate-specific antigen. *Forensic Sci Int* 50 (1):125.

Stubbings, N. A., and P. J. Newall. 1985. An evaluation of gamma-glutamyl transpeptidase (GGT) and p30 determinations for the identification of semen on postcoital vaginal swabs. *J Forensic Sci* 30 (3):604.

Ward, A. M., J. W. Catto, and F. C. Hamdy. 2001. Prostate-specific antigen: Biology, biochemistry and available commercial assays. *Ann Clin Biochem* 38 (Pt 6):633.

Willott, G. M. and J. E. Allard. 1982. Spermatozoa: Their persistence after sexual intercourse. *Forensic Sci Int* 19 (2):135.

Willott, G. M., and M. A. Crosse. 1986. The detection of spermatozoa in the mouth. *J Forensic Sci Soc* 26 (2):125.

Yokota, M. et al. 2001. Evaluation of prostate-specific antigen (PSA) membrane test for forensic examination of semen. *Leg Med (Tokyo)* 3 (3):171.

Yoshida, K. et al. 2003. Quantification of seminal plasma motility inhibitor/semenogelin in human seminal plasma. *J Androl* 24 (6):878.

Yu, H., and E. P. Diamandis. 1995. Prostate-specific antigen in milk of lactating women. *Clin Chem* 41 (1):54.
Zarghami, N. et al. 1997. Prostate-specific antigen in serum during the menstrual cycle. *Clin Chem* 43 (10):1862.

## Study Questions

1. An alternate light source is useful when searching for:
   (a) Semen stains
   (b) Saliva stains
   (c) Blood stains
   (d) Both A and B

2. Which of the following is used to test for the presence of semen?
   (a) Acid phosphatase
   (b) Fluorescin
   (c) TMB
   (d) Luminol

3. Which of the following is true?
   (a) Semen can be identified by the presence of spermatozoa.
   (b) Semen contains seminal vesicle-specific antigen.
   (c) Semen contains prostate-specific antigen.
   (d) All of the above.

4. Prostate-specific antigen can be present:
   (a) In cases of oligospermia (low sperm count)
   (b) In cases of aspermia (absence of sperm)
   (c) When a suspect uses a birth control tool (condom)
   (d) When evidence is collected by using a sexual assault evidence kit

5. Spermatazoa can be present:
   (a) In cases of oligospermia
   (b) In cases of aspermia
   (c) When a suspect uses a birth control tool (condom)
   (d) When evidence is collected by using a sexual assault evidence kit

# Identification of Saliva and Other Biological Fluids

# 9

## 9.1 Identification of Saliva

The identification of saliva may be important in the investigation of criminal and civil cases. For example, saliva evidence can be important in cases in which oral copulation may have occurred. Additionally, the cells in saliva contain DNA and are often used for forensic DNA testing.

### 9.1.1 Biological Characteristics of Saliva

The human salivary glands produce 1.0 to 1.5 liters of saliva daily. About 70% of saliva is produced from the submandibular salivary glands, 25% from the parotids, and 5% from the sublingual salivary glands (Figure 9.1). Although a continuous basal level of saliva secretion is maintained, a large amount of saliva is produced during eating. Saliva is mostly water containing small amounts of electrolytes, buffers, glycoproteins, antibodies, and enzymes. The process of digesting carbohydrates in the diet begins in the oral cavity when amylase in the saliva breaks down carbohydrates such as starch. Thus, detecting amylase indicates the presence of saliva.

#### *9.1.1.1 Amylases*

Amylases are enzymes that cleave polysaccharides such as starches that are composed of D-glucose units connected by $\alpha 1 \rightarrow 4$ linkages. Starches contain two types of glucose polymers: ***amylose*** and ***amylopectin*** (Figure 9.2) Amylose consists of long, unbranched chains of glucose residues connected by $\alpha 1 \rightarrow 4$ linkages. Amylopectin is highly branched and consists of linear chains of glucose residues connected by $\alpha 1 \rightarrow 4$ linkages with the branch points connected by $\alpha 1 \rightarrow 6$ linkages. Both linear amylose and amylopectin can be hydrolyzed by amylase by cleaving the chains at alternate $\alpha 1 \rightarrow 4$ linkages. Amylase cleaves off one maltose (two glucose units) at a time. However, the $\alpha 1 \rightarrow 6$ linkages at the branch points are not cleaved by the amylase.

Two types of amylases can be characterized. β-***amylases*** found in plant and bacterial sources cleave only at the terminal reducing end of a polysaccharide chain. The end of a chain with a free anomeric carbon (not involved in a glycosidic bond) is called the reducing end. Human α-***amylases*** cleave at $\alpha 1 \rightarrow 4$ linkages randomly along the polysaccharide chain.

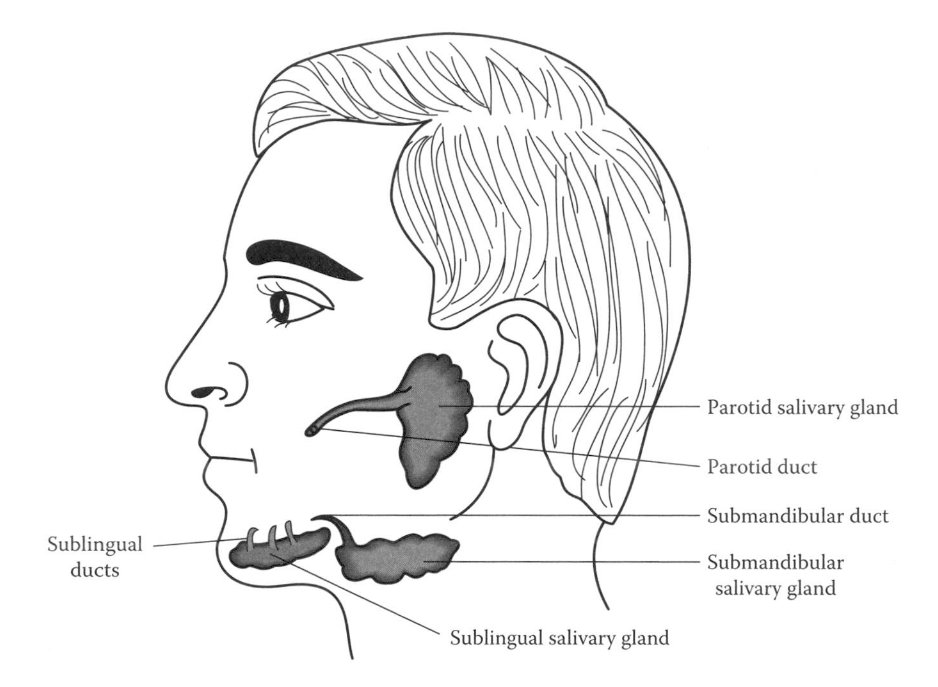


**Figure 9.1** Human salivary glands.

Human α–amylases have two major isoenzymes (multiple forms that differ in their amino acid sequences). Human ***salivary α-amylase*** (HSA) is encoded by the *Amy1* locus, synthesized at the salivary glands and secreted into the oral cavity. Human ***pancreatic α-amylase*** (HPA), encoded by the *Amy2* locus, is synthesized by the pancreas and secreted into the duodenum through the pancreatic duct. The amino acid sequences of the HSA and HPA are highly homologous. However, HSA is inactivated by acids in the stomach, while most HPA is inactivated in the lower portions of the intestine, and some amylase activity remains in the feces.

Amylase activity is found in various biological fluids including semen, tears, milk, perspiration, and vaginal secretions. Most amylase present in normal serum consists of HPA and HSA. The amylases are small molecules and can pass through the glomeruli of the kidney. Thus, the amylase present in urine is derived from plasma. Amylase can be inactivated under boiling temperatures and strong acidic and alkaline conditions. Based on various studies, its stability varies from a few weeks to several months.

### 9.1.2 Analytical Techniques for Identification of Saliva

Visual examination methods such as the use of alternate light sources can be helpful in searching for saliva stains to facilitate further analysis. The

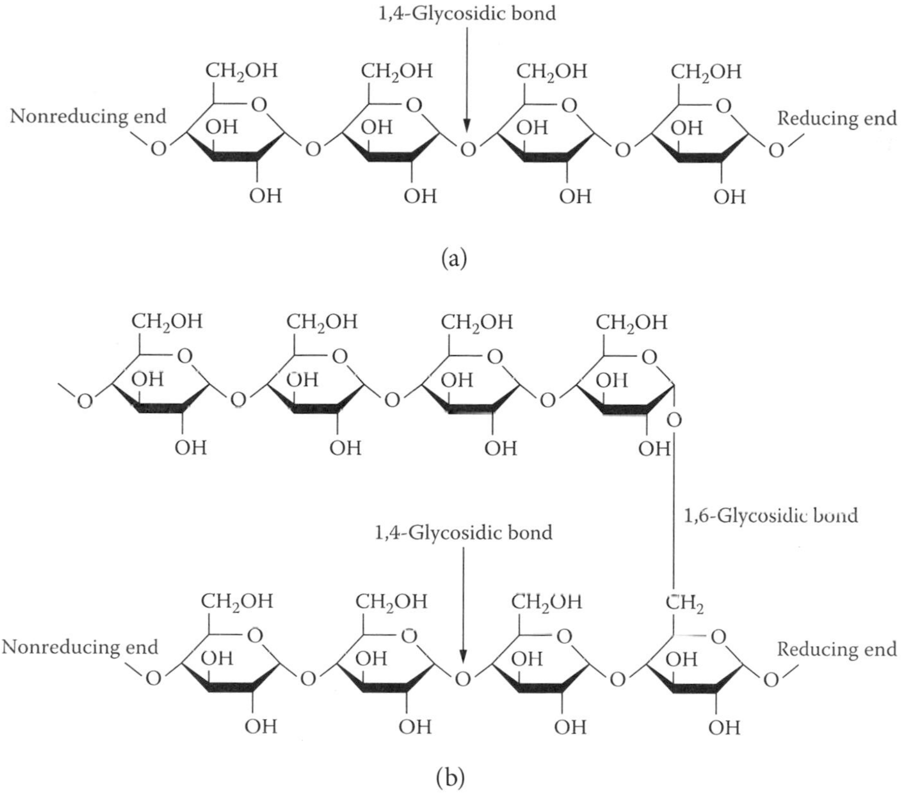

**Figure 9.2** Chemical structures of polysaccharides found in starch. (a) amylose. (b) amylopectin.

identification of saliva is largely based on the presence of amylase in a sample. The amylase assays fall into two categories. The first category measures the enzymatic activity of amylase. Most of these assays measure total amylase and cannot distinguish HSA from other amylases including HPA and non-human amylases such as those from plants, animals, and microorganisms. The second category consists of more confirmatory than enzymatic assays and includes direct detection of HSA proteins and RNA-based assays.

### 9.1.2.1 *Presumptive Assays*

**9.1.2.1.1 Visual Examination** The lighting techniques used to search for semen stains can be employed in searching for saliva stains. For example, a 470-nm excitation wavelength can be used with colored goggles or interference filters of higher wavelengths (555 nm) to allow visualization of fluorescence. However, the fluorescence of a saliva stain is usually less intense than that of a seminal stain. Microscopic examination with proper histological staining can also be performed to identify the buccal epithelial cells that indicate a saliva stain.

#### 9.1.2.1.2 Determination of Amylase Activity

**Starch–Iodine Assay** — Iodine ($I_2$) is used to test for the presence of starch. The amylose in starch reacts strongly with iodine to form a dark blue complex, while amylopectin develops reddish-purple colors. In the presence of amylases, starch is broken down to mono or disaccharides. Consequently, such colors do not develop when iodine is added.

A common configuration of the method is the radial diffusion assay (Figure 9.3). An agar gel containing starch is prepared. A sample well is created by punching a hole in the gel and an extract of the questioned sample is placed into the well. If amylase is present in the sample, it diffuses from the sample well and hydrolyzes the starch in the gel. The gel is then stained using an iodine solution. A clear area in the gel indicates amylase activity and the size of the clear area is proportional to the amount of amylase in the sample. A linear standard curve (in log scale) can be prepared using a series of standard amylases with known concentrations. The amount of amylase can be quantified by comparing the results with the standard curve. The assay is not specific to HSA and can produce false positive results.

**Colorimetric Assays** — Dye-labeled amylase substrates such as dye-conjugated amylose or amylopectin are employed. These substrates are not

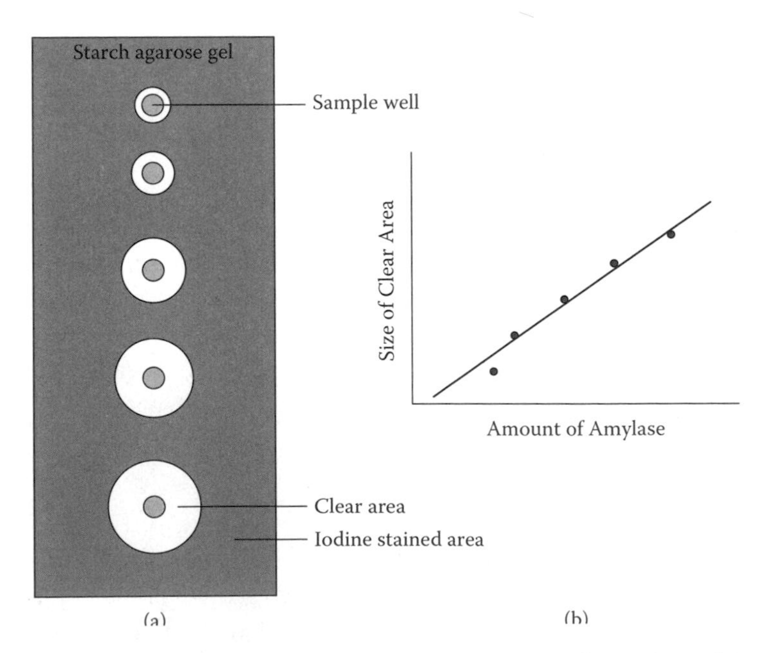


**Figure 9.3** Radial diffusion assay for identification of amylase. (a) Known amounts of amylase standards are applied to the well and allowed to diffuse. (b) The size of clear area arising from amylase activity is plotted. The standard curve can be used to determine the amount of amylase in a questioned sample.

soluble in water. In the presence of amylase, the dye-containing moieties are cleaved and are soluble in water to produce a color. The degree of coloration, which can be measured colorimetrically by spectrophotometric methods, is proportional to the amount of amylase in the sample. Most of these assays are not HSA-specific. Although their specificity can be tested by using inhibitors that preferentially inhibit HSA, the assays are considered presumptive; they are not conclusive for the specific presence of saliva in a sample.

While many substrates are available, the one usually used in forensic laboratories is Phadebas (Pharmacia) reagent. Produced in a tablet form, it is used to detect α-amylase in specimens for clinical diagnostic purposes. A small portion (approximately 3 $mm^2$) of a sample is cut and placed in a tube and incubated for 5 min at 37°C. One Phadebas tablet is added to each tube and mixed. Samples are then incubated for 15 min at 37°C and the reaction is stopped at an alkaline pH by addition of sodium hydroxide. The amylase substrate is an insoluble blue dye conjugated to starch. Amylase hydrolyzes the substrate to generate a blue color that can be measured at 620 nm using a spectrophotometer. The optical density of the supernatant is read and can be converted to amylase units by comparing to a standard curve.

The amylase assay can also be used for amylase mapping as a method of searching for possible saliva stains (Figure 9.4). These assays are based on the principle that amylase is water-soluble and can be transferred from evidence to filter paper and then analyzed via colorimetric assay. This procedure is also referred to as a ***press test***. The sensitivity of the method is similar to that of test tube method.

The substrate can be prepared by evenly spraying the Phadebas reagent on a sheet of filter paper and allowing it to air dry. The dried substrate containing paper can be used immediately or stored until needed. To perform amylase mapping, a piece of paper is placed over the entire area to be tested (the item to be tested must be fairly flat to ensure good contact with the paper). The paper is dampened slightly by spraying with distilled water. An outline may be drawn on the paper to aid in the location of stains. A piece of plastic wrap is placed on top to prevent the paper from drying during the assay and a weight is applied to ensure good contact of substrate and evidence. The test is observed every minute for the first 10 min and every 5 min thereafter, up to 40 min when a positive reaction should appear as a light blue area.

SALIgAE® (Abacus Diagnostics), another commercially available colorimetric assay kit, has been validated by some laboratories. Its manufacturer also produces the SALIgAE spray kit that can be used for amylase mapping.

### *9.1.2.2 Confirmatory Assays*

#### 9.1.2.2.1 Identification of Human Salivary α-Amylase

**Immunochromatographic Assays** — Commercially produced immunochromatographic kits include the RSID®-Saliva (Independent Forensics). A

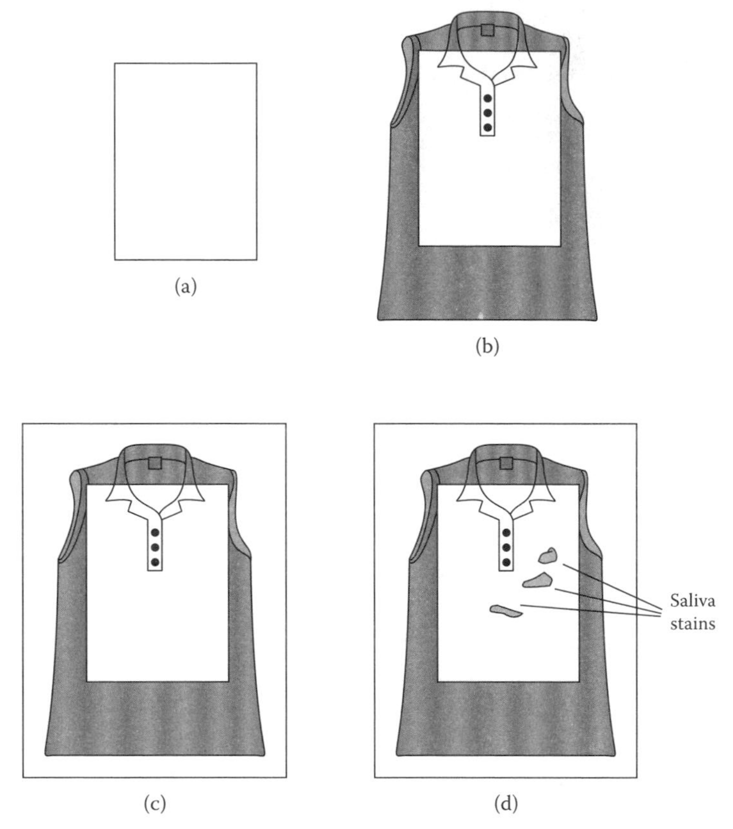


**Figure 9.4** Amylase mapping for saliva stains. (a) Amylase substrate is sprayed on a sheet of filter paper. (b) Substrate-containing paper is placed over the area to be tested. The orientation of the filter paper is marked to aid in locating the stain. (c) The filter paper is dampened by spraying with water and plastic wrap is placed on top to prevent the paper from drying. (d) Blue color indicates a positive reaction.

labeled monoclonal anti-HSA antibody is contained in a sample well, and a second monoclonal anti-HSA antibody is immobilized onto a test zone of a membrane and an antiglobulin that recognizes the antibody is immobilized onto a control zone (Figure 9.5).

A sample can be prepared by cutting out a small portion of a stain or a swab. Each sample is extracted for 1 to 2 h in 200 to 300 μl of an extraction buffer. Approximately 10% of the extract is removed and mixed with a running buffer. The assay is carried out by loading an extracted sample into the sample well where the antigen in the sample binds to the labeled anti-HSA antibody in the well to form a labeled antibody–antigen complex. The complex then diffuses across the membrane to the test zone, where the solid phase anti-HSA antibody binds with the labeled complex to form a labeled antibody–antigen–antibody sandwich.

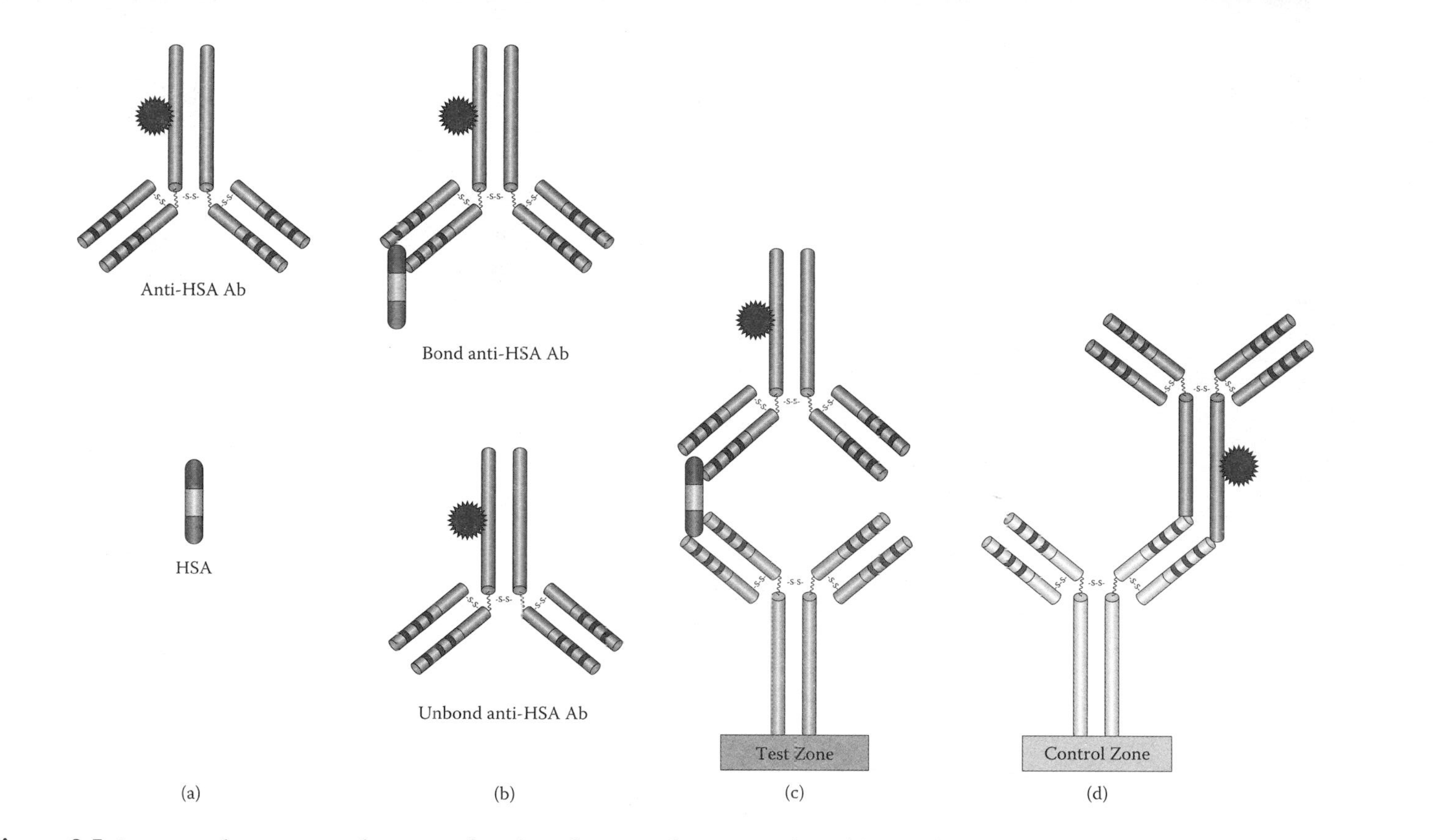


**Figure 9.5** Immunochromatographic assay for identification of HSA in saliva. (a) Sample containing amylases is loaded in a sample well. (b) Antigen binds to a labeled anti-HSA Ab to form a labeled Ab–HSA complex. (c) At the test zone, the labeled complex binds to an immobilized antihuman HSA Ab to form a labeled Ab–HSA–Ab sandwich. (d) At the control zone, the labeled anti-HSA Ab binds to an immobilized antiglobulin and is captured at the control zone. Ab = antibody. HSA = human salivary α-amylase.

The presence of antigen in the sample results in a pink vertical line at the test zone. In the control zone, unbound labeled anti-HSA antibody binds to the solid phase antiglobulin. The labeled antibody–antiglobulin complex at the control zone also results in a pink vertical line. The test is considered valid only if the line in the control zone is observed. The presence of HSA results in a pink line at both the test zone and the control zone while the absence of HSA results in a pink line in the control zone only. A result can be read after 10 min.

The validation studies revealed that the sensitivity of the RSID®-Saliva kit can be as low as 1 µl of saliva. Additionally, the assay was responsive to samples extracted from saliva stains on both smooth and porous surfaces. Species specificity studies have shown that there is no cross reactivity with various animal species, including monkeys (tamarin and callimico), and body fluid specificity studies have also shown that the kit is not responsive to human blood, semen, or urine. The high-dose hook effect, which creates an artifact that may cause false negative results as described previously (see Chapter 5), was not observed when up to 50 µl of saliva was tested. This method is rapid, specific and sensitive and can be used in both laboratory and field analysis.

**Enzyme-Linked Immunosorbent Assay (ELISA)** — This method can be used to detect and quantify a sample with the use of an anti-HSA antibody. The most common configuration in forensic serology is the antibody–antigen–antibody sandwich (Figure 9.6). ELISA utilizes reporting enzymes to produce colorimetric or fluorometric signals. The intensity of the signal can be detected spectrophotometrically and is proportional to the amount of bound antigen. The amount of HSA can be quantified by comparison with a standard of known concentration. This method is specific and highly sensitive in detecting HSA, but it is time-consuming. Chapter 5 discusses the ELISA principle in further detail.

**9.1.2.2.2 RNA-Based Assays** RNA-based assays have been developed recently for the identification of saliva. They are based on the expression of certain genes in certain cell or tissue types. Thus, the techniques used in the identification of saliva are based on the detection of specific types of mRNA expressed exclusively in certain cells in the oral cavity. These assays utilize reverse transcriptase polymerase chain reaction (RT-PCR) methods to detect gene expression levels of mRNAs for saliva identification. Table 9.1 summarizes the tissue-specific genes utilized for saliva identification. Compared to conventional assays used for saliva identification, RNA-based assays present higher specificity and are amenable to automation. However, one limitation is that the RNA is unstable because of degradation by endogenous ribonucleases.

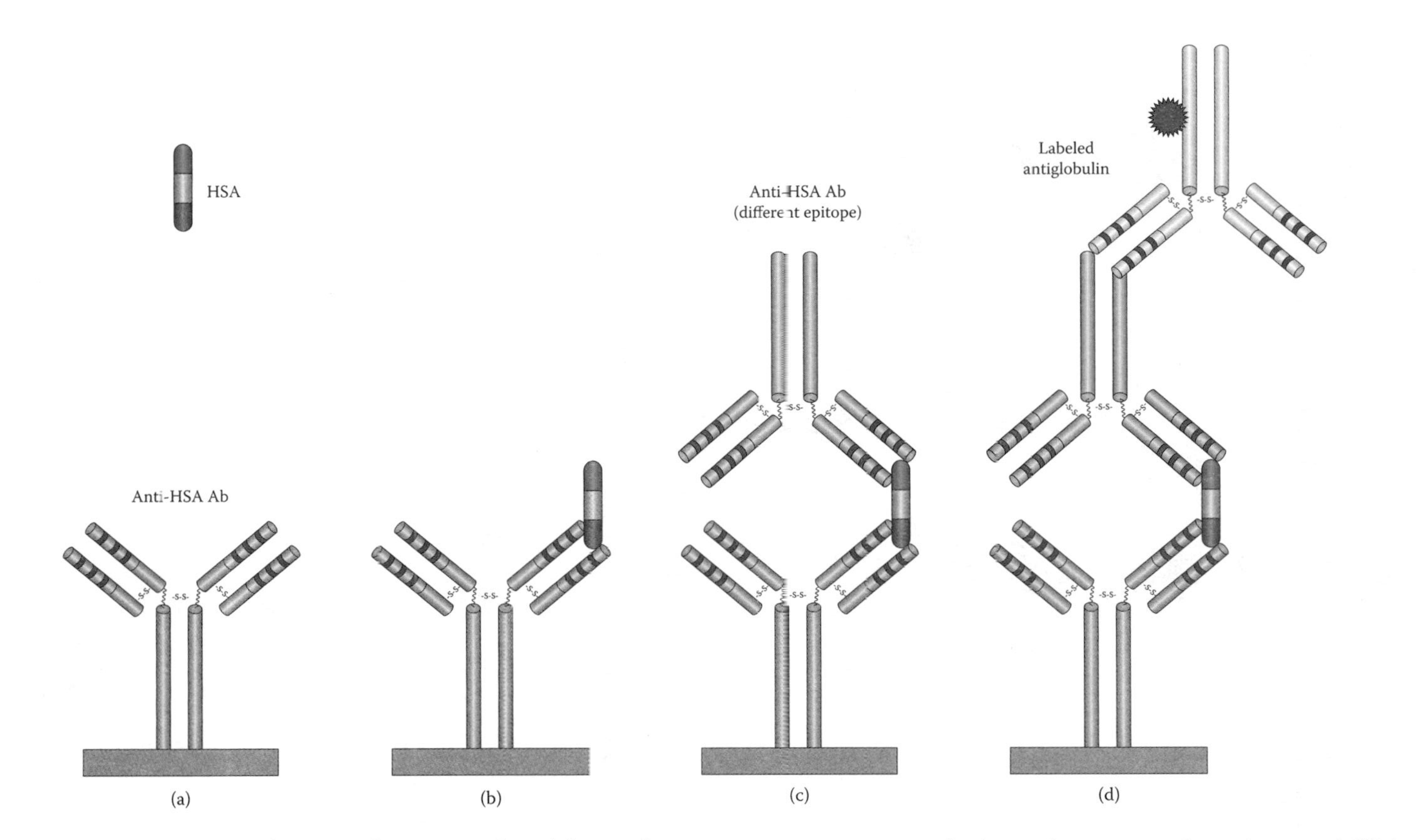


**Figure 9.6** ELISA for identification of HSA in saliva. (a) Sample containing antigen is applied to polystyrene tubes where anti-HSA Ab is immobilized. (b) Antigen binds to immobilized Ab to form HSA-Ab complex. (c) Second anti-HSA Ab, specific for a different epitope of HSA, is added to form Ab–HSA–Ab sandwich. (d) Labeled antiglobulin then binds to the sandwich. The bound antiglobulin can be detected by various reporting schemes. Ab = antibody. HSA = human salivary α-amylase.

**Table 9.1 Application of RT-PCR Assay for Saliva Identification**

| Gene Symbol | Gene Product | Description | Further Reading |
|---|---|---|---|
| *HTN3* | Histatin 3 | Histidine-rich protein involved in nonimmune host defense in oral cavity | Sabatini et al., 1993 |
| *STATH* | Statherin | Inhibitor of precipitation of calcium phosphate salts in oral cavity | Sabatini et al., 1990 |

*Source:* Adapted from Jussola, J., and J. Ballantyne. 2005. *Forensic Sci Int* 152:1.

## 9.2 Other Biological Fluids

Other biological fluids such as vaginal secretions, menstrual blood, urine, feces, and vomitus may require forensic analysis. Although they are less frequently encountered than blood, semen, and saliva, such biological fluids can have probative value. The identification of vaginal secretions can be important for sexual assault cases involving vaginal rape with a foreign object. The identification of menstrual blood can be important to distinguish menstruation from vaginal trauma in sexual assault cases. Urine evidence can be useful for investigating assault involving urination. Fecal samples can be important for sexual assault cases in which anal sodomy occurred. The identification of vomitus may assist in crime scene reconstruction. Table 9.2 summarizes the methods of identification of these biological fluids.

**Table 9.2 Assays for Identification of Miscellaneous Biological Fluids**

| Biological Fluid | Identification | Assay | Further Reading |
|---|---|---|---|
| Vaginal secretion | Locating stains | Lighting techniques (i.e., alternate light source) | Gaensslen, 1983 |
| | Glycogenated nucleated squamous epithelial cells (NSEC); cell type found in vaginal secretions | Lugol's iodine assay or periodic acid Schiff reagent (color assays) | Peabody et al., 1981 |
| | Vaginal acid phosphatase (VAP) | Electrophoresis; can resolve differences between VAP and seminal acid phosphatase | Ablett, 1983 |
| | *HBD1* (β-defensin 1); antimicrobial peptide expressed in vaginal epithelium | RNA-based assay; utilizes RT-PCR to detect tissue-specific gene expression for biological fluid identification | Juusola and Ballantyne, 2005 |
| | *MUC4* (mucin 4); major membrane protein of endocervix that protects epithelial surfaces of reproductive tract against pathogens and controls sperm entry into uterus | RNA-based assay; utilizes RT-PCR to detect tissue-specific gene expression for biological fluid identification | Nussbaumer et al., 2006 |
| Menstrual blood | Glycogenated nucleated squamous epithelial cells (NSEC); cell type found in vaginal secretions<br>Lactate dehydrogenase; isoenyme found in menstrual blood | Lugol's iodine assay or periodic acid Schiff reagent (color assays) | Peabody et al., 1981<br>Thompson and Belschner, 1997 |
| | *MMP7* (matrix metalloproteinase); can be found in menstrual blood but not in circulatory blood; expressed in endometrial tissue, involved in tissue remodeling during menstruation | RNA-based assay; utilizes RT-PCR to detect tissue-specific gene expression for biological fluid identification | Juusola and Ballantyne, 2005 |

**Table 9.2 Assays for Identification of Miscellaneous Biological Fluids** (continued)

| Biological Fluid | Identification | Assay | Further Reading |
|---|---|---|---|
| Urine | Locating stains | Lighting techniques(i. e., alternate light source) | Gaensslen, 1983 |
| | Urea; constituent of urine | Microscopic crystal test; microscopic examination of urea nitrate crystals converted from urea | Nickolls, 1956 |
| | | Color test; use of urease to catalyze breakdown of urea and release ammonia and carbon dioxide; ammonia detected using DMAC (p-dimethylamino cinnamaldehyde) | Rhodes and Thornton, 1976 |
| | | Gas chromatography-mass spectrometry (GC-MS), thin-layer chromatography | Lum, 1991 |
| | Creatinine; constituent of urine | Jaffe color test using picric acid to convert creatinine to colored creatinine picrate | Gaensslen, 1983 |
| | | Color test with potassium ferricyanide | Gaensslen, 1983 |
| | | Color test with o-nitrobenzaldehyde | Gaensslen, 1983 |
| | | Chromatography; GC-MS | Lum, 1991 |
| | Tamm Horsfall glycoprotein (THG); abundant urinary protein | Immunological assays: RIA, ELISA, crossed-over electrophoresis, and Ouchterlony assay | Tsutsumi et al., 1988 |

**Table 9.2 Assays for Identification of Miscellaneous Biological Fluids** (continued)

| Biological Fluid | Identification | Assay | Further Reading |
|---|---|---|---|
| Feces | Fecal urobilinoids; colored metabolites excreted into the feces | Conversion of urobilinoids to urobilin zinc salt that fluoresces under light (450 nm) | Lloyd and Weston, 1982 |
| | Undigested or partially digested plant cells | Microscopic examination | Bock et al., 1988 |
| Vomitus | Food particles | Microscopic examination | Bock et al., 1988 |
| | Pepsin; proteolytic enzyme secreted from gastric gland | Enzymatic assay | Yanada et al., 1992 |
| Sweat | Sweat-specific antigen | Immunological assays using monoclonal antibody | Sagawa et al., 2002 |

*Source:* Adapted from Greenfield, A., and M. M. Sloan. 2005. Identification of biological fluids and stains, in *Forensic Science: An Introduction to Scientific and Investigative Techniques*. James, S. and J. J. Nordby, Eds. Boca Raton, FL: CRC Press; Jones, E. L. 2005. The identification of semen and other body fluids, in *Forensic Science Handbook*. Saferstein, R., Ed. Englewood Cliffs, NJ: Pearson Prentice Hall; Shaler, R.C. 2002. Modern forensic biology, in *Forensic Science Handbook*. Saferstein, R., ed. Upper Saddle River, NJ: Pearson Education Inc.

# Bibliography

## Saliva


Aps, J. K., and L. C. Martens. 2005. The physiology of saliva and transfer of drugs into saliva. *Forensic Sci Int* 150 (2-3):119.

Auvdel, M. J. 1986. Amylase levels in semen and saliva stains. *J Forensic Sci* 31 (2):426.

_____. 1988. Comparison of laser and high-intensity quartz arc tubes in the detection of body secretions. *J Forensic Sci* 33 (4):929.

Barni, F. et al. 2006. Alpha-amylase kinetic test in bodily single and mixed stains. *J Forensic Sci* 51 (6):1389.

Culliford, B. J. 1964. Precipitin reactions in forensic problems: A new method for precipitin reactions on blood, semen, and saliva stains. *Nature* 201:1092.

DeForest, P., R. Gaensslen, and H. C. Lee. 1983. *Forensic Science: An Introduction to Criminalistics*. New York: McGraw-Hill.

DeLeo, D. et al. 1985. A sensitive and simple assay of saliva on stamps. *Z Rechtsmed* 95 (1):27.

Eckersall, P. D. et al. 1981. The production and evaluation of an antiserum for the detection of human saliva. *J Forensic Sci Soc* 21 (4):293.

Gaensslen, R. E. 1983. *Sourcebook in Forensic Serology, Immunology, and Biochemistry*. Washington, D.C.: U.S. Government Printing Office.

Greenfield, A., and M. Sloan. 2005. Identification of biological fluids and stains, in *Forensic Science: An Introduction to Scientific and Investigative Techniques*. James, S. and J. J. Nordby, Eds. Boca Raton, FL: CRC Press.

Hochmeister, M. N. et al. 1997. High levels of α-amylase in seminal fluid may represent a simple artifact in the collection process. *J Forensic Sci* 42 (3):535.

Jones, E. L. 2005. The identification of semen and other body fluids, in *Forensic Science Handbook*. Saferstein, R., Ed. Englewood Cliffs, NJ: Pearson Prentice Hall.

Jones, E. L. Jr., and J. A. Leon. 2004. Lugol's test reexamined again: Buccal cells. *J Forensic Sci* 49 (1):64.

Juusola, J., and J. Ballantyne. 2003. Messenger RNA profiling: A prototype method to supplant conventional methods for body fluid identification. *Forensic Sci Int* 135 (2):85.

_____. 2005. Multiplex mRNA profiling for the identification of body fluids. *Forensic Sci Int* 152 (1):1.

Keating, S. M., and D. F. Higgs. 1994. The detection of amylase on swabs from sexual assault cases. *J Forensic Sci Soc* 34 (2):89.

Lee, H. C. 1982. Identification and grouping of blood stains. *Forensic Science Handbook*. Saferstein, R., Ed. Englewood Cliffs, NJ: Prentice Hall.

Martin, N. C., N. J. Clayson, and D. G. Scrimger. 2006. The sensitivity and specificity of red-starch paper for the detection of saliva. *Sci Justice* 46 (2):97.

Miller, D. W., and J. C. Hodges. 2005. Validation of Abacus SALIgAE test for forensic identification of saliva. http://www.dnalabsinternational.com/SalivaValidation.pdf.

Nussbaumer, C., E. Gharehbaghi-Schnell, and I. Korschineck. 2006. Messenger RNA profiling: A novel method for body fluid identification by real-time PCR. *Forensic Sci Int* 157 (2-3):181.

Pretty, I. A. 2006. The barriers to achieving an evidence base for bitemark analysis. *Forensic Sci Int* 159 Suppl 1:S110.

Quarino, L. et al. 1993. Differentiation of α-amylase from various sources: An approach using selective inhibitors. *J Forensic Sci Soc* 33 (2):87.

Quarino, L. et al. 2005. An ELISA method for the identification of salivary amylase. *J Forensic Sci* 50 (4):873.

Rushton, C. et al. 1979. The distribution and significance of amylase-containing stains on clothing. *J Forensic Sci Soc* 19 (1):53.

Sabatini, L. M., Y. Z. He, and E. A. Azen. 1990. Structure and sequence determination of the gene encoding human salivary statherin. *Gene* 89 (2):245.

Searcy, R. L. et al. 1965. The interaction of human serum protein fractions with the starch–iodine complex. *Clin Chim Acta* 12 (6):631.

Shaler, R. C. 2002. Modern forensic biology, in *Forensic Science Handbook*. Saferstein, R., Ed. Upper Saddle River, NJ: Person Education Inc.

Soukos, N. S. et al. 2000. A rapid method to detect dried saliva stains swabbed from human skin using fluorescence spectroscopy. *Forensic Sci Int* 114 (3):133.

Sweet, D. et al. 1997. An improved method to recover saliva from human skin: the double swab technique. *J Forensic Sci* 42 (2):320.

Vandenberg, N., and R. A. van Oorschot. 2006. The use of Polilight in the detection of seminal fluid, saliva, and bloodstains and comparison with conventional chemical-based screening tests. *J Forensic Sci* 51 (2):361.

Wawryk, J., and M. Odell. 2005. Fluorescent identification of biological and other stains on skin by the use of alternative light sources. *J Clin Forensic Med* 12 (6):296.

Whitehead, P. H., and A. E. Kipps. 1975. The significance of amylase in forensic investigations of body fluids. *Forensic Sci* 6 (3):137.

____. 1975. A test paper for detecting saliva stains. *J Forensic Sci Soc* 15 (1):39.

Willott, G. M. 1974. An improved test for the detection of salivary amylase in stains. *J Forensic Sci Soc* 14 (4):341.

Willott, G. M., and M. Griffiths. 1980. A new method for locating saliva stains: Spotty paper for spotting spit. *Forensic Sci Int* 15 (1):79.

## Other Biological Fluids

Ablett, P. J. 1983. The identification of the precise conditions for seminal acid phosphatase (SAP) and vaginal acid phosphatase (VAP) separation by isoelectric focusing patterns. *J Forensic Sci Soc* 23 (3):255.

Bock, J. H., M. A. Lane, and D. O. Norris. 1988. *Identifying Plant Food Cells in Contents for Use in Forensic Investigations: A Laboratory Manual.* Washington, D.C.: National Institute of Justice.

Brunzel, N. A. 2004. *Fundamentals of Urine and Body Fluid Analysis*, 2nd ed. Philadelphia: W.B. Saunders.

Gaensslen, R. E. 1983. Identification of urine, in *Sourcebook in Forensic Serology, Immunology, and Biochemistry.* Washington, D.C.: U.S. Government Printing Office.

Hausmann, R., and B. Schellmann. 1994. Forensic value of the Lugol's staining method: Further studies on glycogenated epithelium in the male urinary tract. *Int J Legal Med* 107 (3):147.

Jones, E. L. 2005. The identification of semen and other body fluids, in *Forensic Science Handbook.* Saferstein, R., Ed. Englewood Cliffs, NJ: Pearson Prentice Hall.

Juusola, J., and J. Ballantyne. 2005. Multiplex mRNA profiling for the identification of body fluids. *Forensic Sci Int* 152 (1):1.

Lloyd, J. B. and N. T. Weston. 1982. A spectrometric study of the fluorescence detection of fecal urobilinoids. *J Forensic Sci* 27 (2):352.

London Metropolitan Police Forensic Science Laboratory. 1978. *Biology Methods and Manual.*

Lum, P. 1991. Seven month old substituted urine sample. *Cal Dept Justice Tie-Line* 16:79.

Nickolls, L. C. 1956. Urine, in *The Scientific Investigation of Crime.* London: Butterworth.

Nussbaumer, C., E. Gharehbaghi-Schnell, and I. Korschineck. 2006. Messenger RNA profiling: A novel method for body fluid identification by real-time PCR. *Forensic Sci Int* 157 (2–3):181.

Peabody, A. J., R. M. Burgess, and R. E. Stockdale. 1981. A Re-examination of Lugol's iodine test. Home Office Central Research Establishment.

Poon, H. H. L. 1984. Identification of human urine by immunological techniques. *Can Soc Forensic Sci J* 17:81.

Randall, B. 1988. Glycogenated squamous epithelial cells as a marker of foreign body penetration in sexual assault. *J Forensic Sci* 33 (2):511.

Rhodes, E. F., and J. I. Thornton. 1976. DMAC test for urine stains. *J Police Sci Admin* 4:88.

Sagawa, K. et al. Production and characterization of a monoclonal antibody for sweat-specific protein and its application for sweat identification. *Int J Legal Med* 117 (2):90.

Taylor, M. C., and J. S. Hunt. 1983. Forensic identification of human urine by radioimmunoassay for Tamm-Horsfall urinary glycoprotein. *J Forensic Sci Soc* 23 (1):67.

Thompson, J., and K. Belschner. 1997. Differentiation of various menstrual blood stains using LDH. *Cal Dept Justice Justice Tie-Line* 21:31-39.

Tsutsumi, H. et al. 1988. Identification of human urinary stains by enzyme-linked immunosorbent assay for human uromucoid. *J Forensic Sci* 33 (1):237.

Yamada, S. et al. 1992. Vomit identification by a pepsin assay using a fibrin blue-agarose gel plate. *Forensic Sci Int* 52 (2):215.

## Study Questions

1. Where are amylases secreted?
   (a) Salivary glands
   (b) Pancreas
   (c) Both of the above

2. Which of the following reacts with Phadebas reagent?
   (a) Human salivary amylase
   (b) Human pancreatic amylase
   (c) Plant amylase
   (d) Bacterial amylase

3. Which of the following is considered confirmatory for identifying saliva?
   (a) Starch–iodine assay
   (b) Phadebas assay
   (c) Use of antihuman salivary amylase antibody
   (d) RNA-based assay

4. When is identification of saliva in a stain necessary?

5. Which of the following can distinguish human salivary amylase and human pancreatic amylase?
   (a) Starch–iodine assay
   (b) Phadebas assay
   (c) Use of antihuman salivary amylase antibody
   (d) RNA-based assay

# Blood Group Typing and Protein Profiling

# 10

## 10.1 Blood Group Typing

### 10.1.1 Blood Groups

For purposes of this text, blood groups are defined as antigen polymorphisms present on erythrocyte surfaces. Human erythrocyte surface membranes contain a variety of blood group antigens. Transfusion reactions occur when an incompatible type of blood is transfused into an individual, which can lead to severe symptoms or even death. Karl Landsteiner discovered the first blood group known as the ABO system in the early 1900s while studying transfusion and transplantation. The discovery made blood transfusions feasible and Landsteiner was awarded the Nobel Prize in 1930.

The International Society of Blood Transfusion currently recognizes 29 blood group systems that include 245 antigen polymorphisms (Table 10.1). From the 1950s to the 1970s, the structures and biosynthesis activities of many blood group antigens became known. The genes for most of these blood group systems have been identified. The isolation of blood group genes has made it possible to understand the molecular mechanisms of antigenic characteristics of the blood group systems.

The ABO system of antigens in human erythrocytes is the most commonly used blood group system for forensic applications. Forensic laboratories also use others including the Rh, MNS, Kell, Duffy, and Kidd systems.

### 10.1.2 ABO Blood Group System

In the ABO blood group system, two types of antigens designated A and B give rise to four blood types:

- Type A individuals have the A antigen.
- Type B individuals have the B antigen.
- Type AB individuals have both A and B antigens.
- Type O individuals have neither A nor B antigens.

In addition to blood, the antigens may be found in many tissues and other biological fluids, including the salivary glands and saliva, pancreas, kidney, liver, lungs, testes, semen, and amniotic fluid.

**Table 10.1 Blood Group Systems**

| No. | Name | Symbol | No. of Antigens | Gene Names | Chromosomal Location |
|---|---|---|---|---|---|
| 001 | ABO | ABO | 4 | *ABO* | 9q34.2 |
| 002 | MNS | MNS | 43 | *GYPA, GYPB, GYPE* | 4q31.21 |
| 003 | P | P1 | 1 | | 22q11.2-qter |
| 004 | Rh | RH | 49 | *RHD, RHCE* | 1p36.11 |
| 005 | Lutheran | LU | 20 | *LU* | 19q13.32 |
| 006 | Kell | KEL | 25 | *KEL* | 7q34 |
| 007 | Lewis | LE | 6 | *FUT3* | 19p13.3 |
| 008 | Duffy | FY | 6 | *FY* | 1q23.2 |
| 009 | Kidd | JK | 3 | *SLC41A1* | 18q12.3 |
| 010 | Diego | DI | 21 | *SLC4A1* | 17q21.31 |
| 011 | Yt | YT | 2 | *ACHE* | 7q22.1 |
| 012 | Xg | XG | 2 | *XG, MIC2* | Xp22.33, Yp11.3 |
| 013 | Scianna | SC | 5 | *ERMAP* | 1p34.2 |
| 014 | Dombrock | DO | 5 | *DO* | 12p12.3 |
| 015 | Colton | CO | 3 | *AQP1* | 7p14.3 |
| 016 | Landsteiner-Wiener | LW | 3 | *ICAM4* | 19p13.2 |
| 017 | Chido/Rodgers | CH/RG | 9 | *C4A, C4B* | 6p21.3 |
| 018 | H | H | 1 | *FUT1* | 19q13.33 |
| 019 | Kx | XK | 1 | *XK* | Xp21.1 |
| 020 | Gerbich | GE | 8 | *GYPC* | 2q14.3 |
| 021 | Cromer | CROM | 12 | *DAF* | 1q32.2 |
| 022 | Knops | KN | 8 | *CR1* | 1q32.2 |
| 023 | Indian | IN | 2 | *CD44* | 11p13 |
| 024 | Ok | OK | 1 | *BSG* | 19p13.3 |
| 025 | Raph | RAPH | 1 | *CD151* | 11p15.5 |
| 026 | John Milton Hagen | JMH | 1 | *SEMA7A* | 15q24.1 |
| 027 | I | I | 1 | *GCNT2* | 6p24.2 |
| 028 | Globoside | GLOB | 1 | *B3GALT3* | 3q26.1 |
| 029 | Gill | GIL | 1 | *AQP3* | 9p13.3 |

*Source:* Adapted from Daniels, G. L. et al. 2004. *Vox Sang* 87 (4):304.

#### *10.1.2.1 Biosynthesis of Antigens*

All individuals generate the ***O antigen***, also known as the ***H antigen***. The O antigen is synthesized by ***fucosyltransferase***, a fucose transferase coded by the *FUT* genes, that adds a fucose on the end of a glycolipid (in erythrocytes) or glycoprotein (in tissues). An additional monosaccharide (Figure 10.1) is then transferred to the O antigen by a ***transferase*** encoded by the *ABO* locus. The specificity of this enzyme determines the ABO blood type (Figure 10.2):

- The *A* allele produces the A-transferase that transfers N-acetylgalactosamine to the O antigen and thus synthesizes the A antigen.
- The *B* allele produces the B-transferase that transfers galactose to the O antigen and thus synthesizes the B antigen.
- The *O* allele has a mutation (small deletion) that eliminates transferase activity and no modification of the O antigen occurs.

As a result, the A and B antigens differ in their terminal sugar molecules. Subgroups of blood types A and B have been described. The most important

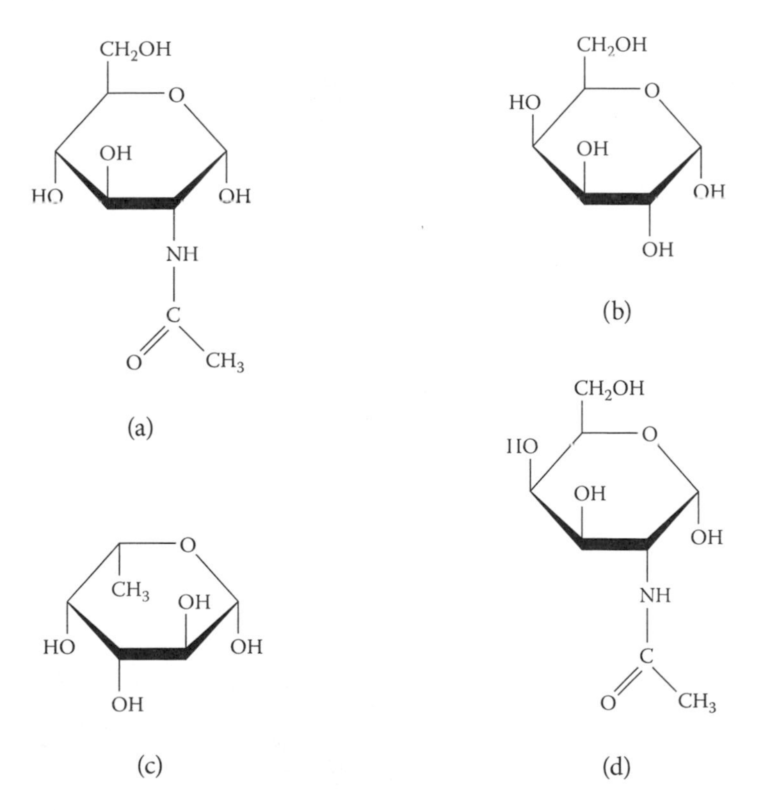


**Figure 10.1** Chemical structures of (a) N-acetylglucosamine; (b) galactose; (c) fucose; and (d) N-acetylgalactosamine.

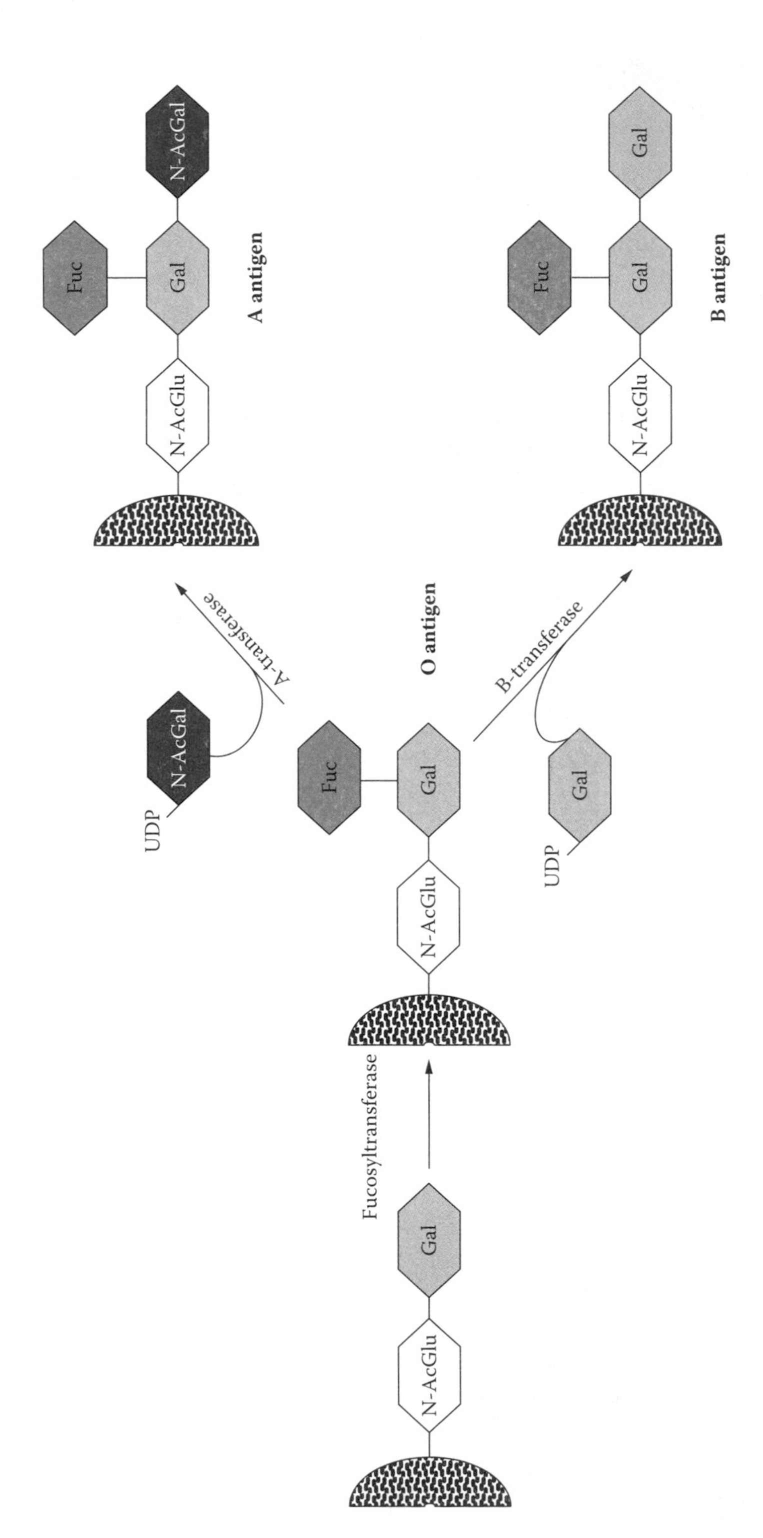


**Figure 10.2** Biosynthesis of ABO antigens. **O antigen** biosynthesis is catalyzed by fucosyltransferase. **A antigen** biosynthesis is catalyzed by the A-transferase that transfers the N-acetylgalactosamine from the donor, uridine diphosphate (UDP)-N-acetylgalactosamine to the O antigen. **B antigen** biosynthesis is catalyzed by the B-transferase that transfers the galactose from UDP-galactose to the O antigen. N-AcGlu = N-acetytglucosamine. Gal = galactose. Fuc = fucose. N-AcGal = N-acetylgalactosamine.

are the $A_1$ and $A_2$ antigens. Both $A_1$ and $A_2$ (and $A_1B$ and $A_2B$) cells react with anti-A antibodies. However, $A_1$ cells react more strongly than $A_2$ cells. The apparent difference between $A_1$ and $A_2$ is that each $A_1$ cell contains more copies of the A antigen than $A_2$ cells.

#### *10.1.2.2 Molecular Basis of ABO System*

A- and B-transferases are encoded by a single gene, *ABO,* on chromosome 9. The *ABO* gene (approximately 20 kilobases) is organized into seven exons. Most of its coding regions are located in exons 6 and 7 of the *ABO* locus including the domain responsible for catalytic activity (Figure 10.3). The products of the *A* and *B* alleles differ by four amino acid substitutions (Table 10.2). However, amino acid residues at positions 266 and 268 are the most important in determining whether a gene product is an A-transferase or B-transferase.

The $A^1$ allele and $A^2$ allele differ in a single nucleotide deletion before the translation stop codon. The resulting $A^2$ reading-frame shift abolishes the

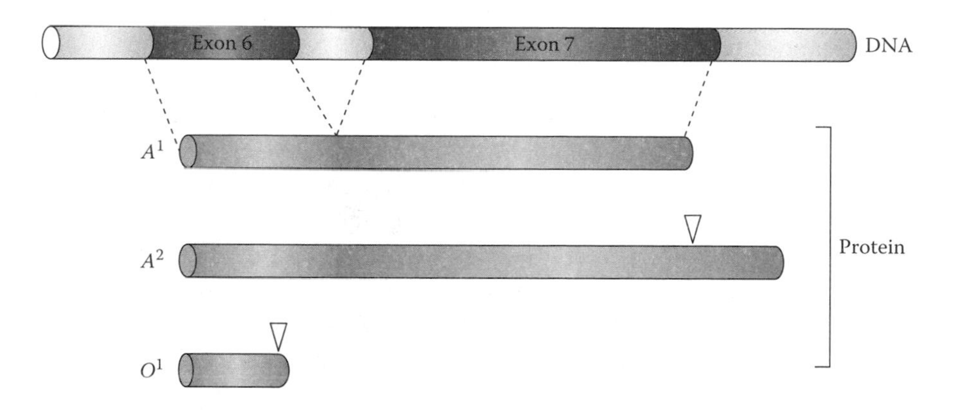


**Figure 10.3** Structure of ABO gene and variants. Exons 6 and 7 are shown. The deletion mutation in $A^2$ and $O^1$ variants is indicated by an inverted triangle and leads to their $A_2$ and O phenotypes, respectively.

**Table 10.2 Amino Acid Substitutions at Four Positions in Human ABO Variants** $A^1$, $B$, **and** $O^2$

| ABO Variant | Amino Acid Position | | | |
|---|---|---|---|---|
| | 176 | 235 | 266 | 268 |
| $A^1$ | Arginine | Glycine | Leucine | Glycine |
| $B$ | Glycine | Serine | Methionine | Alanine |
| $O^2$ | Glycine | Glycine | Leucine | Arginine |

stop codon, yielding a product with an extra 21-amino acid residue at the C-terminus.

Subgroups of blood types O have also been reported. The sequence of the $O^{1}$ allele has a deletion of a single nucleotide at exon 6. This nucleotide deletion leads to a reading-frame shift generating a truncated protein, which lacks the catalytic domain. While the $O^{1y}$ allele also has a single nucleotide deletion, it differs from $O^{1}$ by nine nucleotides within the coding sequence. $O^{1}$ and $O^{1y}$ have identical phenotypes. There is also an $O^{2}$ allele that is inactivated by a substitution mutation at glycine (position 268) by arginine. Additionally, a few dozen other rare *O* alleles that yield inactive products have been documented.

#### 10.1.2.3 Secretors

In addition to erythrocytes, A, B, and O antigens are widely distributed in tissues and are present in body secretions. They are known as ***secretors***. As described earlier, the O antigen is the substrate for the A- and B-transferase because the A- and B-transferase can only utilize a fucosylated substrate. The O antigen is synthesized by fucosylation of the terminal galactosyl residue catalyzed by the fucosyltransferase which is encoded by *FUT* genes.

Chromosome 19 contains two homologous genes: *FUT1* and *FUT2*. *FUT1* is expressed in tissues of mesodermal origin (embryonic tissues that serve as precursors of hemopoietic tissues, muscle, skeleton, and internal organs) and is responsible for the synthesis of O antigen in erythrocytes. *FUT2* is expressed in tissues of endodermal origin (embryonic tissues that are precursors of the gut and other internal organs); it is responsible for the synthesis of O antigen in secretions. About 20% of Caucasian individuals (called ***non-secretors***) are homozygous for a nonsense mutation in *FUT2* at amino acid position 143, resulting in a truncated protein.

Secretions of type A or B non-secretors contain no A or B antigens despite containing active A- or B-transferases (Figure 10.4). However, nonsecretors have O antigens on erythrocytes synthesized by *FUT1*, and thus have A or B antigens in blood. Homozygous *FUT1* mutations produce very rare erythrocyte O-deficient phenotypes in which the erythrocytes express no O and thus no A or B, regardless of *ABO* genotype. Individuals with such phenotypes can be either secretors or non-secretors. The non-secretors who also have no O (or A or B) in their secretions are known as Bombay ($O_h$) phenotypes.

#### 10.1.2.4 Inheritance of A and B Antigens

*A* and *B* alleles are dominant. For *AO* and *BO* heterozygotes, the corresponding transferase synthesizes the A or B antigen. *A* and *B* alleles are codominant in *AB* heterozygotes because both transferase activities are expressed. The *OO* homozygote produces neither transferase activity and therefore lacks both antigens. The inheritance of *A* and *B* alleles obeys Mendelian principles.

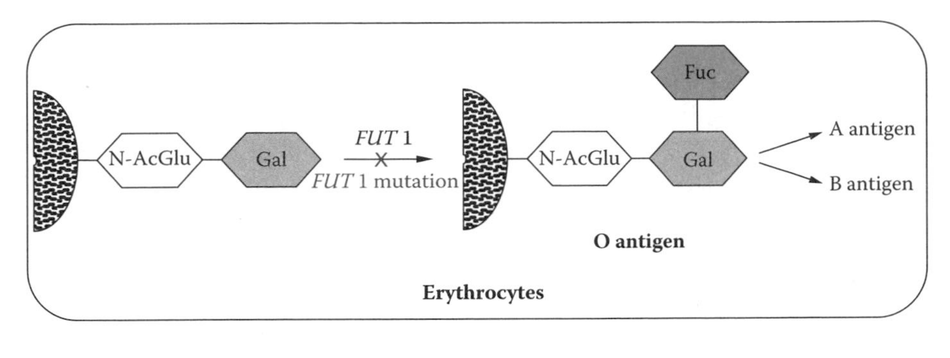


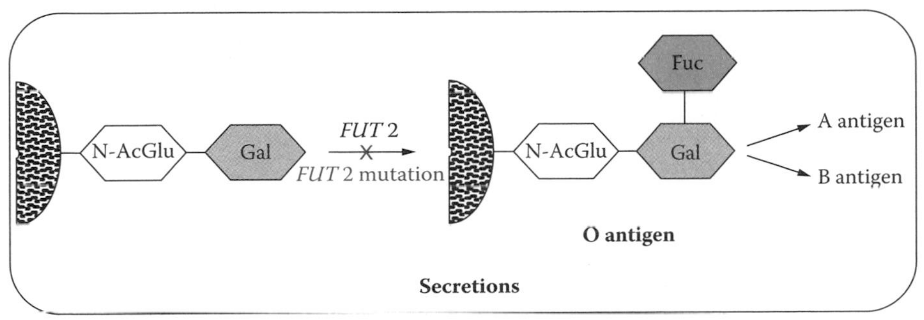


**Figure 10.4** Tissue-specific O antigen biosynthesis in erythrocytes is catalyzed by the *FUT1* gene product; in secretions it is catalyzed by the *FUT2* gene product. The mutations abolishing the biosynthesis of O antigens are indicated. The *FUT2* mutation produces a nonsecretor phenotype.

For example, an individual with type B blood may have inherited a *B* allele from each parent or a *B* allele from one parent and an *O* allele from the other; thus, an individual whose phenotype is B may have the *BB* (homozygous) or *BO* (heterozygous) genotype. Conversely, if the blood types of the parents are known, the possible genotypes of their children can be determined. When both parents are type B (heterozygous), they may produce children with genotype *BB* (B antigens from both parents), *BO* (B antigen from one parent, O from the other heterozygous parent), or *OO* (O antigens from parents who are both heterozygous). Thus, blood group typing can be used for paternity testing.

### 10.1.3 Forensic Applications of Blood Group Typing

The application and the usefulness of blood typing to forensic identification are based on the ability to group individuals into four different types with the ABO blood system, allowing individuals to be identified. For example, if one crime scene blood sample is type B and a suspect has type A, the crime scene sample must have a different origin. However, if both the sample and the suspect are type A, the sample may have come from the same origin or from a different origin that happened to be type A.

Unfortunately, the probability that any two randomly chosen individuals having an identical blood type is very high. Approximately 42% of Caucasians have type A blood. The frequency of other blood types within the ABO system is shown in Figure 10.5. Multiple blood group systems were utilized to decrease the probability of a coincident match.

The A and B antigens are very stable and can be identified in dried blood even after many years. They can also be found in semen and other biological fluids of secretors. Approximately 80% of Caucasians are secretors. Thus, in sexual assault cases, for example, the ABO type of a semen sample can be examined to identify a perpetrator.

## 10.1.4 Blood Group Typing Techniques

The most common assays used in forensic serology involve agglutination and include the Lattes crust and absorption–elution assays.

### *10.1.4.1 Lattes Crust Assay*

In the early 1900s, Karl Landsteiner used his blood and blood obtained from his laboratory coworkers to test the effects of sera on erythrocytes. He discovered that naturally occurring antibodies in sera caused agglutination of certain erythrocytes and the agglutination patterns observed were designated A, B, and O. Each pattern indicated the presence or absence of a particular antigen on erythrocytes.

Infants rapidly develop antibodies against antigens that are not present in their own cells. Thus, type A individuals develop anti-B antibodies, type B individuals develop anti-A antibodies, type O individuals develop both types of antibodies, and type AB individuals do not develop anti-A or anti-B antibodies. When the plasma of a type A individual is mixed with type B

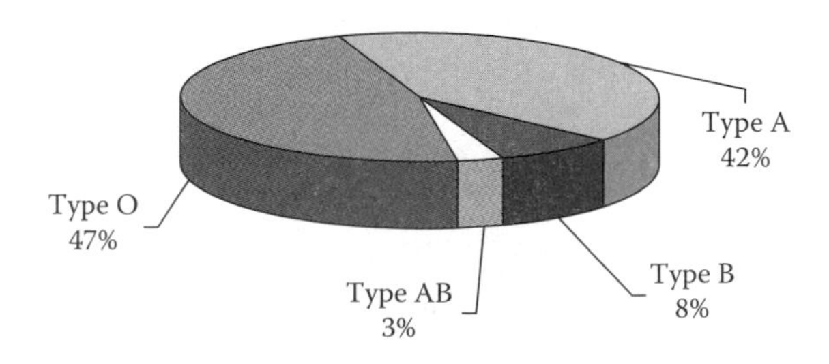


**Figure 10.5** Frequency distributions of ABO types observed in American Caucasians. Different human populations may exhibit different frequencies of the four blood types.

cells, the anti-B antibodies from the type A individual cause the type B cells to agglutinate. This result forms the basis for blood group typing.

The Lattes crust assay relies upon the principles of Landsteiner's experiments. It is an agglutination-based assay that utilizes the A, B, and O indicator cells to test the agglutination reaction with its corresponding naturally occurring serum antibodies in a questioned sample. The procedure for the Lattes crust assay is described in Box 1 and illustrated in Figure 10.6. Typical results are summarized in Table 10.3 and illustrated in Figure 10.7. Type A blood contains naturally occurring anti-B antibody and agglutinates only with B cells. Likewise, type B blood agglutinates only with A cells; type O blood agglutinates with both A and B cells; and type AB blood will not agglutinate with any cells.

The Lattes crust assay is simple and rapid. However, one limitation is that the assay is not very sensitive and requires a large quantity of blood. It is not reliable for testing old stains.

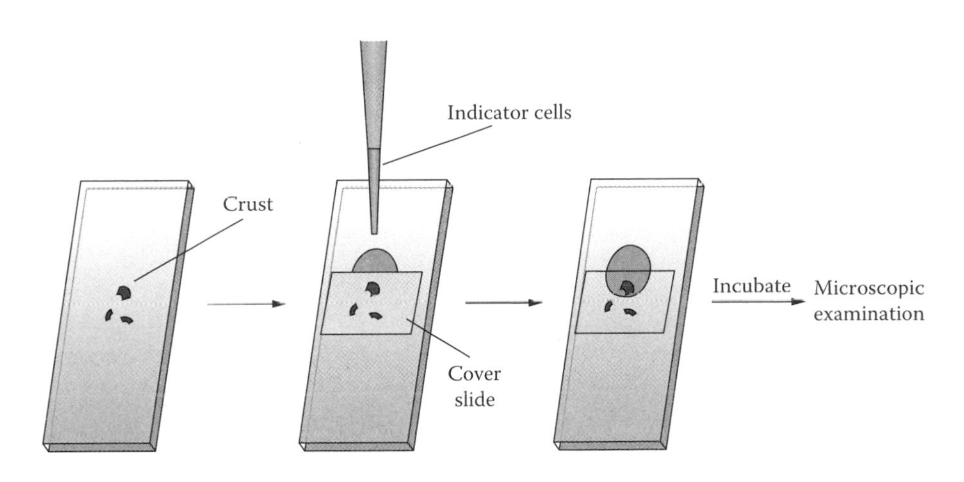


**Figure 10.6** Lattes crust assay.

**Table 10.3 Representative Results of Lattes Crust Assay**

| Blood Type | Serum Antibody | Agglutination Reaction Observed |
|---|---|---|
| A | Anti-B | B cells |
| B | Anti-A | A cells |
| O | Anti-A, anti-B | A cells, B cells |
| AB | None | None |

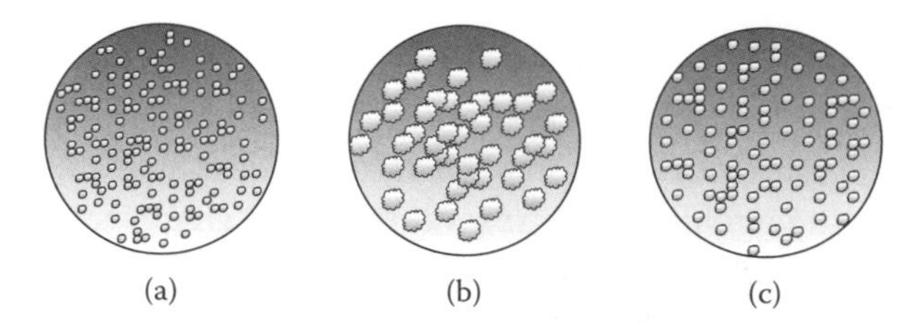


**Figure 10.7** Microscopic observation of Lattes crust assay results. (a) Indicator cells added before incubation. (b) Strong agglutination: large clumps observed after incubation. (c) Negative agglutination: cloudy background may be observed after incubation.

**Box 1**

**Lattes Crust Assay Procedure**

1. Place small quantities of crusts of dried blood from a specimen on a piece of microscopic slide and place a cover slide over the crusts. Prepare such slides for testing A, B, and O cells.
2. Prepare cell suspension with saline (0.85% NaCl in phosphate buffer, pH 7.4) for the A, B, and O cells.
3. Apply few drops of the A cell suspension and allow the cells to diffuse under the cover slip. Repeat this step for B cells and O cells.
4. Incubate the slides in a moisture chamber at room temperature for 2 hr.
5. Examine results under a microscope.

### *10.1.4.2 Absorption–Elution Assay*

The absorption–elution assay is highly sensitive and can be used for testing dried blood stains. It indirectly detects the presence of antigens. Recall that successful agglutination reactions usually require intact cells. The agglutination reaction is, therefore, difficult to carry out because blood cells lyse when they are dry. However, with this method, the lysed cells containing antigen are immobilized on a solid phase (Figure 10.8).

At low temperatures, the antigen is allowed to bind to its corresponding antibody: anti-A antibodies, anti-B antibodies, or anti-O lectins. (The anti-O lectin is isolated from plants reacts strongly with the O antigen present in type O blood, but has some cross-reaction with the A antigen). The excess unbound antibodies are removed by washing and the bound antibodies are then eluted at higher temperatures (recall that antigen–antibody binding can

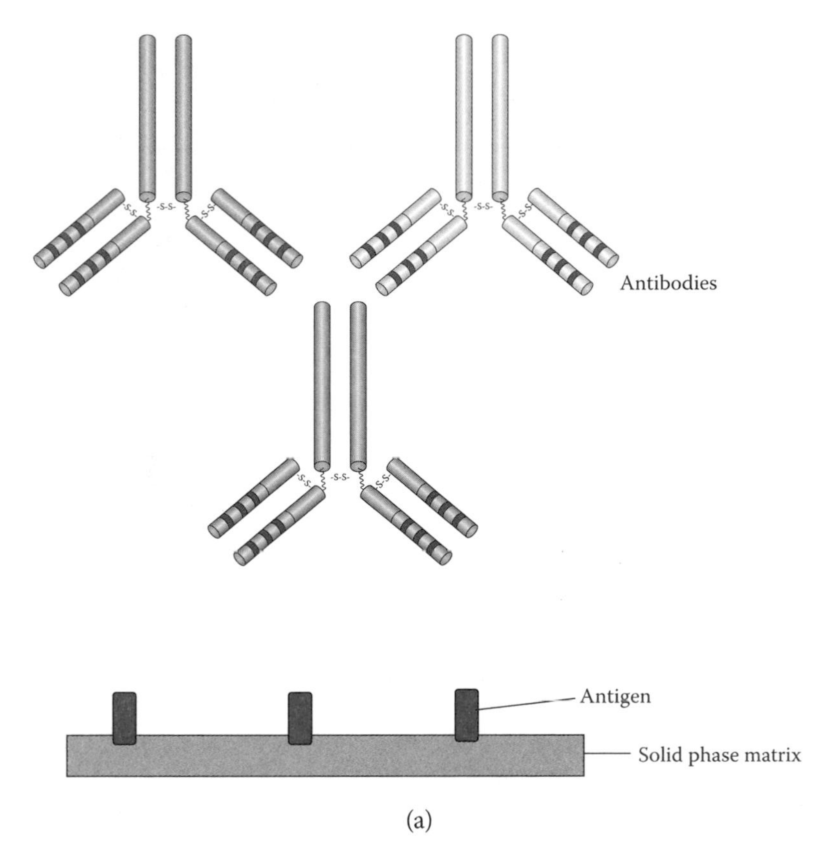


**Figure 10.8** Absorption–elution assay. (a) Antigens are immobilized on a solid phase matrix. (b) Antibodies are added and absorbed. (c) Unbound antibodies are washed away. (d) The bound antibody is eluted. (e) The eluted antibody is tested with indicator cells. (f) The eluted antibody is identified by a positive agglutination reaction.

be affected by temperature). The eluted antibodies can then be identified by an agglutination assay using A, B, and O indicator cells.

Typical results of an absorption–elution assay are summarized in Table 10.4. The blood stains containing A antigen can bind to anti-A antibodies. The eluted anti-A antibody can form agglutination with A cells. Likewise, for type B blood, the eluted anti-B antibody can form agglutination with B cells; for type AB blood, the eluted antibodies can form agglutination with both A and B cells; and with type O blood, the eluted antibodies can form agglutination with O cells.

## 10.2 Forensic Protein Profiling

Because of the limitations of blood group systems, inherited protein polymorphic markers have been utilized to decrease the chances of matches

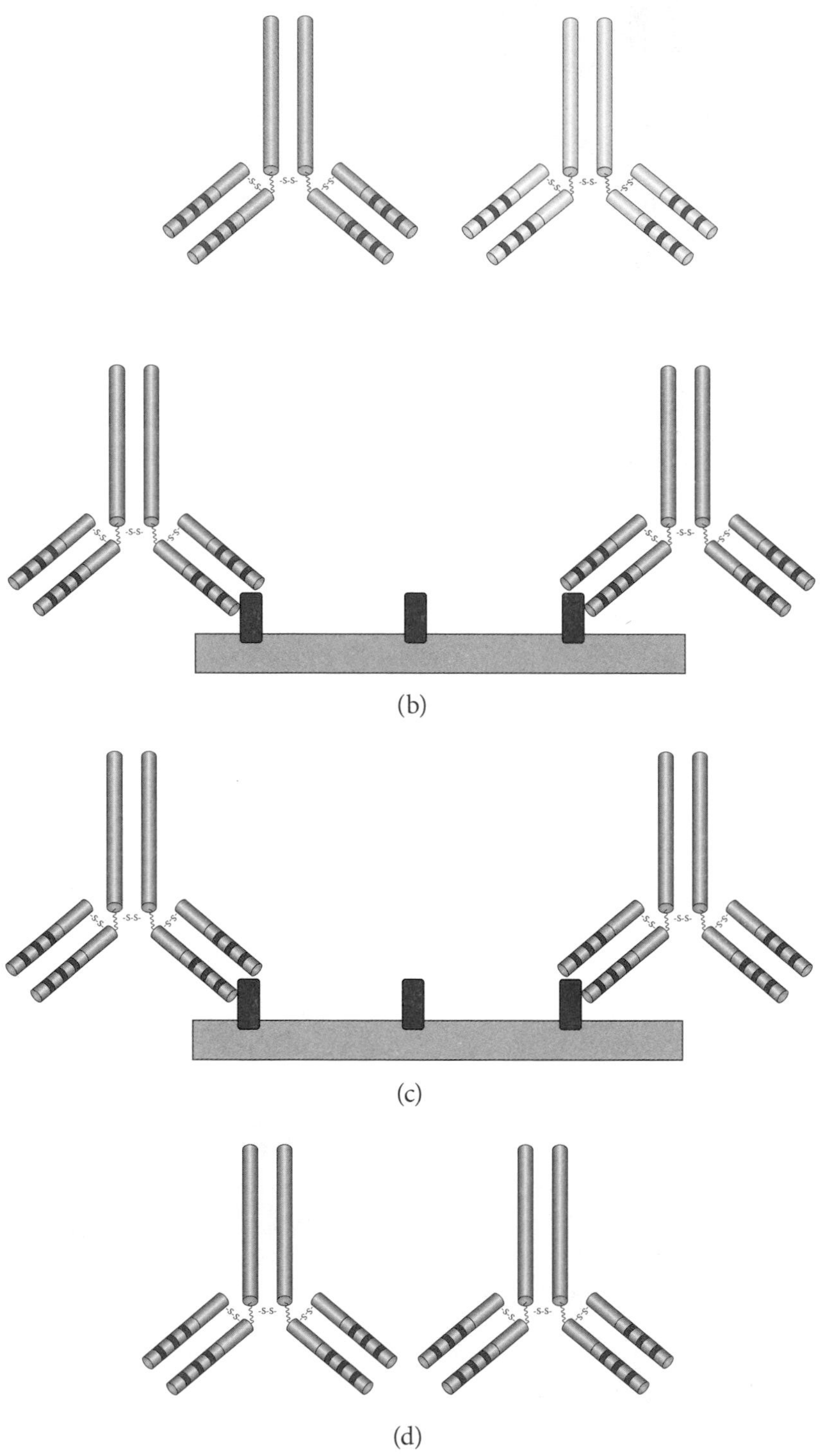


**Figure 10.8** (continued)

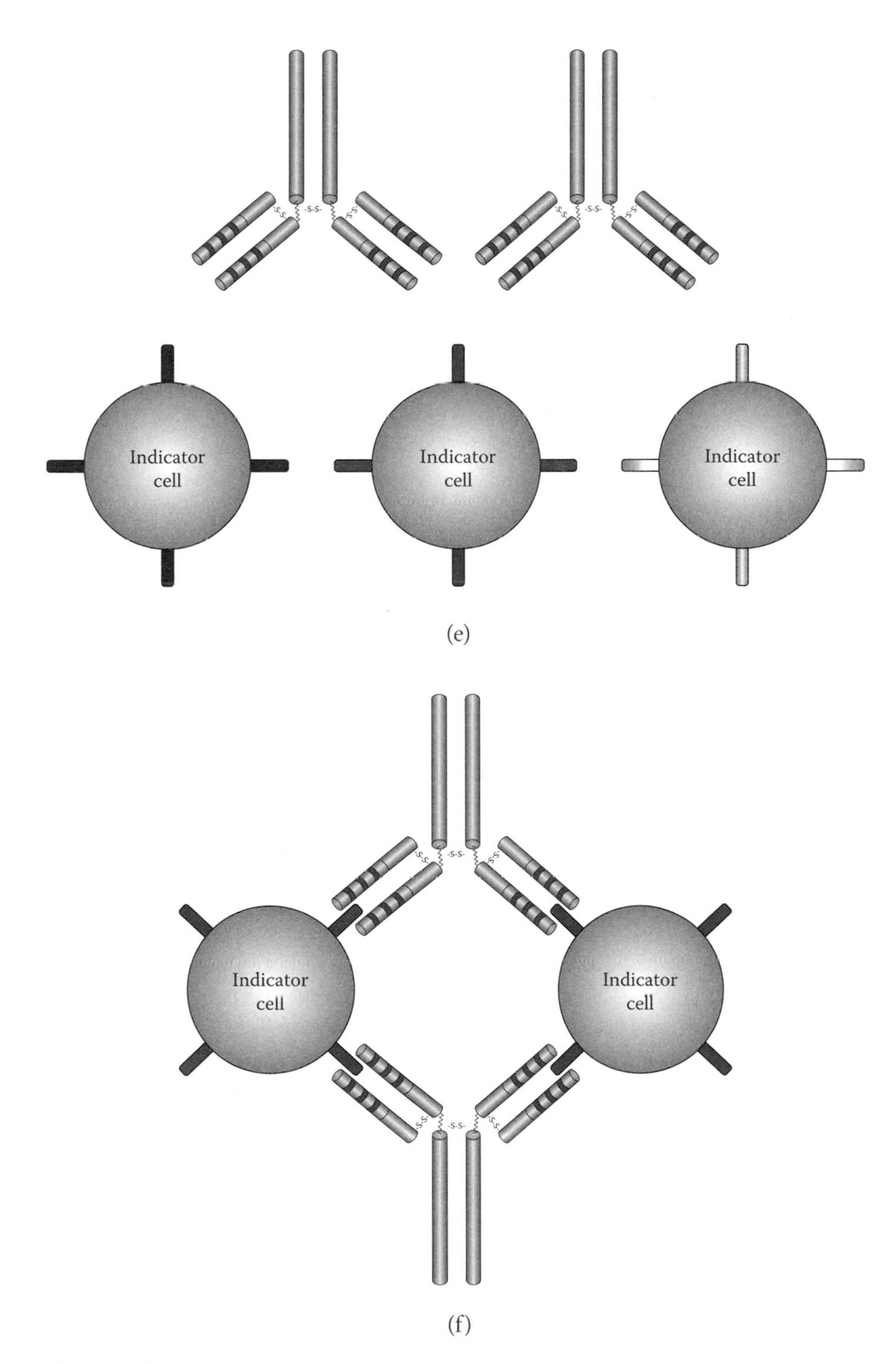


**Figure 10.8** (continued)

**Table 10.4 Representative Results of Absorption–Elution Assay**

| Blood Type (Stain) | Antibody Bound and Eluted | Agglutination Reaction Observed |
|---|---|---|
| A | Anti-A | A cells |
| B | Anti-B | B cells |
| O | Anti-O | O cells |
| AB | Anti-A, anti-B | A cells, B cells |

between two unrelated individuals. The amino acid sequences of many proteins vary in the human population. An estimated 20 to 30% of the proteins in humans are polymorphic. Some of the variations in amino acid sequences affect the function of proteins but many of them exert little or no effect on protein function. Thus, the population can be divided into groups based on protein polymorphisms. A combination of the blood group systems and protein polymorphic markers can be used for criminal investigations and paternity testing. The chance that results for two unrelated persons would match was decreased to one in several hundred through use of both techniques.

### 10.2.1 Methods

Identification of protein polymorphisms is performed through electrophoretic separation based on the molecular weights (*M*r) and charges of the protein variants.

#### *10.2.1.1 Matrices Supporting Protein Electrophoresis*

Electrophoresis of proteins is generally carried out in a support material, also called the matrix, that acts as a molecular sieve for the separation of different sized macromolecules. Thus, the larger the size of the protein, the smaller its electrophoretic mobility. The matrix also reduces the effects of diffusion and convection on the macromolecules.

Historically, protein profiling for forensic application employs two types of matrices including papers such as cellulose acetate and gels composed of starch, agar, agarose, or polyacrylamide. The first polymorphic protein marker, phosphoglucomutase, was characterized by starch gel electrophoresis. However, the agarose and polyacrylamide became more commonly used in electrophoresis due to good reproducibility and reliability (Table 10.5).

#### *10.2.1.2 Separation by Molecular Weight*

An electrophoretic method is frequently employed to resolve various proteins based on their molecular weights. Native, also known as non-denaturing, electrophoresis can be used to provide information about the functions

**Table 10.5 Properties of Matrices Supporting Protein Electrophoresis**

| Supporting Matrix | Pore Size | EEO | Reproducibility | Strength | | Toxicity |
|---|---|---|---|---|---|---|
| Cellulose acetate | Large | High | Poor | Good | Simple | Nontoxic |
| Starch | Large | High | Poor | Fragile | Simple | Nontoxic |
| Agar | Large | High | Poor | Fragile | Simple | Nontoxic |
| Agarose | Large | Low | Good | Fragile | Simple | Nontoxic |
| Polyacrylamide | Small | Very Low | Good | High; tolerates high electronic field | Complex | Toxic |

EEO = electroendosmosis occurring when fixed charges of the supporting matrix cause liquid flow toward the electrodes. A matrix with high EEO may affect the mobilities and separation performances of proteins during electrophoresis.

of proteins after separation. Biological activity can be retained for further analysis. However, some proteins are not well separated in electrophoresis in their native form. Thus, it may be necessary to denature the proteins in order to resolve them better during separation. The process is called ***denaturing protein electrophoresis***. The following additives can be used:

**Reducing Agents** — It is common to include reducing agents such as mercaptoethanol (ME), dithiothreitol (DTT), or sodium mercaptoethane sulfonate (MESNA) to denature proteins. Reducing agents cleave the disulfide bonds of proteins. As a result, protein shape becomes unfolded and more linear. These agents can be used during sample preparation and can also be added to the electrophoresis buffer.

**Detergents** — These additions disrupt non-covalent interactions within the structures of native proteins. The procedure is generally performed with sodium dodecylsulfate (SDS), a strong anion detergent that binds to most proteins in amounts proportional to the molecular weight of the protein (approximately one molecule of SDS for two amino acids). The bound SDS contributes a large net negative charge on the protein that masks any surface charges of the native protein. As a result, the charge-to-mass ratio of the protein becomes a constant. As with reducing agents, the various native conformations of proteins change to a more uniformly linear shape when SDS is bound. Thus, the separation is based more upon the sieving mechanism. Electrophoretic mobility in the presence of SDS, therefore, becomes based on *Mr* rather than both *Mr* and the charge. Smaller proteins move through the gel more rapidly than larger proteins.

SDS gel electrophoresis can also be used to determine the *Mr* of an unidentified protein based on its electrophoretic mobility on the gel. Standard marker proteins of known molecular weight are run on the same gel and allow estimation of the *Mr* of an unknown protein. A linear plot of log *Mr* values of marker proteins versus relative migration during electrophoresis allows the molecular weight of the unknown protein to be determined from the graph.

### 10.2.1.3 Separation by Isoelectric Point

The ***isoelectric focusing (IEF)*** technique can be used to separate proteins according to their ***isoelectric points (pI)***. The pI is the pH value at which the net electric charge of an amino acid is zero. All proteins are composed of amino acids and each has its own characteristic pI at which its net electric charge is zero and does not migrate in an electric field.

In IEF electrophoresis, a pH gradient is created between the electrodes and a protein sample is placed in a well on the gel. With an applied electric field, proteins enter the gel and migrate until they reach a pH equivalent to their pI values at which they lose mobility (Figure 10.9).

A pH gradient in the gel is established by employing materials such as ampholytes and immobilines that are dispersed in the spaces between gel molecules. Ampholytes are synthetic compounds with weak ionizable groups that can act as either acids or bases. A mixture of slightly different

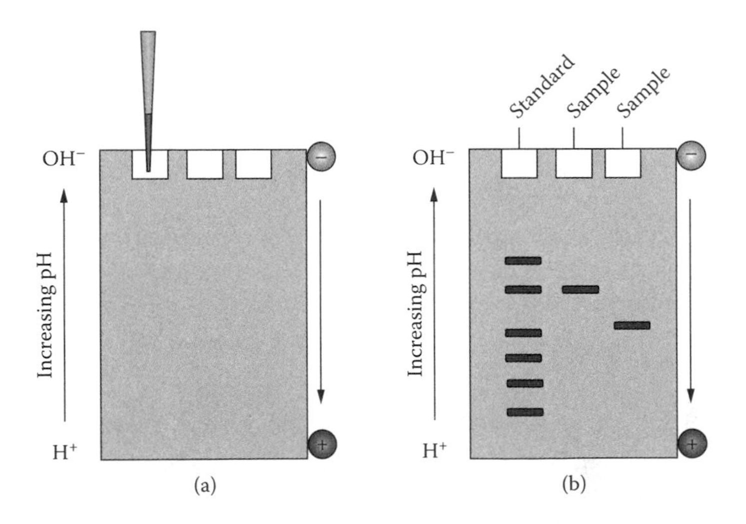


**Figure 10.9** Isoelectric focusing. (a) A pH gradient is established by allowing low molecular weight organic acids and bases to distribute themselves in an electric field across a gel. A sample containing protein mixture is loaded into a sample well. (b) During the electrophoresis, each protein migrates until it matches its pI. Proteins with different pI are separated.

ampholytes can allow for the creation of a gradual pH gradient between the electrodes. Immobilines are modified acrylamide monomers that act as both acids and bases. The gradient is generated using a gradient-forming device that changes the proportion of the immobilines added to the gel matrix mixture as it is loaded into the gel casting apparatus.

IEF, based on molecular charge, is capable of producing sharper bands than denaturing protein electrophoresis and thus has a higher resolving power. The technique can detect very low quantities of proteins in samples.

## 10.2.2 Erythrocyte Protein Polymorphisms

### *10.2.2.1 Erythrocyte Isoenzymes*

The human erythrocyte contains a number of ***isoenzymes*** (multiple forms of an enzyme that catalyze the same reaction but differ in their amino acid sequences) that show variations from person to person. Thus, individuals can be divided into groups on the basis of the different isoenzymes present in their erythrocytes. The isoenzyme type is also inherited according to Mendelian principles.

Polymorphism of erythrocyte phosphoglucomutase (PGM) was first described in the 1960s and later successfully applied to the testing of blood stains. PGM, an important metabolic enzyme, catalyzes the reversible conversion of glucose-1-phosphate and glucose-6-phosphate. The PGM found in erythrocytes is encoded at the *PGM1* locus at chromosome 1. The PGM encoded by *PGM1* can also be found in semen and thus can be utilized for the testing of semen samples in sexual assault cases. The protein polymorphisms of the PGM have two alleles that result in three different phenotypes, depending on the combination of the two alleles. The success in the forensic application of PGM led to the similar use of many other erythrocyte isoenzyme polymorphisms. The most commonly used erythrocyte isoenzyme systems are listed in Table 10.6.

**Table 10.6 Comon Isoenzymes Used for Forensic Protein Profiling**

| Erythrocyte Isoenzyme | Protein Symbol | No. of Alleles |
|---|---|---|
| Phosphoglucomutase | PGM | 2* |
| Erythrocyte Acid Phosphatase | ACP/EAP | 3 |
| Esterase D | ESD | 2 |
| Adenylate Kinase | AK | 2 |
| Glyoxalase I | GLO | 2 |
| Adenosine Deaminase | ADA | 2 |

* Ten alleles can be observed using IEF electrophoresis.

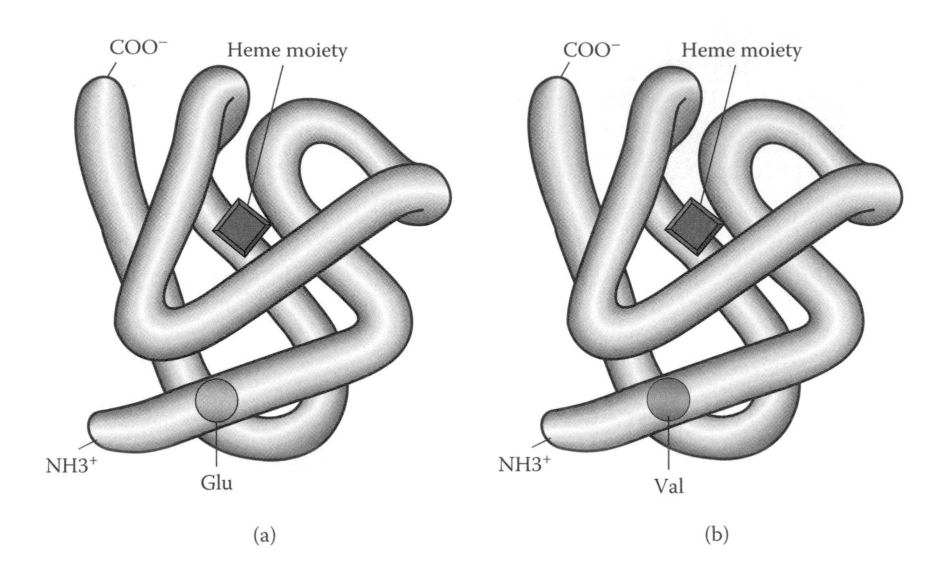


**Figure 10.10** Normal and sickle cell hemoglobin β chains. (a) Normal hemoglobin βchain contains a glutamic acid residue (Glu) at position 6 of the N-terminal of the protein. (b) At position 6 of the sickle cell hemoglobin β chain, the glutamic acid residue is replaced by a valine (Val).

### *10.2.2.2 Hemoglobin*

Recall that the use of hemoglobin (Hb) in screening and confirmatory blood tests was discussed in Chapter 6. Adult human Hb consists of two α chains and two β chains. Each polypeptide chain contains a heme group involved in oxygen binding. A very small portion of blood possesses a form of the human adult Hb consisting of two α chains and two δ chains.

More than 200 Hb variants have been identified and can be useful as markers for forensic applications. In particular, two types of human Hb variants are important in forensic testing: fetal Hb and sickle cell Hb (Hb S, the factor responsible for sickle cell disease [Figure 10.10]). Hb variants can be resolved using electrophoresis (Figure 10.11).

**10.2.2.2.1 Fetal Hemoglobin** Humans have three forms of Hb during their development: embryonic, fetal, and adult Hb. In adults, the Hb tetramer consists of two identical α and two identical β chains. Embryonic erythrocytes contain tetramers that are different from the adult form. Each embryonic Hb consists of two identical α-like chains and two identical β-like chains. The embryonic Hb is gradually replaced during pregnancy (approximately 3 months after conception) by fetal Hb that comprises approximately 70% of the Hb in fetal blood. The fetal Hb has two identical α chains and two identical γ chains.

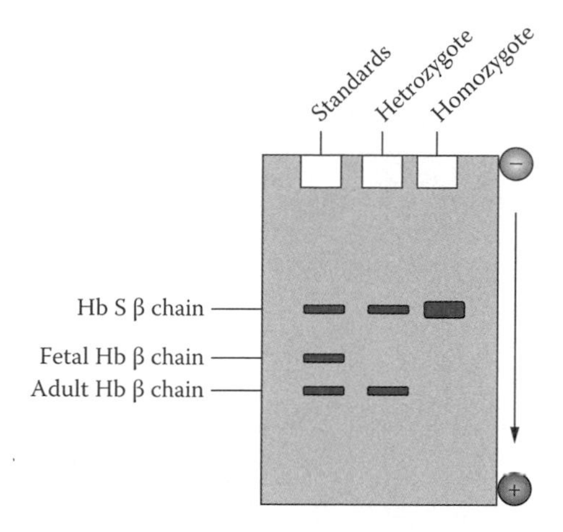


**Figure 10.11** Hemoglobins resolved by isoelectric focusing electrophoresis.

The embryonic and fetal Hbs have higher affinities to oxygen required to provide the embryo and fetus with sufficient amounts of oxygen from maternal blood. Fetal Hb is replaced by adult Hb after birth. These Hbs are encoded by their corresponding genes located at the globin gene clusters. Detection of fetal Hb in a blood stain via electrophoresis can provide important evidence in cases of infanticide and concealed birth.

**10.2.2.2.2 Hemoglobin S** Hb S polymorphism has forensic importance in identifying individuals. The Hb S polymorphism is observed in high frequencies among those of African heritage and some Hispanic populations. Such a protein polymorphic marker can provide investigational leads for the indication of the ethnic origin of a perpetrator. Hb S transports oxygen much less efficiently than normal Hb. Individuals who are homozygous for Hb S usually die early after suffering from sickle cell anemia and a variety of health-related problems. However, a heterozygous individual (an individual with a copy of the wild type Hb gene from one parent and a copy of Hb S gene from the other) will survive. This condition is known as the sickle cell trait.

In the 1950s, Vernon Ingram of Cambridge University discovered the molecular mechanism of the Hb S defect. His work revealed that the Hb S bears a mutation that changes the glutamic acid (in wild type) to a valine at the sixth amino acid from the N-terminal end of the β chain. This substitution of amino acids causes a major change in the structure of the β chain that in turn results in sickle cell anemia.

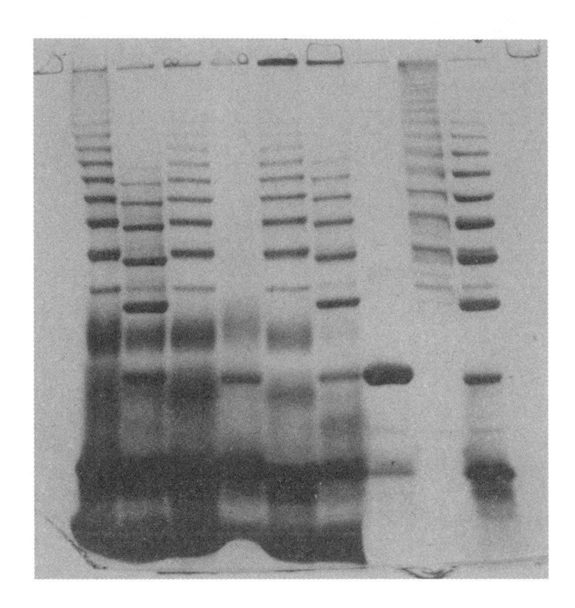

**Figure 10.12** Polyacrylamide gel electrophoresis of haptoglobin proteins. From left to right: Hp2, Hp2-1, Hp2, Hp1, Hp2, Hp2-1, Hp1, Hp2, Hp2-1. A node is at bottom (Source: James, S., and J. J. Nordby. 2005. *Forensic Science: An Introduction to Scientific and Investigative Techniques*. Boca Raton, FL: CRC Press. With permission.)

### 10.2.3 Serum Protein Polymorphisms

The serum portion of blood consists of a large number of proteins. The work on serum proteins for forensic purposes started in the 1950s when variations in serum proteins were found useful for distinguishing individuals. Over the years, a number of serum proteins were characterized and applied for forensic testing. Haptoglobin (Hp) was the most widely used of the polymorphic serum proteins in forensic biology (Figure 10.12). Haptoglobin is a protein that binds and transports Hb from the blood stream to the liver for the recycling of the iron contained in the Hb.

Immunoglobulin accounts for approximately 15% of serum protein and has been found to be highly variable. Two immunoglobulin proteins became utilized for forensic application. The γ chain protein ($G_m$) is the heavy chain of immunoglobulin G and the κ chain protein ($K_m$) is the light chain of all immunoglobulins. Table 10.7 lists common serum group systems. All exhibit genetic variations and can be detected in blood stains. The variants of these proteins can be determined by electrophoresis or serological methods.

**Table 10.7 Serum Proteins Used for Forensic Protein Profiling**

| Serum Protein | Protein Symbol | Gene Symbol | Chromosomal Location | No. of Amino Acids | No. of Alleles |
|---|---|---|---|---|---|
| Haptoglobin | Hp | *HP* | 16q22.1 | 328/387 | 2 |
| Group-specific component | Gc | *GC* | 4q12-13 | 458 | 2 |
| Transferrin | Tf | *TF* | 3q22.1 | 679 | 3 |
| Protease inhibitor (α1-antitrypsin) | Pi | *PI* | 14q32.1 | 394 | Many |

*Source:* Adapted from Yuasa, I., and K. Umetsu. 2005. *Leg Med* 7:251.

# Bibliography

## Blood Groups


Achermann, F. J. et al. 2005. Soluble type A substance in fresh-frozen plasma as a function of ABO and secretor genotypes and Lewis phenotype. *Transfus Apher Sci* 32 (3):255.

Allen, F. H., Jr., and R. E. Rosenfield. 1972. Review of Rh serology: Eight new antigens in nine years. *Haematologia (Budap)* 6 (1):113.

Anstee, D. J., and M. J. Tanner. 1975. Separation of ABH, I, Ss antigenic activity from the MN-active sialoglycoprotein of the human erythrocyte membrane. *Vox Sang* 29 (5):378.

Bargagna, M., and M. Pereira. 1967. A study of absorption–elution as a method of identification of Rhesus antigens in dried bloodstains. *J Forensic Sci Soc* 7 (3):123.

Bargagna, M., M. Sabelli, and C. Giacomelli. 1982. The detection of Rh antigens (D,C,c,E,e) on bloodstains by a micro-elution technique using low ionic strength solution (LISS) and papain-treated red cells. *Forensic Sci Int* 19 (2):197.

Bassler, G. 1986. Determination of the Lewis blood group substances in stains of forensically relevant body fluids. *Forensic Sci Int* 30 (1):29.

Boorman, K. E., B. E. Dodd, and P. J. Lincoln. 1977. *Blood Group Serology*, 5th ed., Vol. 4. London: Churchill Livingstone.

Brauner, P. 1996. DNA typing and blood transfusion. *J Forensic Sci* 41 (5):895.

Brewer, C. A., P. L. Cropp, and L. E. Sharman. 1976. A low ionic strength, hemagglutinating Autoanalyzer for Rhesus typing of dried bloodstains. *J Forensic Sci* 21 (4):811.

Busuttil, A. et al. 1993. Assessment of Lewis blood group antigens and secretor status in autopsy specimens. *Forensic Sci Int* 61 (2-3):133.

Camp, F. R., Jr. 1980. Forensic serology in the United States. I. Blood grouping and blood transfusion: Historical aspects. *Am J Forensic Med Pathol* 1 (1):47.

Cartron, J. P. et al. 1974. 'Weak A' phenotypes. Relationship between red cell agglutinability and antigen site density. *Immunology* 27 (4):723.

Cayzer, I., and P. H. Whitehead. 1973. The use of sensitized latex in the identification of human bloodstains. *J Forensic Sci Soc* 13 (3):179.

Chatterji, P. K. 1978. A simplified mixed agglutination technique for ABO grouping of dried bloodstains using cellulose acetate sheets. *J Forensic Sci Soc* 17 (2–3):143.

Chester, M. A., and M. L. Olsson. 2001. The ABO blood group gene: A locus of considerable genetic diversity. *Transfus Med Rev* 15 (3):177.

Daniels, G. 2002. *Human Blood Groups*, 2nd ed. Oxford: Blackwell Science.

Daniels, G. L. et al. 2004. Blood group terminology 2004 from the International Society of Blood Transfusion Committee on Terminology for Red Cell Surface Antigens. *Vox Sang* 87 (4):304.

De Soyza, K., and D. G. Garland. 1988. Studies and observations on Lewis grouping of body fluids and stains. *Forensic Sci Int* 38 (1–2):129.

DeForest, P., R. E. Gaensslen, and H. C. Lee. 1983. Blood, in *Forensic Science: An Introduction to Criminalistics.* New York: McGraw-Hill.

Denomme, G. A. 2004. The structure and function of the molecules that carry human red blood cell and platelet antigens. *Transfus Med Rev* 18 (3):203.

Erskine, A. G. 1973. *The Principles and Practices of Blood Grouping.* St. Louis: Mosby.

Fiori, A., and P. Benciolini. 1972. The ABO grouping of stains from body fluids. *Z Rechtsmed* 70 (4):214.

Fiori, A., M. Marigo, and P. Benciolini. 1963. Modified absorption–elution method Siracusa for ABO and MN grouping of blood-stains. *J Forensic Sci* 8 (3):419.

Gaensslen, R. E. et al. 1985. Evaluation of antisera for bloodstain grouping. II. Ss, Kell, Duffy, Kidd, and Gm/Km. *J Forensic Sci* 30 (3):655.

Gardas, A., and J. Koscielak. 1974. Megaloglycolipids: unusually complex glycosphingolipids of human erythrocyte membrane with A, B, H and I blood group specificity. *FEBS Lett* 42 (1):101.

Grunbaum, B. W. 1976. Some new approaches to the individualization of fresh and dried bloodstains. *J Forensic Sci* 21 (3):488.

Haak, W., J. Burger, and K. W. Alt. 2004. ABO genotyping by PCR-RFLP and cloning and sequencing. *Anthropol Anz* 62 (4):397.

Hakomori, S., and A. Kabata. 1974. Blood group antigens, in *The Antigens*. New York: Academic Press.

Hamaguchi, H., and H. Cleve. 1972. Solubilization of human erythrocyte membrane glycoproteins and separation of the MN glycoprotein from a glycoprotein with I, S, and A activity. *Biochim Biophys Acta* 278 (2):271.

Harrington, J. J. et al. 1990. Chemically sensitized erythrocytes for hemagglutination reactions. *J Forensic Sci* 35 (5):1115.

Harrington, J. J., R. Martin, and L. Kobilinsky. 1988. Detection of hemagglutinins in dried saliva stains and their potential use in blood typing. *J Forensic Sci* 33 (3):628.

Henry, S. et al. 1996. Molecular basis for erythrocyte $Le^{(a+\ b+)}$ and salivary ABH partial secretor phenotypes: Expression of a FUT2 secretor allele with an A→T mutation at nucleotide 385 correlates with reduced α-(1,2) fucosyltransferase activity. *Glycoconj J* 13 (6):985.

Hughes-Jones, N. C., and B. Gardner. 1971. The Kell system studied with radioactively-labelled anti-K. *Vox Sang* 21 (2):154.

Hughes-Jones, N. C., B. Gardner, and P. J. Lincoln. 1971. Observations of the number of available c,D, and E antigen sites on red cells. *Vox Sang* 21 (3):210.

Kabat, E. A. 1956. *The Blood Group Substances*. New York: Academic Press.

Kaneko, M. et al. 2003. Molecular characterization of a human monoclonal antibody to B antigen in ABO blood type. *Immunol Lett* 86 (1):45.

Kelly, R. J. et al. 1994. Molecular basis for H blood group deficiency in Bombay (Oh) and para-Bombay individuals. *Proc Natl Acad Sci USA* 91 (13):5843.

Kelly, R .J. et al. 1995. Sequence and expression of a candidate for the human secretor blood group α-(1,2)fucosyltransferase gene (FUT2): Homozygosity for an enzyme-inactivating nonsense mutation commonly correlates with the nonsecretor phenotype. *J Biol Chem* 270 (9):4640.

Kimura, A. et al. 1991. ABO blood grouping of semen from mixed body fluids with monoclonal antibody to tissue-specific epitopes on seminal ABO blood group substance. *Int J Legal Med* 104 (5):255.

Kimura, A. et al. 2005. Blood group A glycosphingolipid accumulation in the hair of patients with α-N-acetylgalactosaminidase deficiency. *Life Sci* 76 (16):1817.

Kimura, H., and S. Matsuzawa. 1991. Lewis blood group determination in bloodstains by planimetric measurement of eluted monoclonal antibodies. *J Forensic Sci* 36(4):999.

Kind, S. S. 1960. Absorption–elution grouping of dried blood smears. *Nature* 185:397.

_____. 1960. Absorption–elution grouping of dried blood-stains on fabrics. *Nature* 187:789.

Kind, S. S., and R. M. Cleevely. 1970. The fluorescent antibody technique: Its application to the detection of blood group antigens in stains. *J Forensic Med* 17(3):121.

Kobilinsky, L., and J .J. Harrington. 1988. Detection and use of salivary hemagglutinins for forensic blood grouping. *J Forensic Sci* 33 (2):396.

Koda, Y. et al. 1997. Missense mutation of FUT1 and deletion of FUT2 are responsible for Indian Bombay phenotype of ABO blood group system. *Biochem Biophys Res Commun* 238 (1):21.

Koda, Y., M. Soejima, and H. Kimura. 2001. The polymorphisms of fucosyltransferases. *Leg Med (Tokyo)* 3 (1):2.

Korchagina, E. Y. et al. 2005. Design of the blood group AB glycotope. *Glycoconj J* 22 (3):127.

Ladd, C. et al. 1996. A PCR-based strategy for ABO genotype determination. *J Forensic Sci* 41 (1):134.

Lappas, N. T., and M. E. Fredenburg. 1981. The identification of human bloodstains by means of a micro-thin-layer immunoassay procedure. *J Forensic Sci* 26 (3):564.

Lattes, L. 1932. *The Individuality of Blood*. Translated by L. W. H. Bertie. London: Oxford University Press.

Lee, H. C. 1982. Identification and grouping of bloodstains, in *Forensic Science Handbook*, Saferstein, R., Ed. Englewood Cliffs: Prentice Hall.

Lee, H. C. 1991. et al. Genetic markers in human bone: II. Studies on ABO (and IGH) grouping. *J Forensic Sci* 36 (3):639.

Lee, J. C. et al. 1996. ABO genotyping by mutagenically separated polymerase chain reaction. *Forensic Sci Int* 82 (3):227.

Lincoln, P. J., and B. E. Dodd. 1975. The application of a micro-elution technique using anti-human globulin for the detection of the S, s, K, Fya, Fyb and Jka antigens in stains. *Med Sci Law* 15 (2):94.

Liu, Y. H. et al. 1998. Distribution of H type 1 and of H type 2 antigens of ABO blood group in different cells of human submandibular gland. *J Histochem Cytochem* 46 (1):69.

Martin, P. D. 1978. A manual method for the detection of Rh antigens in dried bloodstains. *J Forensic Sci Soc* 17 (2–3):139.

McDowall, M. J., P. J. Lincoln, and B. E. Dodd. 1978. Increased sensitivity of tests for the detection of blood group antigens in stains using a low ionic strength medium. *Med Sci Law* 18 (1):16.

_____. 1978. Observations on the use of an Autoanalyzer and a manual technique for the detection of the red cell antigens C, D, E, c, K and S in bloodstains. *Forensic Sci* 11 (2):155.

Miller, C. H. et al. 2003. Measurement of von Willebrand factor activity: Relative effects of ABO blood type and race. *J Thromb Haemost* 1 (10):2191.

Mollicone, R., A. Cailleau, and R. Oriol. 1995. Molecular genetics of H, Se, Lewis and other fucosyltransferase genes. *Transfus Clin Biol.*

Moureau, P. 1963. Determination of blood groups in bloodstains, in *Methods of Forensic Science*. Lundquist, F., Ed. New York: Interscience.

Nickolls, L. C., and M. Pereira. 1962. A study of modern method of grouping dried bloodstains. *Med Sci Law* 2:172.

Ohmori, T. et al. 2003. Monoclonal antibodies against blood group A secretors and nonsecretors saliva. *Hybrid Hybridomics* 22 (3):183.

Okiura, T. et al. 2003. A-elute alleles of the ABO blood group system in Japanese. *Leg Med (Tokyo)* 5 Suppl 1:S207.

Oriol, R. 1995. ABO, Hh, Lewis, and secretion: Serology, genetics, and tissue distribution, in *Blood Cell Chemistry*. Cartron, J.P. and P. Rouger, Eds. New York: Plenum Press.

Oriol, R., J. J. Candelier, and R. Mollicone. 2000. Molecular genetics of H. *Vox Sang* 78 Suppl 2:105.

Outteridge, R. A. 1962. Absorption–elution grouping of blood stains: Modification and development. *Nature* 194:385.

_____. 1962. Absorption–elution method of grouping blood stains. *Nature* 195:818.

_____. 1965. Recent advances in the grouping of dried blood and secretion stains, in *Methods of Forensic Science*. Curry, A.S., Ed. New York: Interscience.

Painter, T. J., W. M. Watkins, and W. T. Morgan. 1965. Serologically active fucose-containing oligosaccharides isolated from human blood-group A and B substances. *Nature* 206 (984):594.

Prokop, O., and G. Uhlenbruck. 1969. *Human Blood and Serum Groups*, 2nd ed. Translated by J. Raven. New York: Wiley Interscience.

Reid, M. E., and C. Lomas-Francis. 2004. *The Blood Group Antigen Facts Book*, 2nd ed. London: Academic Press.

Reid, M. E., and N. Mohandas. 2004. Red blood cell blood group antigens: structure and function. *Semin Hematol* 41 (2):93.

Rosenfield, R. E., F. H. Allen, Jr., and P. Rubinstein. 1973. Genetic model for the Rh blood group system. *Proc Natl Acad Sci USA* 70 (5):1303.

Sasaki, M., and H. Shiono. 1996. ABO genotyping of suspects from sperm DNA isolated from postcoital samples in sex crimes. *J Forensic Sci* 41 (2):275.

Schleyer, F. 1962. Investigation of biological stains with regard to species origin, in *Methods of Forensic Science*. Lundquist, F., Ed. New York: Interscience.

Shaler, R. C., A. M. Hagins, and C. E. Mortimer. 1978. MN determination in bloodstains: Selective destruction of cross-reacting activity. *J Forensic Sci* 23 (3):570.

Smeets, B., H. van de Voorde, and P. Hooft. 1991. ABO bloodgrouping on tooth material. *Forensic Sci Int* 50 (2):277.

Snyder, L. H. 1973. *Blood Groups*. Minneapolis: Burgess Publishing.

Springer, G. F., and P. R. Desai. 1975. Human blood-group MN and precursor specificities: Structural and biological aspects. *Carbohydr Res* 40 (1):183.

Storry, J. R., and M. L. Olsson. 2004. Genetic basis of blood group diversity. *Br J Haematol* 126 (6):759.

Styles, W. M., B. E. Dodd, and R. R. Coombs. 1963. Identification of human bloodstains by means of mixed antiglobulin reaction on separate cloth fibrils. *Med Sci Law* 20:257.

Watkins, W. M. 1966. Blood-group substances. *Science* 152 (3719):172.

______. 2001. Commemoration of the centenary of the discovery of the ABO blood group system. *Transfus Med* 11:239.

Wynbrandt, F., and W. J. Chisum. 1971. Determination of the ABO blood group in hair. *J Forensic Sci Soc* 11 (3):201.

Xingzhi, X. et al. 1993. ABO blood grouping on dental tissue. *J Forensic Sci* 38 (4):956.

Yamamoto, F. et al. 1990. Molecular genetic basis of the histo-blood group ABO system. *Nature* 345 (6272):229.

## Protein Profiling

Allen, R. C., R. A. Harley, and R. C. Talamo. 1974. A new method for determination of α-1-antitrypsin phenotypes using isoelectric focusing on polyacrylamide gel slabs. *Am J Clin Pathol* 62 (6):732.

Bark, J. E., M. J. Harris, and M. Firth. 1976. Typing of common phosphoglucomutase variants using isoelectric focusing: a new interpretation of the phosphoglucomutase system. *J Forensic Sci Soc* 16 (2):115.

Blake, N. M., and K. Omoto. 1975. Phosphoglucomutase types in the Asian-Pacific area: A critical reveiw including new phenotypes. *Ann Hum Genet* 38 (3):251.

Budowle, B. 1984. A method to increase the volume of sample applied to isoelectric focusing gels. *Forensic Sci Int* 24 (4):273.

______. 1987. A method for subtyping group-specific component in bloodstains. *Forensic Sci Int* 33 (3):187.

Budowle, B., and P. Eberhardt. 1986. Ultrathin-layer polyacrylamide gel isoelectric focusing for the identification of hemoglobin variants. *Hemoglobin* 10 (2):161.

Budowle, B., and E. Scott. 1985. Transferrin subtyping of human bloodstains. *Forensic Sci Int* 28 (3-4):269.

Budowle, B., S. Sundaram, and R. E. Wenk. 1985. Population data on the forensic genetic markers: Phosphoglucomutase-1, esterase D, erythrocyte acid phosphatase and glyoxylase I. *Forensic Sci Int* 28 (2):77.

Burdett, P. E. 1981. Isoelectric focusing in agarose: Phosphoglucomutase (PGM locus 1) typing. *J Forensic Sci* 26 (2):405.

Burdett, P. E., and P. H. Whitehead. 1977. The separation of the phenotypes of phosphoglucomutase, erythrocyte acid phosphatase, and some haemoglobin variants by isoelectric focusing. *Anal Biochem* 77 (2):419.

Carracedo, A., and L. Concheiro. 1982. The typing of α-1-antitrypsin in human bloodstains by isoelectric focusing. *Forensic Sci Int* 19 (2):181.

Carracedo, A. et al. 1983. Silver staining method for the detection of polymorphic proteins in minute bloodstains after isoelectric focusing. *Forensic Sci Int* 23 (2–3):241.

Constans, J., and M. Viau. 1977. Group-specific component: Evidence for two subtypes of the Gc1 gene. *Science* 198 (4321):1070.

Constans, J. et al. 1978. Analysis of the Gc polymorphism in human populations by isoelectrofocusing on polyacrylamide gels: Demonstration of subtypes of the Gc allele and of additional Gc variants. *Hum Genet* 41 (1):53.

Cox, D. W., S. Smyth, and G. Billingsley. 1982. Three new rare variants of α-1-antitrypsin. *Hum Genet* 61 (2):123.

Divall, G. B. 1981. Studies on the use of isoelectric focusing as a method of phenotyping erythrocyte acid phosphatase. *Forensic Sci Int* 18 (1):67.

______. 1984. The esterase D polymorphism as revealed by isoelectric focusing in ultra-thin polyacrylamide gels. *Forensic Sci Int* 26 (4):255.

Divall, G. B., and M. Ismail. 1983. Studies and observations on the use of isoelectric focusing in ultra-thin polyacrylamide gels as a method of typing human red cell phosphoglucomutase. *Forensic Sci Int* 22 (2-3):253.

Dorrill, M., and P. H. Whitehead. 1979. The species identification of very old human blood-stains. *Forensic Sci Int* 13 (2):111.

Dykes, D. D., and H. F. Polesky. 1984. Review of isoelectric focusing for Gc, PGM1, Tf, and Pi subtypes: Population distributions. *Crit Rev Clin Lab Sci* 20 (2):115.

Frants, R. R., and A. W. Eriksson. 1976. A-1-Antitrypsin: Common subtypes of Pi M. *Hum Hered* 26 (6):435.

Gorg, A., W. Postel, and R. Westermeier. 1978. Ultrathin-layer isoelectric focusing in polyacrylamide gels on cellophane. *Anal Biochem* 89 (1):60.

Grunbaum, B. W., and P. L. Zajac. 1977. Rapid phenotyping of the group-specific component by immunofixation on cellulose acetate. *J Forensic Sci* 22 (3):586.

______. 1978. Phenotyping of erythrocyte acid phosphatase in fresh blood and in bloodstains on cellulose acetate. *J Forensic Sci* 23 (1):84.

Hoste, B. 1979. Group-specific component (Gc) and transferrin (Tf) subtypes ascertained by isoelectric focusing: a simple nonimmunological staining procedure for Gc. *Hum Genet* 50 (1):75.

Itoh, Y., and S. Matsuzawa. 1990. Detection of human hemoglobin A (HbA) and human hemoglobin F (HbF) in biological stains by microtiter latex agglutination-inhibition test. *Forensic Sci Int* 47 (1):79.

Jones, D. A. 1972. Blood samples: Probability of discrimination. *J Forensic Sci Soc* 12(2):355.

Khalap, S., and G. B. Divall. 1979. Gm(5) grouping of dried bloodstains. *Med Sci Law* 19 (2):86.

Kido, A. et al. 1984. A stability study on Gc subtyping in bloodstains: Comparison by two different techniques. *Forensic Sci Int* 26 (1):39.

Kimura, H. et al. 1983. The typing of group-specific component (Gc protein) in human blood stains. *Forensic Sci Int* 22 (1):49.

Kipps, A. E. 1979. Gm and Km typing in forensic science-a methods monograph. *J Forensic Sci Soc* 19 (1):27.

Kipps, A. E., V. E. Quarmby, and P. H. Whitehead. 1978. The detection of mixtures of blood and other body secretions in stains. *J Forensic Sci Soc* 18 (3-4):189.

Kueppers, F., and B. Harpel. 1979. Group-specific component (Gc) 'subtypes' of Gc1 by isoelectric focusing in US blacks and whites. *Hum Hered* 29 (4):242.

Kuhnl, P., U. Schmidtmann, and W. Spielmann. 1977. Evidence for two additional common alleles at the PGM1 locus (phosphoglucomutase: E.C. 2.7.5.1): Comparison by three different techniques. *Hum Genet* 35 (2):219.

Kuhnl, P., and W. Spielmann. 1978. Transferrin: evidence for two common subtypes of the TfC allele. *Hum Genet* 43 (1):91.

____. 1979. A third common allele in the transferrin system, TfC3, detected by isoelectric focusing. *Hum Genet* 50 (2):193.

Lamm, L. U. 1970. Family studies of red cell acid phosphatase types: Report of a family with the D variant. *Hum Hered* 20 (3):329.

Lincoln, P. J., and B. E. Dodd. 1973. An evaluation of factors affecting the elution of antibodies from bloodstains. *J Forensic Sci Soc* 13 (1):37.

Markert, C. L. 1968. The molecular basis for isozymes. *Ann N Y Acad Sci* 151 (1):14.

Miscicka, D., T. Dobosz, and S. Raszeja. 1977. Determination of phenotypes of phosphoglucomutase (PGM1) in bloodstains by cellulose acetate electrophoresis. *Z Rechtsmed* 79 (4):297.

Murch, R. S., and B. Budowle. 1986. Applications of isoelectric focusing in forensic serology. *J Forensic Sci* 31 (3):869.

Neilson, D. M. et al. 1976. Simultaneous electrophoresis of peptidase A, phosphoglucomutase, and adenylate kinase. *J Forensic Sci* 21 (3):510.

Olaisen, B. et al. 1981. The ESD polymorphism: further studies of the ESD2 and ESD5 allele products. *Hum Genet* 57 (4):351.

Randall, T., W. A. Harland, and J. W. Thorpe. 1980. A method of phenotyping erythrocyte acid phosphatase by iso-electric focusing. *Med Sci Law* 20 (1):43.

Rees, B., T. J. Rothwell, and J. Bonnar. 1974. Correlation of phosphoglucomutase isoenzymes in blood and semen. *Lancet* 1 (7861):783.

Sensabaugh, G. F. 1982. Biochemical markers for individuality, in *Forensic Science Handbook*. Saferstein, R., Ed. Englewood Cliffs, NJ: Prentice Hall.

Shaler, R. C. 1982. Interpretation of Gm testing results: Two case histories. *J Forensic Sci* 27 (1):231.

Sonneborn, H. H. 1972. Comments on the determination of isoenzyme polymorphism (ADA, AK, 6-PGD, PGM) by cellulose acetate electrophoresis. *Humangenetik* 17(1):49.

Sorensen, S. A. 1974. Agarose gel electrophoresis of the human red cell acid phosphatase. *Vox Sang* 27 (6):556.

____. 1975. Report and characterization of a new variant, EB, of human red cell acid phosphatase. *Am J Hum Genet* 27 (1):100.

Stedman, R. 1972. Human population frequencies in twelve blood grouping systems. *J Forensic Sci Soc* 12 (2):379.

Stolorow, M. D., and B. G. Wraxall. 1979. An efficient method to eliminate streaking in the electrophoretic analysis of haptoglobin in bloodstains. *J Forensic Sci* 24(4):856.

Sutton, J. G., and R. Burgess. 1978. Genetic evidence for four common alleles at the phosphoglucomutase-1 locus (PGM1) detectable by isoelectric focusing. *Vox Sang* 34(2):97.

Teige, B. et al. 1988. Forensic aspects of haptoglobin: Electrophoretic patterns of haptoglobin allotype products and an evaluation of typing procedure. *Electrophoresis* 9 (8):384.

Turowska, B., and M. Bogusz. 1978. The rare silent gene Po of human red cell acid phosphatase in a second family in Poland. *Forensic Sci* 11 (3):175.

Twibell, J., and P. H. Whitehead. 1978. Enzyme typing of human hair roots. *J Forensic Sci* 23 (2):356.

Whitehead, P. H. et al. 1970. The examination of bloodstains by Laurell electrophoresis (antigen–antibody crossed electrophoresis). *J Forensic Sci Soc* 10 (2):83.

Whitehead, P. H., L. A. King, and D. J. Werrett. 1979. New information from bloodstains. *Naturwissenschaften* 66 (9):446.

Wraxall, B. G., and E. G. Emes. 1976. Erythrocyte acid phosphatase in bloodstains. *J Forensic Sci Soc* 16 (2):127.

Wrede, B., E. Koops, and B. Brinkmann. 1971. Determination of three enzyme polymorphisms in one electrophoretic step on horizontal polyacrylamide gels. *Humangenetik* 13 (3):250.

Yuasa, I., and K. Umetsu. 2005. Molecular aspects of biochemical markers. *Leg Med* 7:251.

## Study Questions

1. What is the molecular basis for designating an individual a nonsecretor?

2. Which serum antibody does a type B individual have?
   (a) Anti-A
   (b) Anti-B
   (c) Anti-A and anti-B
   (d) Neither

3. Which of the following performs better when testing an aged blood stain? Why?
   (a) Lattes crust assay
   (b) Absorption–elution assay

4. What is the difference between the agglutination and hemagglutination assays?

5. Which of the following antigen(s) does a type O individual have?
   (a) A antigen
   (b) B antigen
   (c) A and B antigen
   (d) Neither A nor B antigen

# Basic DNA Techniques

# IV

# Introduction to Human Genome 11

## 11.1 Structure and Properties of DNA

### 11.1.1 Nucleotides and Polynucleotides

***Deoxyribonucleic acid (DNA)*** is a linear polynucleotide consisting of four types of monomeric nucleotides. Each ***nucleotide*** contains three components: a deoxyribose, a nitrogenous base, and a phosphate group (Figure 11.1). The four bases for DNA are adenine (A), cytosine (C), guanine (G), and thymine (T) (Figure 11.2). The deoxyribose is attached to the nitrogen of a base. The phosphate group is attached to the deoxyribose (Figure 11.3). In a ***polynucleotide***, individual nucleotides are linked by phosphodiester bonds (Figure 11.4).

### 11.1.2 Double Helix

DNA is a double helical molecule as described by Watson and Crick who received the 1962 Nobel Prize for their work. Figure 11.5 shows the double helix, also called the B-form of DNA characterized by right-handed turns and two grooves that spiral along the length of the helix. The ***major groove*** is relatively wide and deep; the ***minor groove*** is narrow and more shallow.

The double helix is stabilized by chemical interactions. Base pairing of the two strands involves the formation of hydrogen bonds that provide weak electrostatic attractions between electronegative atoms. An A always pairs with a T (two hydrogen bonds), and a C pairs with a G (three hydrogen bonds). See Figure 11.6. Base stacking involves hydrophobic interactions between adjacent base pairs and provides stability to the double helix.

### 11.1.3 Denaturation and Renaturation of DNA

Double stranded DNA is maintained by hydrogen bonding between the bases of complementary pairs. ***Denaturation*** occurs when the hydrogen bonds of DNA are disrupted and the strands are separated. A ***melting curve*** can be obtained from measuring DNA denaturation by slowly heating a solution of DNA. As shown in Figure 11.7, an increasing temperature increases the percentage of the DNA that is denatured. The temperature at which 50% of DNA strands is denatured is defined as the ***melting temperature*** ($Tm$). The value of $Tm$ is affected by the salt concentration of the solution, but can also be

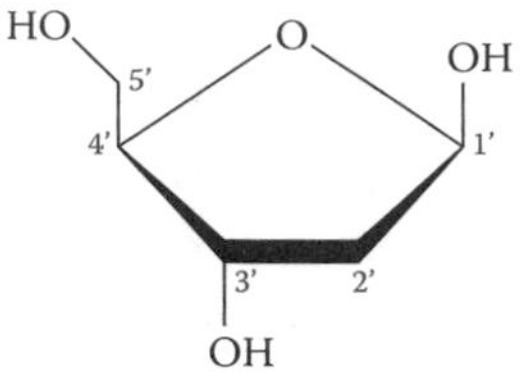

**Figure 11.1** Deoxyribose. A pentose type of sugar composed of five numbered carbon atoms. Note that the hydroxyl group attached to the 2′ carbon on ribose is replaced by a hydrogen group in deoxyribose.

affected by nucleotide content, high pH, and length of the molecule.

Nucleotide content affects the value of *Tm* because GC pairs are joined by three hydrogen bonds while the AT pairs are joined only by two. Increasing the GC content of a DNA molecule will increase the *Tm*. Excessively high pH causes the hydrogen bonds to break and the paired strands to separate. Finally, the length of the molecule also affects the *Tm* simply because a longer molecule of DNA requires more energy to break more bonds than a shorter molecule.

The single strands in a solution of denatured DNA can, under certain conditions, re-form into double-stranded DNA. The process is called ***renaturation***, or reannealing, and two requirements must be met for it to occur. First, the salt concentration must be high enough to neutralize the negative charges of the phosphate groups that would otherwise cause the complementary strands to repel one another. Also, the temperature must be high enough to sufficiently disrupt hydrogen bonds that form at random between the bases within the same strand, but not so high as to disrupt the base pairs

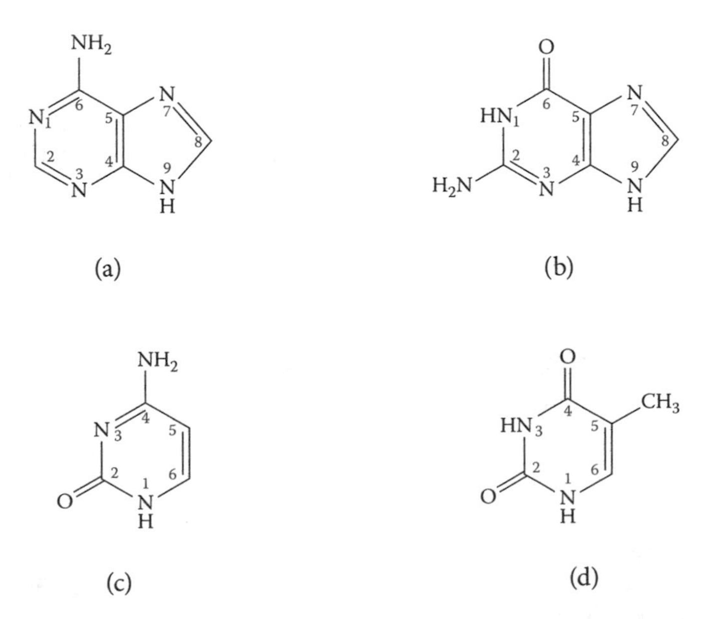


**Figure 11.2** Chemical structures of nitrogenous bases: (a) adenine; (b) guanine; (c) cytosine; and (d) thymine.

between the complementary strands. A temperature approximately 20°C below *Tm* is usually optimal.

### 11.1.4 Organization of DNA into Chromosomes

The nuclear ***chromosomes*** of humans consist of complexes of DNA, histone proteins, and nonhistone chromosomal proteins. Each chromosome consists of one linear double stranded DNA molecule. The large amounts of DNA present in the human chromosome are compacted by their association with histones into nucleosomes and even further compacted by higher levels of folding of the nucleosomes into ***chromatin*** fibers. Each chromosome contains a large number of looped domains of chromatin fibers attached to a protein scaffold.

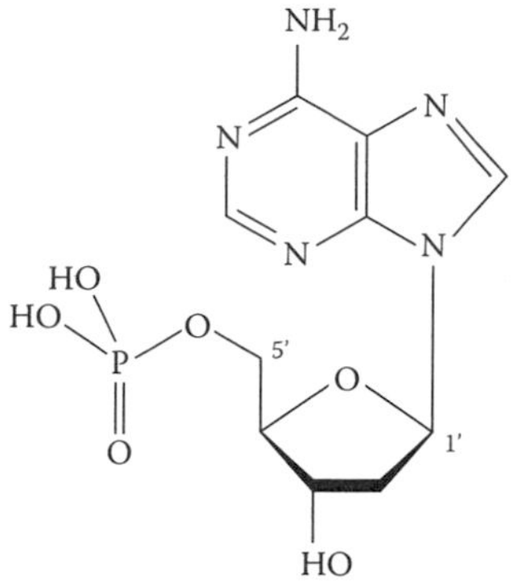

**Figure 11.3** Structure of nucleotide. Each nucleotide in a DNA polymer is made up of three components: a deoxyribose, a nitrogenous base (adenine is shown as an example), and a phosphate group. The 1′ carbon of the deoxyribose is attached to the nitrogen of a nitrogenous base. A phosphate group is attached to the 5′ carbon of the deoxyribose.

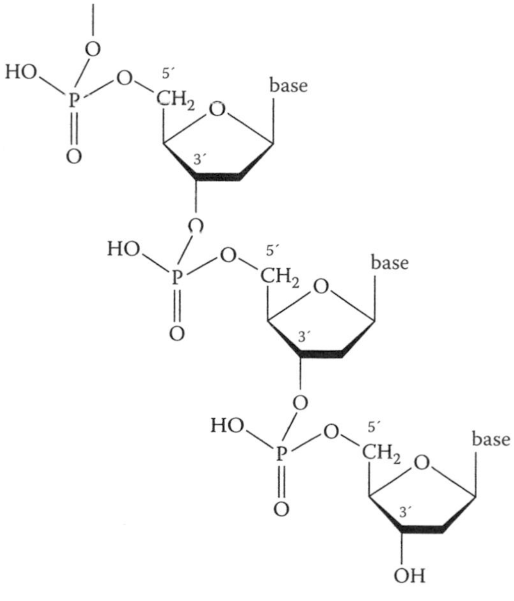

**Figure 11.4** DNA polynucleotide chain. Individual nucleotides are linked by phosphodiester bonds between their 5′ and 3′ carbons.

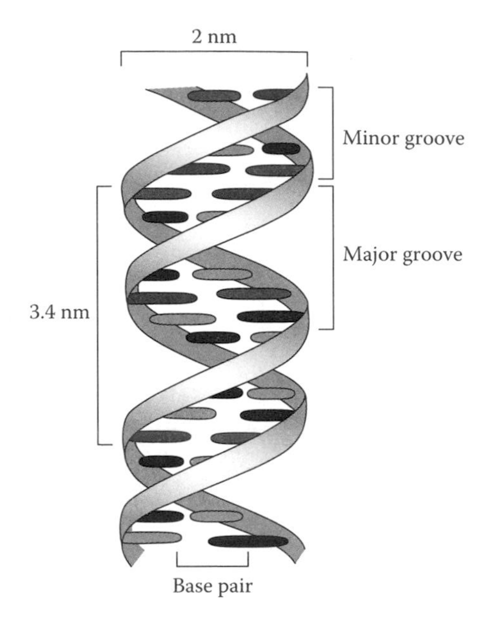


**Figure 11.5** DNA double helix. With a helical diameter of 2 nm, each turn of the helix takes 3.4 nm which corresponds to 10 base pairs per turn. The major and minor grooves are shown.

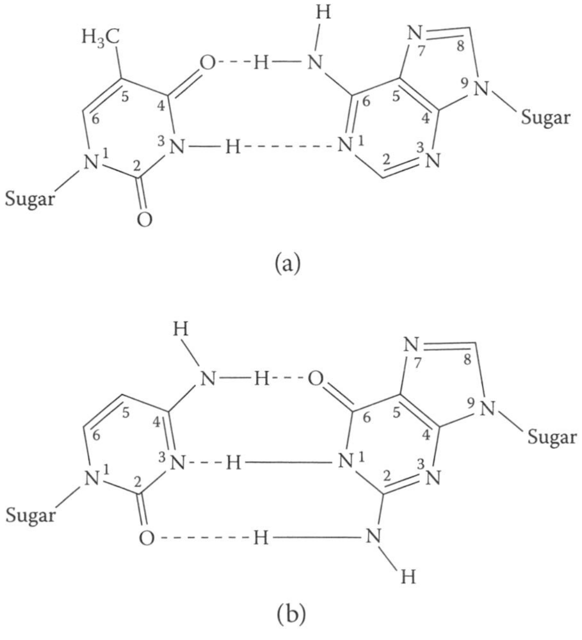

H
$H_3C$ N 7 8
O- - -H—N
5 4 6 5 9 N
6 4 Sugar
3 N—H- - - - - - -N 1 3
N 1 2 2 N
Sugar
O

(a)

H
N
N—H- - -O 7 8
5 4 6 5 9 N
6 4 Sugar
3 N- - -H———N 1 3
N 1 2 2 N
Sugar
O- - - - - - -H———N
H

(b)

**Figure 11.6** Base pairing between two DNA stands. (a) Adenine (right) pairs with a thymine (left). (b) A cytosine (left) pairs with a guanine (right).

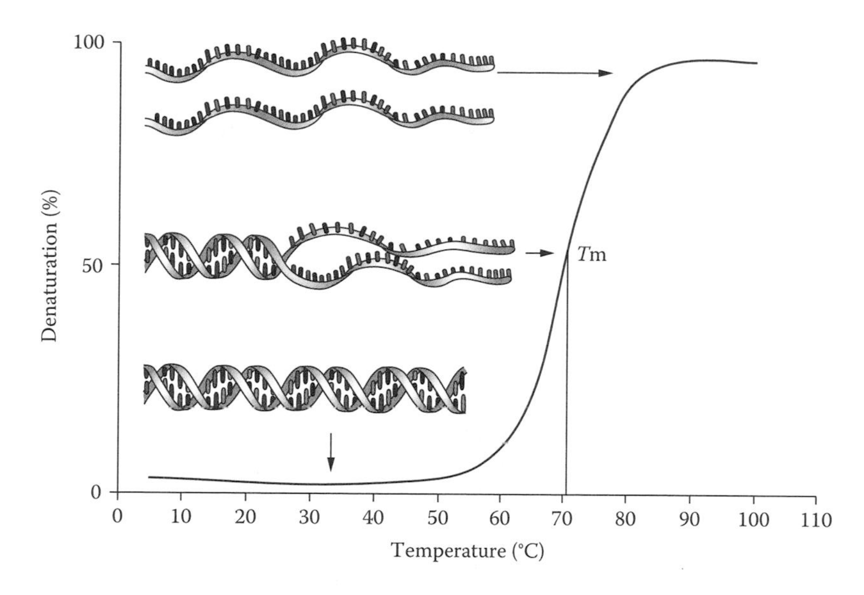


**Figure 11.7** Melting curve of DNA. The degree of DNA denaturation is increased by increasing the temperature. *Tm* and possible shapes of a DNA molecule are shown.

The degree of DNA packing varies throughout the cell cycle. During the metaphase of mitosis and meiosis of the cell cycle, chromatin is the most condensed. Two forms of chromatin have been defined on the basis of chromosome-staining properties. ***Euchromatin*** regions are areas of chromosomes that undergo normal chromosome condensation and decondensation during the cell cycle. The intensity of staining of euchromatin is the darkest in metaphase and the lightest in the synthesis (S) phase. Euchromatic regions account for most of the genome and lack repetitive DNA. Usually, the genes within the euchromatin can be expressed. ***Heterochromatin*** comprises the chromosomal regions that usually remain condensed throughout the cell cycle. It contains repetitive DNA and can be found at centromeres, much of the Y chromosome long arm, and the short arms of the acrocentric chromosomes (chromosomes with centromeres near one end). Genes within heterochromatic DNA are usually inactive.

## 11.1.5 Human Chromosomes

Each human chromosome has a short arm, designated p (for petit), and a long arm, designated q (for queue) separated by a centromere (Figure 11.8). ***Centromeres*** are the DNA sequences found near the points of attachment of mitotic or meiotic spindle fibers. The centromere region of each chromosome is responsible for accurately segregating the replicated chromosomes to

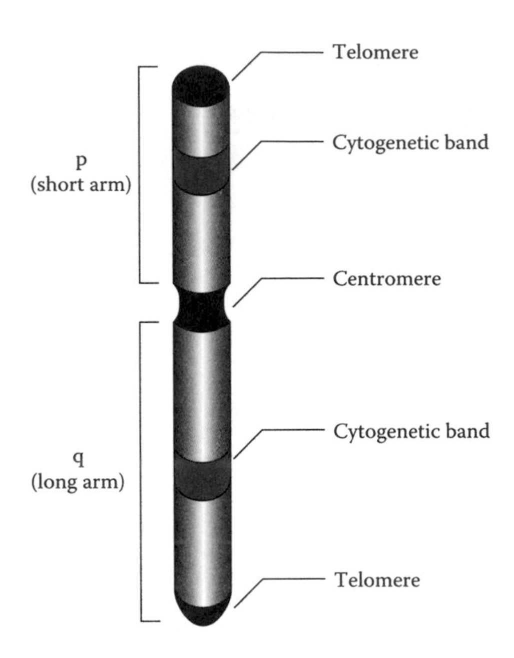


**Figure 11.8** Chromosome structure with cytogenetic banding pattern shown after chemical staining. The short arm is designated p; the long arm is q. The centromere and telomere are also shown.

daughter cells during cell divisions. The ends of the chromosome are called ***telomeres*** and they help stabilize the chromosome and play a role in the replication of DNA in the chromosome.

Chemical staining of metaphase chromosomes results in an alternating dark and light banding pattern (cytogenetic banding) that can be seen under a microscope. Each chromosome arm is divided into regions based on the cytogenic bands. This process is known as ***cytogenetic mapping***. The cytogenetic bands are labeled p1, p2, p3, q1, q2, q3, and so on counting from the centromere out toward the telomeres. At higher resolutions, sub-bands can be observed. For example, the cytogenetic map location of a gene termed *AMELY* (amelogenin, Y-linked) is Yp11.2, which indicates its location on chromosome Y, p arm, band 11, sub-band 2. The visually distinct banding pattern gives each chromosome a unique appearance. Recently, the cytogenic map has been integrated with the human genome sequence to allow the determination of the positions of cytogenetic bands within the DNA sequence.

Spermatozoa and ova formed by germ cells are called ***gametes***. In humans, each gamete is haploid, containing 22 ***autosomes*** (chromosomes other than sex chromosomes) plus one ***sex chromosome***. In ova, the sex chromosome is always an X while in spermatozoa it may be an X or a Y. After fertilization, the zygote becomes diploid as a result of fusion of the haploid spermatozoon and ovum. Most other cells of the body, known as ***somatic cells*** are diploid. This means they have two copies of each autosome plus two sex chromosomes, XX for females or XY for males. This results in a total of 46 chromosomes per diploid cell. The two chromosomes of a pair in a diploid cell are ***homologous chromosomes.*** One of the homologous chromosomes is inherited from the sperm and the other from the ovum.

Although most somatic cells are diploid, exceptions exist. Some differentiated cells such as red blood cells and platelets have no nuclei and are designated ***nulliploid.*** A few other cells have more than two sets of chromosomes as a result of DNA replication without cell division and are referred to as ***polyploid.*** The regenerating cells of the liver and other tissues are naturally tetraploid, while the giant megakaryocytes of the bone marrow may contain 8, 16, or even 32 copies of chromosomes.

Chromosomes are identified on the basis of size and the positions of the centromeres and cytogenetic banding patterns. This allows chromosomes to become easily identifiable. The chromosome constitution is described as a ***karyotype*** and can be displayed as a ***karyogram*** that includes the total number of chromosomes and the sex chromosome constitution (Figure 11.9). Chromosomes are numbered in order of size, with chromosome 1 as the largest (except chromosome 21 is smaller than 22). In cases of chromosomal abnormality, the karyotype can also reflect the type of abnormality and allow visualization of the affected chromosome bands.

## 11.2 Human Nuclear Genome

The human genome contains the biological information needed to allow humans to perform biological functions. It consists of the ***nuclear genome*** and the ***mitochondrial genome*** (Chapter 23 discusses the mitochondrial genome). The human nuclear genome, a set of 23 chromosomes, contains approximately 3.2 billion base pairs (bp). The Human Genome Project was initiated in 1990 to sequence the entire human nuclear genome. About 2.6 billion bp (more than 80%) of the human genome sequence were completed in 2001 including the most important parts of the genome. The genome contains genes and intergenic non-coding sequences.

### 11.2.1 Genes and Related Sequences

It is estimated that the human genome has about 30,000 to 40,000 genes that encode the information for the synthesis of proteins. The functions of about half these genes have been identified. Most encode the proteins responsible

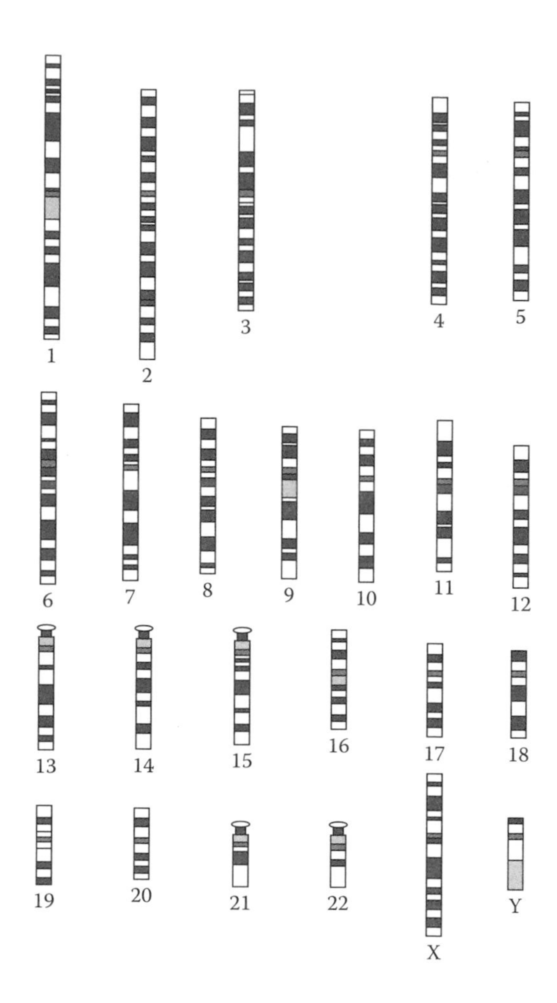


**Figure 11.9** Human karyogram. The chromosomes are numbered. Cytogenic patterns show alternating dark and light bands (Source: www.genome.gov).

for the maintenance of the genome, functioning of cells, the immune response, and the structural proteins of cells.

Most human genes are discontinuous. The coding regions of genes are called ***exons*** and are separated by ***introns***. During gene expression, the primary messenger RNA transcript (mRNA), consisting of both the exons and introns is produced. The mRNA is a template for protein synthesis in which the sequence is based on a complementary strand of DNA. Through the process of splicing, the introns are removed and the exons are joined, producing the spliced mRNA form that can be used for protein synthesis via the translation process. Other gene-related sequences include those responsible for gene transcription such as ***promoter sequences***; those responsible for gene activation such as ***cis-regulatory sequences*** (or enhancers); and ***untranslated sequences*** that are transcribed but do not encode proteins. Figure 11.10 depicts the features of a representative human gene.

## 11.2.2 Intergenic Noncoding Sequences

More than 60% of the human genome sequence consists of intergenic noncoding sequences located between genes. The functions of these sequences are yet to be discovered. The intergenic noncoding sequences contain large quantities of various types of ***repetitive DNA*** that falls into two categories: genome-wide or interspersed repeats and tandem repeats (Figure 11.11)

### *11.2.2.1 Interspersed Repeats*

***Interspersed repeats*** are randomly located throughout the human genome. Several human types have been characterized based on their characteristic sequences, namely SINEs, LINEs, LTR elements, and DNA transposons. Some of these sequences can be further divided into subtypes. For example, SINEs, the most abundant types, have three subtypes designated Alu elements, MIR, and MIR3. It is believed that these interspersed repeats are derived from a transposable element, a mobile segment of DNA, that changes location by transposition. These elements duplicate themselves during transposition and propagate throughout the genome. The transposition of LINEs, SINEs and LTR elements requires an RNA intermediate; DNA transposons do not.

### *11.2.2.2 Tandem Repeats*

***Tandem repeats*** are repeat units placed next to each other in an array. One type is called ***satellite DNA*** because of the observation of satellite bands

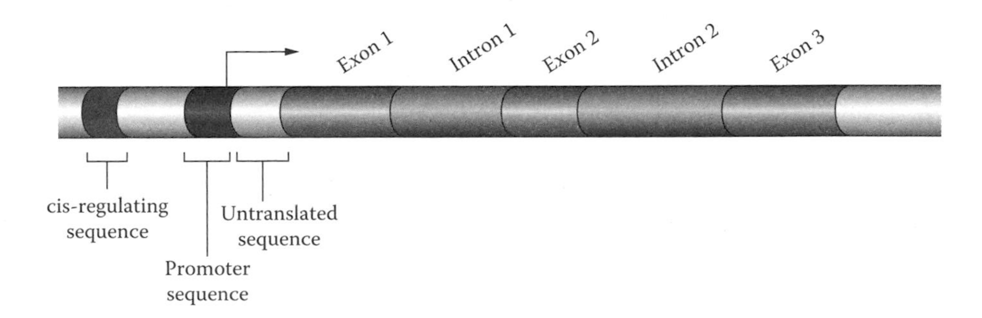


**Figure 11.10** Gene structure. Transcription, which can be regulated by the cis-regulatory sequence, is initiated at the transcription start site (arrow) near the promoter. The exons, noncoding introns, and the untranslated sequence are also shown.

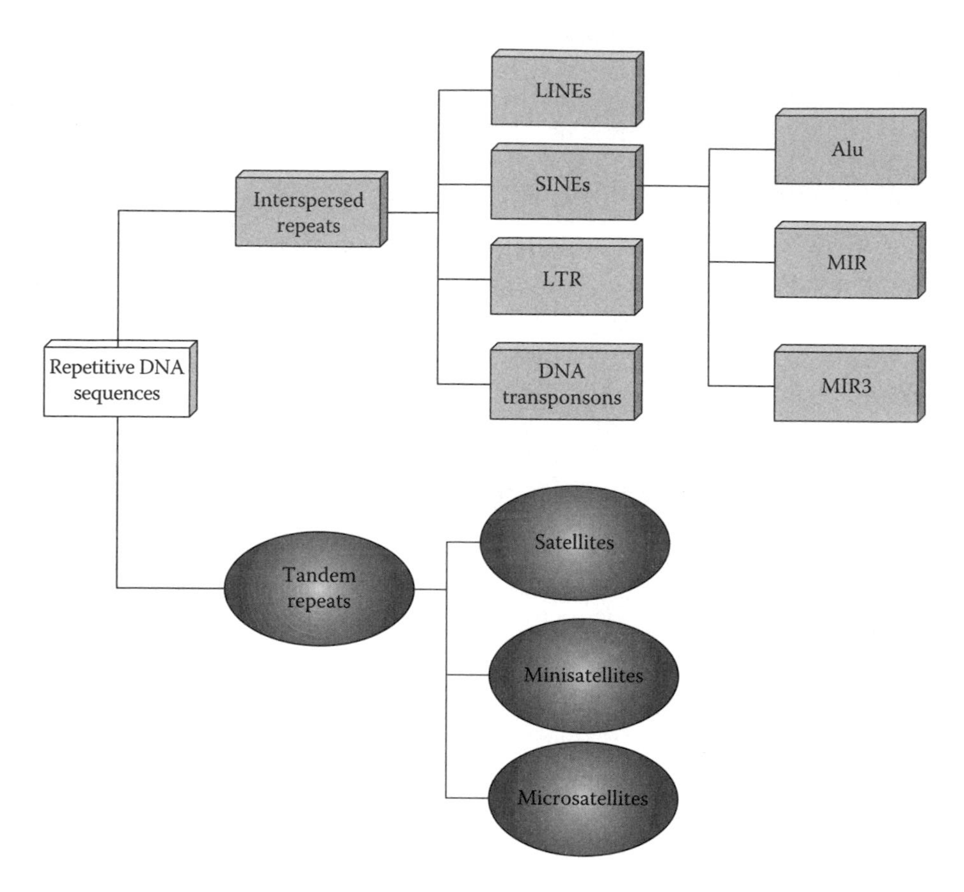


**Figure 11.11** Classes of repetitive DNA sequences.

containing DNA with tandem repeats during density gradient centrifugation. Satellite DNA can be found at centromeres and telomeres consisting of fragments composed of long stretches of tandem repeats. Minisatellites and microsatellites are two other types of shorter tandem repeats. ***Minisatellites***, also known as ***variable number tandem repeats (VNTRs)*** form arrays of tandem repeats with repeat unit length from several to hundreds of base pairs. In a ***microsatellite***, also known as a ***short tandem repeat (STR)*** or ***simple sequence repeat (SSR)***, the repeating unit length can be 2 to 6 bp long.

## 11.3 Human DNA Polymorphisms

Most of individual human genome sequences are very similar. However, variations in sequences do occur. The differences between individual genomes that occur at DNA level are called ***DNA polymorphisms***. In particular, a DNA polymorphism with alternative forms of a chromosomal locus that differ in nucleotide sequence is known as a ***sequence polymorphism***. A DNA polymorphism that differs in the numbers of tandem repeat units is known as a ***length polymorphism***. A DNA polymorphism can occur anywhere in a genome including genes and other chromosomal locations. Many DNA polymorphisms are useful for genetic mapping studies and hence are called ***DNA markers***. DNA polymorphisms form the basis of forensic DNA profiling. The focus of this text is on the human genome, but polymorphisms also occur in the genomes of other organisms.

Most DNA polymorphisms are ***single nucleotide polymorphisms (SNPs)*** involving a single base pair change or a point mutation. Over 1.4 million SNPs have been identified. SNPs arise by spontaneous mutation. Most SNPs occur in noncoding regions of the genome, although some appear in coding regions as well.

Other forms of DNA polymorphisms are tandem repeats such as microsatellites and minisatellites. Although their functions are unknown, microsatellites and minisatellites are very useful for forensic DNA analysis. Many are highly polymorphic, and the number of repeat units varies greatly among different individuals of a population. It is unlikely that two unrelated individuals will have exactly the same combination of microsatellite or minisatellite polymorphisms if sufficient markers are examined. Thus, a resulting genetic profile can be used for human identification.

Alternative forms of DNA polymorphisms are called ***alleles***. The same allele present in both homologous chromosomes is referred to as ***homozygous***. Two different alleles present in homologous chromosomes are referred to as ***heterozygous***. A combination of alleles at a given locus is a ***genotype***. In forensic analysis, the genotype for a panel of analyzed loci is called the ***DNA profile***.

## Bibliography

### Books

Brown, T. A. 2002. *Genomes*, 2nd ed. New York: John Wiley & Sons, Inc.

Lewin, B. 2004. *Genes*, Vol. 8. Upper Saddle River, NJ: Pearson Education.

Nelson, D., and M. Cox. 2004. *Lehninger's Principles of Biochemistry*, 4th ed. New York: W.H. Freeman.

Watson, J. et al. 2004. *Molecular Biology of the Gene*, 5th ed. San Francisco: Pearson Education.

## DNA

Dickerson, R. E. 1983. The DNA helix and how it works. *Sci Amer* 249:17.
Sachidanandam, R. et al. 2001. A map of human genome sequence variation containing 1.42 million single nucleotide polymorphisms. *Nature* 409 (6822):928.
Watson, J. D., and F. H. Crick. 1953. Genetical implications of the structure of deoxyribonucleic acid. *Nature* 171 (4361):964.
_____. 1953. The structure of DNA. *Cold Spring Harb Symp Quant Biol* 18:123.
Wilkins, M. H., A. R. Stokes, and H. R. Wilson. 1953. Molecular structure of deoxypentose nucleic acids. *Nature* 171 (4356):738.
Wu, R. 1978. DNA sequence analysis. *Annu Rev Biochem* 47:607.

## Human Genome

Altshuler, D. et al. 2000. An SNP map of the human genome generated by reduced representation shotgun sequencing. *Nature* 407 (6803):513.
Blackburn, E. H. 1984. The molecular structure of centromeres and telomeres. *Annu Rev Biochem* 53:163.
Breathnach, R., and P. Chambon. 1981. Organization and expression of eucaryotic split genes coding for proteins. *Annu Rev Biochem* 50:349.
Britten, R. J., and E. H. Davidson. 1971. Repetitive and non-repetitive DNA sequences and a speculation on the origins of evolutionary novelty. *Q Rev Biol* 46 (2):111.
Davidson, E. H., and R. J. Britten. 1973. Organization, transcription, and regulation in the animal genome. *Q Rev Biol* 48 (4):565.
Dib, C. et al. 1996. A comprehensive genetic map of the human genome based on 5,264 microsatellites. *Nature* 380 (6570):152.
Donis-Keller, H. et al. 1987. A genetic linkage map of the human genome. *Cell* 51(2):319.
Hogenesch, J. B. et al. 2001. A comparison of the Celera and Ensembl predicted gene sets reveals little overlap in novel genes. *Cell* 106 (4):413.
Lander, E. S. et al. 2001. Initial sequencing and analysis of the human genome. *Nature* 409 (6822):860.
Maki, H. 2002. Origins of spontaneous mutations: Specificity and directionality of base substitution, frameshift, and sequence substitution mutageneses. *Annu Rev Genet* 36:279.
Venter, J. C. et al. 2001. The sequence of the human genome. *Science* 291 (5507):1304.
White, R. et al. Construction of linkage maps with DNA markers for human chromosomes. *Nature* 313 (5998):101.

## Study Questions

1. The melting temperature of double stranded DNA can be affected by:
    (a) pH
    (b) Ionic strength
    (c) Length of DNA molecule
    (d) All of the above

2. At melting temperature, ____________________.
   (a) 100% of DNA strands are denatured
   (b) 75% of DNA strands are denatured
   (c) 50% of DNA strands are denatured
   (d) 25% of DNA strands are denatured

3. Given that a haploid cell has 23 chromosomes with approximately $3 \times 10^9$ bp of DNA, how much DNA by weight does a haploid cell contain? (An average base pair is 618 g/mol and 1 mole contains $6.02 \times 10^{23}$ molecules.)

4. One ng of DNA can be obtained from approximately:
   (a) 100 - 200 diploid cells
   (b) 300 - 400 diploid cells
   (c) 600 - 700 diploid cells
   (d) 900 - 1000 diploid cells

5. During capillary electrophoresis, ____________________.
   (a) The DNA fragment migrates from cathode to anode
   (b) The DNA fragment migrates from anode to cathode

6. Which of the following is true?
   (a) GC pairs have three hydrogen bonds.
   (b) GC pairs have two hydrogen bonds.

7. Which of the following types of biological evidence contains haploid cells?
   (a) Hair
   (b) Blood
   (c) Sperm
   (d) Saliva

# DNA Extraction 12

## 12.1 Basic Principles

Finding a suitable DNA isolation system to satisfy forensic DNA testing needs is important for the successful completion of testing. This chapter introduces basic techniques for DNA extraction used in forensic laboratories (Table 12.1). It is important to note that extraction procedures may vary according to the type of biological evidence such as cell types, substrates, and the quantity of biological evidence collected. The method of choice is often the one that yields sufficient quantity, good quality, and high purity of DNA. An insufficient quantity of DNA may result in a partial DNA profile or the failure to obtain a profile.

Poor quality DNA, for example, degraded DNA, may also result in a partial DNA profile or a failure to obtain a profile. Low purity of DNA may cause interference during subsequent DNA testing. For example, DNA polymerase inhibitors interfere with DNA amplification. Additional considerations for selecting proper DNA extraction methods include adaptability to automation, throughput, simplicity, risk of contamination, and cost effectiveness.

The most common methods involve isolating total cellular DNA, which is suitable for most forensic DNA analysis. DNA isolation methods can differ in many respects because the biochemical compositions of tissues vary. Many specific DNA isolation procedures for any sample of interest may be found in the literature. Nevertheless, certain basic principles, reagents, and isolation procedures apply to various types of samples. The most common DNA extraction protocols include certain components discussed below.

### 12.1.1 Cell and Tissue Disruption

In most DNA isolation protocols, enzymatic digestions, such as with proteinase K, have been used for cell and tissue disruption. Tissues can also be disrupted by boiling and alkali treatment. Materials such as bone and teeth can be frozen in liquid nitrogen and then ground to a fine powder with a mortar or cryogenic grinder such as the SPEX CertiPrep® freezer mill.

### 12.1.2 Lysis of Membranes and Organelles

The lysis of membranes and release of DNA from nuclei or mitochondria are performed during or immediately after tissue disruption. The lysis buffer usually consists of (1) detergents such as anionic compounds, sarkosyl, and sodium dodecyl sulfate (SDS) to destroy membranes, denature proteins, and dissociate proteins from the DNA; (2) a buffer system, such as Tris-HCl to maintain the pH in a range that avoids the activities of degrading enzymes; (3) high salt concentrations to dissociate nuclear proteins such as histones from the DNA; (4) reducing agents such as mercaptoethanol or dithiothreitol (DTT) to inhibit oxidization processes that can damage DNA; and (5) chelating agents such as ethylenediaminetetraacetic acid (EDTA) or Chelex® to capture divalent metal ions that are cofactors of endogenous DNases that catalyze the hydrolysis of DNA.

### 12.1.3 Removal of Proteins and Cytoplasmic Contaminants

Most cell constituents interfering with DNA isolation reside in the cytoplasm. Proteins are dissociated from the DNA by the action of detergents and high salt concentrations in the lysis buffer. Dissolved proteins are then usually removed by one or more rounds of extraction with phenol, phenol–chloroform mixtures, or chloroform–isoamyl alcohol mixtures, followed by centrifugation to separate the phases. Another strategy to remove cytoplasmic contaminants is to employ the reversible binding of DNA to a solid matrix such as silica that selectively binds DNA in chaotropic salt solutions. The proteins and cytoplasmic contaminants can be removed by washing steps.

Over the years, high throughput procedures have been developed to process large numbers of samples in parallel. Some of these methods are adapted to automated DNA extraction platforms.

### 12.1.4 Contamination

Contamination usually is caused by the introduction of exogenous DNA. It can occur between a person and a sample, between samples, between samples with amplified DNA, or through contamination from other organisms. To prevent the occurrence of contamination, certain procedures should be followed. The evidence samples should be processed and extracted in separate areas (or at different times) from the DNA amplification area. Evidence and reference samples should be processed separately in different rooms to avoid between-sample contamination. In situations where space is limited, the evidence sample should be processed before the reference sample.

Solutions and test tubes used for extraction should be DNA-free and aerosol-resistant pipette tips should be used during the extraction process. Additionally, the levels of contamination should be monitored by using ***reagent blanks*** (reagents but no samples) that monitor contamination from the extraction.

### 12.1.5 Storage of DNA Solutions

Highly purified and high molecular weight DNA is usually stored in TE buffer (10 mM Tris-HCl, 1 mM EDTA, pH 8.0). EDTA is usually included in storage solutions to chelate magnesium ions and thereby inhibit magnesium-dependent DNases. Such a DNA solution may be stored at 4°C or –20°C. For long-term storage, –80°C is recommended. Frequent freezing or thawing cycles should be avoided because the fluctuations may cause breaks of single- and double-stranded DNA. Additionally, DNA preparations containing impurities such as preparations generated via the Chelex method are much less stable. Certain substances such as heavy metals and free radicals should also be avoided because they can cause breakage of phosphodiester bonds in the molecules. Furthermore, contamination of nucleases may lead to subsequent degradation of DNA.

## 12.2 Methods of DNA Extraction

### 12.2.1 Extraction with Phenol-Chloroform

This method is also called organic extraction. Major steps include the following:

**Cell Lysis and Protein Digestion** — These steps can be achieved by digestion with proteolytic enzymes such as proteinase K before extraction with organic solvents.

**Extraction with Organic Solvents** — The removal of proteins is carried out by extracting aqueous solutions containing DNA with a mixture of phenol:chloroform:isoamyl alcohol (25:24:1). Phenol is used to extract the proteins from the aqueous solution. However, although phenol has a slightly higher density than water, it may be difficult to separate from the aqueous phase so chloroform is employed due to its higher density. The phenol–chloroform mixture forms the lower organic phase and is easily separated from the aqueous phase. Isoamyl alcohol is often added to the phenol–chloroform mixture to reduce foaming. During partition, DNA is solubilized in the aqueous phase while lipids are solubilized in the organic phase. Proteins are located at the interface between the two phases (Figure 12.1).

**Concentrating DNA** — Two common methods for concentrating DNA are ethanol precipitation and ultrafiltration. In the first method, the DNA is precipitated from the aqueous solution with ethanol and salts. Ethanol depletes the hydration shell of DNA, thus exposing its negatively charged phosphate groups. Ethanol precipitation can only occur if cations are available in sufficient quantity to neutralize the charge on the exposed phosphate residues. The most commonly used cations are ammonium, lithium, and sodium. Ultrafiltration is an alternative to ethanol precipitation for concentrating DNA solutions. The Microcon® and Centricon® are centrifugal

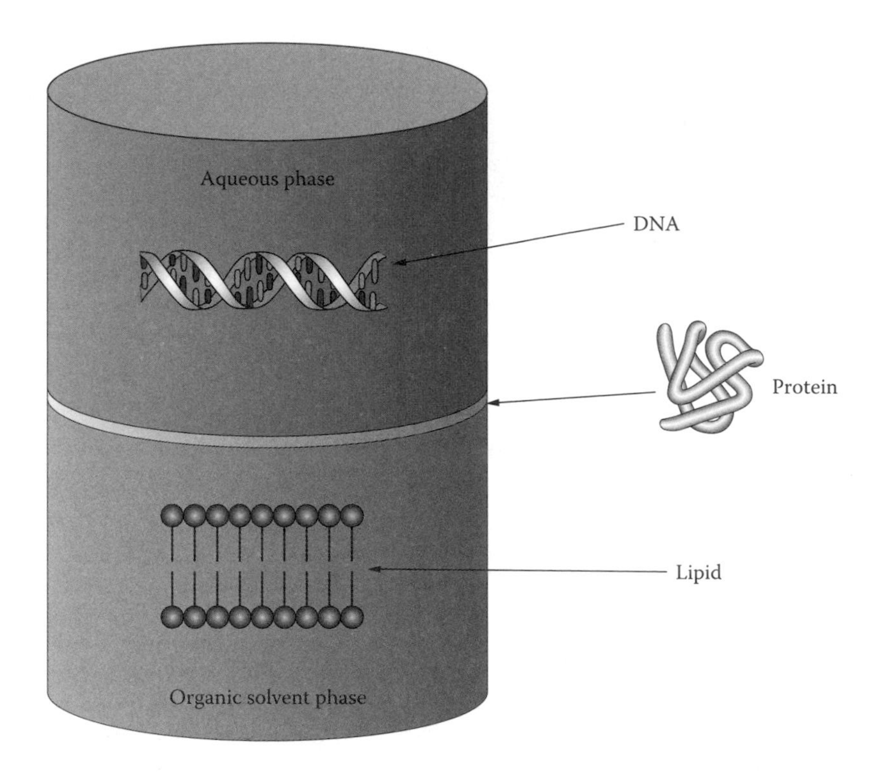


**Figure 12.1** DNA extraction using organic solvent. The DNA is contained in the aqueous phase while cellular materials such as lipids are contained in the organic solvent phase. Proteins remain in the barrier between the two phases.

ultrafiltration devices that can concentrate DNA samples (Figure 12.2). A proper Microcon® unit can be selected with a nucleotide cut-off equal to or smaller than the molecular weight of the DNA fragment of interest. Usually the cut-off size is 100 for forensic DNA samples.

Phenol–chloroform extraction yields large sized, double stranded DNA, and can be used for either restriction fragment length polymorphism (RFLP)- or polymerase chain reaction (PCR)-based analysis. However, the organic extraction method is time-consuming, involves the use of hazardous reagents, and requires transferring samples among tubes.

### 12.2.2 Extraction by Boiling Lysis and Chelation

This technique, also called the Chelex® extraction method was introduced in the early 1990s to forensic DNA laboratories. It usually includes the following steps:

**Washing** — This step removes contaminants and inhibitors that may interfere with DNA amplification. For example, heme should be removed from blood samples because it inhibits DNA amplification.

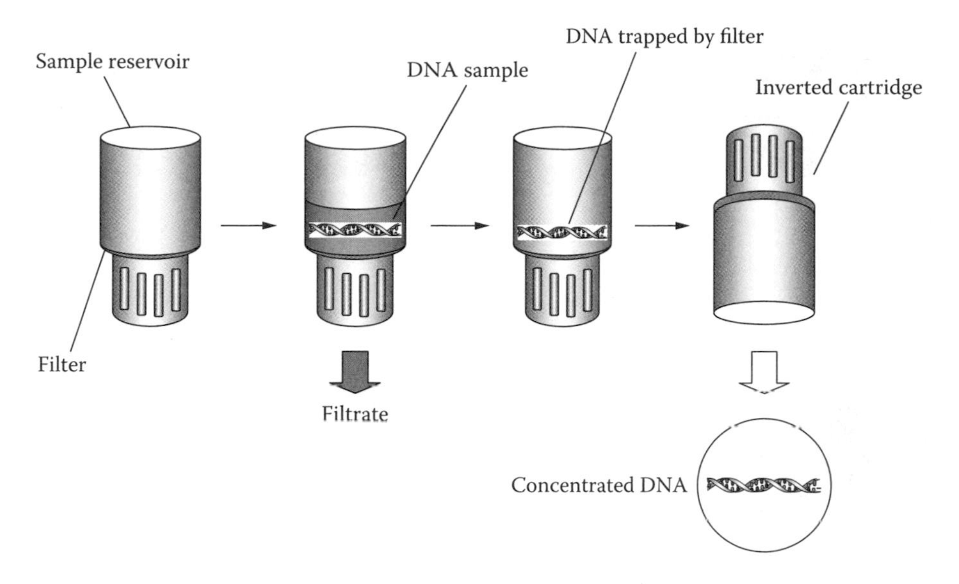


**Figure 12.2** Concentration of DNA solutions using filtration devices. DNA samples are loaded into the reservoir. The liquid is filtered by centrifugation and the DNA becomes trapped in the membrane. The cartridge is then inverted to recover the trapped DNA by centrifugation.

**Boiling** — Cells are suspended in solution and then lysed by heating to boiling temperature to break open the membranes and release the DNA. A chelating resin (Chelex® 100) is employed during the extraction process. It is an ion-exchange resin composed of styrene divinylbenzene copolymers. The paired iminodiacetate ion groups in Chelex® 100 act as chelators by binding to polyvalent metal ions such as magnesium. Magnesium is a cofactor of endogenous DNase. Thus, sequestering magnesium in the solution using Chelex® 100 protects DNA from degradation by DNase (Figure 12.3).

**Centrifugation** —Brief centrifugation is performed to pull the Chelex® 100 resin and cellular debris to the bottom of the tube. The supernatant is used for the DNA analysis. One should avoid carrying the Chelex® 100 resin over into the DNA amplification solutions because magnesium is a necessary cofactor for the DNA polymerase required for amplification.

This method is simple, rapid, and uses only a single tube for extraction, thus reducing the risks of contamination and sample mix-ups. However, the heating step of this method disrupts and denatures proteins and also affects the chromosomal DNA. The resulting DNA extracted from the solution becomes single stranded. Thus, the DNA extracted is not suitable for RFLP analysis because RFLP requires double stranded DNA samples. The DNA obtained by lysis and chelation can only be used for PCR-based DNA analysis.

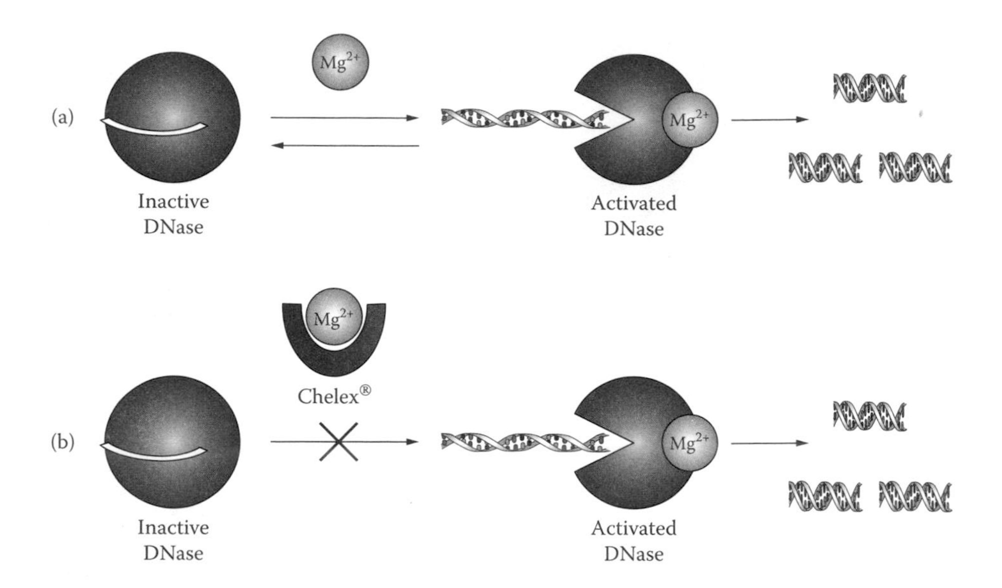


**Figure 12.3** Functioning of Chelex®. (a) Cations such as $Mg^{2+}$ are required for the activity of endogenous DNase which degrades DNA. (b) Chelex® prevents DNA degradation from endogenous DNase by sequestering the $Mg^{2+}$.

### 12.2.3 Silica-Based Extraction

The method of adsorbing DNA molecules to solid silica surfaces has been used for extraction. The method is based on the phenomenon that DNA is reversibly adsorbed to silica in the presence of high concentrations of chaotropic salts (Figure 12.4). Silica (silicon dioxide, $SiO_2$) is the oxide of silicon. Chaotropic salts were originally used to disrupt the three-dimensional structures of proteins.

The specific role of chaotropic agents on the adsorption of DNA to silica is still not well understood. However, it is believed that the adsorption to silica is due to the dehydrating effects caused by chaotropic salts. The dehydration of the phosphodiester backbone of DNA allows the phosphate residues to become available to interact with the silica surface. Common chaotropic salts employed for DNA extraction include guanidinium salts such as guanidinium thiocyanate (GuSCN) and guanidinium hydrochloride (GuHCl). GuSCN is a more potent chaotropic salt and also facilitates cell lysis and DNA adsorption. This technique usually includes the following steps:

**Cell Lysis and Protein Digestion** — This is carried out by proteinase K digestion. The cell membranes are broken open and DNA is released.

**DNA Adsorption onto Silica** — This step employs silica as the stationary phase in a column configuration to bind the DNA contained in the cell

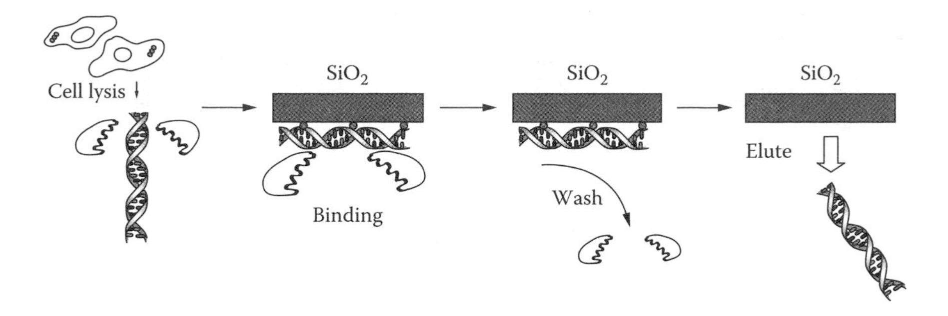


**Figure 12.4** Silica-based DNA extraction. Cells are lysed in the presence of proteinase. The DNA then binds to the silica matrix. A washing step removes unbound cellular materials and salts from the matrix. The purified DNA is then eluted for use in downstream applications.

lysate. Adsorption of the DNA to the silica occurs in the presence of high concentrations of chaotropic agents (some protocols adjust pH conditions to enhance adsorption). Under these conditions, cellular materials and other contaminants that can inhibit DNA amplification reactions are not retained on the column. The adsorbed DNA is double stranded but partial denaturation may still occur.

**Washing** — This step removes chaotropic agents and other contaminants. The adsorbed DNA cannot be eluted from the silica matrix by certain solvents, for example, an ethanol-based wash solution but the chaotropic agents and contaminants can be removed via an ethanol-based wash from the column.

**Elution of DNA** — The adsorbed DNA can be eluted by rehydration with aqueous low-salt solutions. The eluted DNA is double stranded and can be used for a wide variety of applications.

The original device used a silica resin configuration. The protocol for purification by silica resin involved combining the lysate with a resin slurry and using vacuum filtration to wash the adsorbed DNA, followed by centrifugation to elute the purified DNA. Silica resin methods yield high-quality DNA. Recently, however, the silica membrane column device has been employed due to its convenience. The device also adapts to automation, for example, by using 96-well silica membrane plates or a variety of robotic platforms.

Another type of device employs silica-coated paramagnetic particles that adsorb DNA in the solution. A magnet is employed for particle capture instead of centrifugation or vacuum filtration. The solution containing contaminants and cellular materials is discarded. The magnetic particles can then be resuspended during the wash steps, and DNA can be eluted after washing. This device can also be adapted to automated, high throughput methods.

**Table 12.1 Three Basic DNA Extraction Methods**

| Method | Forensic Application | Purity | DNA Strand | DNA Size | Throughput | Note |
|---|---|---|---|---|---|---|
| Solvent-based | RFLP and PCR-based assay | High | Double stranded | Large | Time consuming; difficult to adapt for automation | Uses toxic solvent; multiple transfer steps between tubes |
| Boiling | PCR-based assay | Low | Single stranded | Small | ~ 30 min per sample | No transfer required |
| Silica-based | RFLP and PCR-based assay | High | Double stranded | Large | ~ 1 hr per sample; amenable to automation | Minimum transfer required |

### 12.2.4 Differential Extraction

This method is very useful for the extraction of DNA from biological evidence derived from sexual assault cases, for example, vaginal swabs and body fluid stains. These types of evidence often contain mixtures of sperm cells from a male contributor and epithelial cells from a female victim. Mixtures of individual DNA profiles can complicate data interpretation.

This method selectively lyses the nonsperm and sperm cells in separate steps based on the differences in cell membrane properties of spermatozoa and other types of cells. Thus, the DNA from sperm and nonsperm cell fractions can be isolated.

First, the differential extraction procedure involves preferentially lysing the nonsperm cells with proteolytic degradation using proteinase. Sperm plasma membrane contains proteins cross-linked by disulfide bonds. The membrane exhibits a much higher mechanical stability than nonsperm cells and is thus resistant to proteolytic degradation. The nonsperm DNA is released into the supernatant and the liquid containing it (the nonsperm fraction) is extracted, yielding a fraction that predominantly contains DNA from nonsperm cells.

To lyse the sperm cells, it is necessary to cleave the disulfide bonds in addition to proteolytic digestion. The application of DTT, a reducing agent, is an approach that can be used for cleavage. In the presence of DTT and proteinase K, the sperm plasma membrane is then lysed (Figure 12.5). Subsequently, DNA from the sperm cells can be extracted (Figure 12.6).

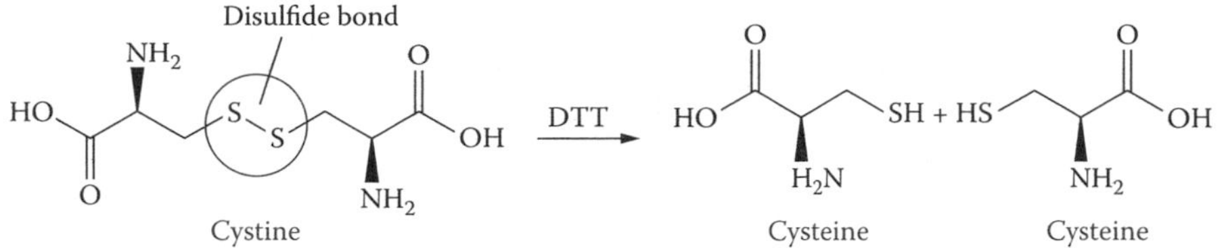

**Figure 12.5** DTT reaction. The breaking of disulfide bonds in cystine residues is carried out by the reduction of DTT.

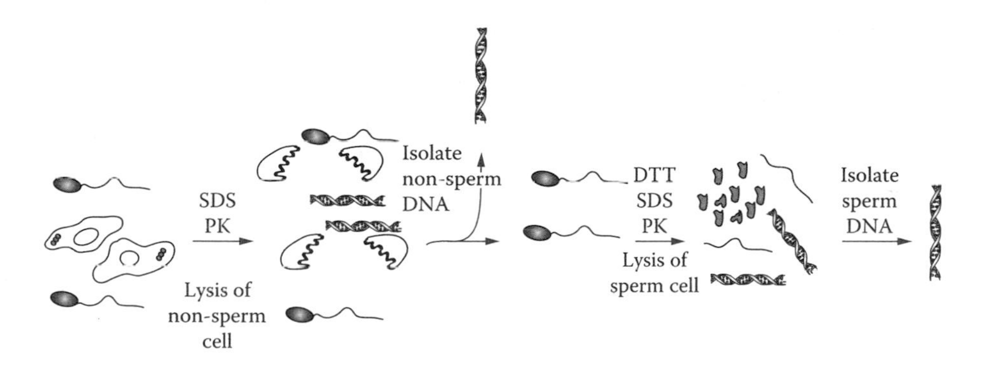


**Figure 12.6** Differential extraction process used to separate male sperm cells from nonsperm cells. Nonsperm cells are lysed in the presence of SDS and proteinase K (PK); sperm cells are resistant to such conditions. The nonsperm cell DNA is extracted. The sperm cells are then lysed separately in the presence of SDS and proteinase K plus DTT to extract the sperm DNA.

The process purifies and enriches each fraction so that DNA profiles from both male and female contributors in a sample can be identified separately. However, the nonsperm cell DNA and sperm cell DNA may not be completely separated, for example, if the sperm cells have already lysed due to poor sample conditions. Some sperm DNA may be present in the nonsperm cell fraction. Additionally, if a mixture has an abundance of nonsperm cells and fewer sperm cells, nonsperm cell DNA may be detected in the sperm fraction.

# Bibliography

## Chelex Method

Crouse, C. A., J. D. Ban, and J. K. D'Alessio. 1993. Extraction of DNA from forensic-type sexual assault specimens using simple, rapid sonication procedures. *Biotechniques* 15 (4):641.

Eminovic, I. et al. 2005. A simple method of DNA extraction in solving difficult criminal cases. *Med Arch* 59 (1):57.
Ginestra, E. et al. 2004. DNA extraction from blood determination membrane card test. *Forensic Sci Int* 146 Suppl:S145.
Hochmeister, M. N. et al. 1991. PCR-based typing of DNA extracted from cigarette butts. *Int J Legal Med* 104 (4):229.
Hoff-Olsen, P. et al. 1999. Extraction of DNA from decomposed human tissue: Evaluation of five extraction methods for short tandem repeat typing. *Forensic Sci Int* 105 (3):171.
Iwasa, M. et al. 2003. Y-chromosomal short tandem repeats haplotyping from vaginal swabs using a chelating resin-based DNA extraction method and a dual-round polymerase chain reaction. *Am J Forensic Med Pathol* 24 (3):303.
Rerkamnuaychoke, B. et al. 2000. Comparison of DNA extraction from blood stain and decomposed muscle in STR polymorphism analysis. *J Med Assoc Thai* 83 Suppl 1:S82.
Sweet, D. et al. 1996. Increasing DNA extraction yield from saliva stains with a modified Chelex method. *Forensic Sci Int* 83 (3):167.
Tsuchimochi, T. et al. 2002. Chelating resin-based extraction of DNA from dental pulp and sex determination from incinerated teeth with Y-chromosomal alphoid repeat and short tandem repeats. *Am J Forensic Med Pathol* 23 (3):268.
Vandenberg, N., R. A. van Oorschot, and R. J. Mitchell. 1997. An evaluation of selected DNA extraction strategies for short tandem repeat typing. *Electrophoresis* 18 (9):1624.
Walsh, P. S., D. A. Metzger, and R. Higuchi. 1991. Chelex 100 as a medium for simple extraction of DNA for PCR-based typing from forensic material. *Biotechniques* 10 (4):506.

## Organic Solvent Method

Andelinovic, S. et al. 2005. Twelve-year experience in identification of skeletal remains from mass graves. *Croat Med J* 46 (4):530.
Anzai-Kanto, E. et al. 2005. DNA extraction from human saliva deposited on skin and its use in forensic identification procedures. *Pesqui Odontol Bras* 19 (3):216.
Cattaneo, C. et al. 1995. A simple method for extracting DNA from old skeletal material. *Forensic Sci Int* 74 (3):167.
Fridez, F., and R. Coquoz. 1996. PCR DNA typing of stamps: evaluation of the DNA extraction. *Forensic Sci Int* 78 (2):103.
Hoff-Olsen, P. et al. 1999. Extraction of DNA from decomposed human tissue: An evaluation of five extraction methods for short tandem repeat typing. *Forensic Sci Int* 105 (3):171.
Kochl, S., H. Niederstatter, and W. Parson. 2005. DNA extraction and quantitation of forensic samples using the phenol-chloroform method and real-time PCR. *Methods Mol Biol* 297:13.
Ma, H. W., J. Cheng, and B. Caddy. 1994. Extraction of high quality genomic DNA from microsamples of human blood. *J Forensic Sci Soc* 34 (4):231.
Yamada, M. et al. 1994. Determination of ABO genotypes with DNA extracted from formalin-fixed, paraffin-embedded tissues. *Int J Legal Med* 106 (6):285.

Ye, J. et al. 2004. A simple and efficient method for extracting DNA from old and burned bone. *J Forensic Sci* 49 (4):754.

## Silica-Based Method

Baker, L. E., W. F. McCormick, and K. J. Matteson. 2001. A silica-based mitochondrial DNA extraction method applied to forensic hair shafts and teeth. *J Forensic Sci* 46 (1):126.
Castella, V. et al. 2006. Forensic evaluation of the QIAshredder/QIAamp DNA extraction procedure. *Forensic Sci Int* 156 (1):70.
Greenspoon, S. A. et al. 1998. QIAamp spin columns as a method of DNA isolation for forensic casework. *J Forensic Sci* 43 (5):1024.
Greenspoon, S. A. et al. 2004. Application of the BioMek 2000 Laboratory Automation Workstation and DNA IQ System to the extraction of forensic casework samples. *J Forensic Sci* 49 (1):29.
Hanselle, T. et al. 2003. Isolation of genomic DNA from buccal swabs for forensic analysis, using fully automated silica-membrane purification technology. *Leg Med (Tokyo)* 5 Suppl 1:S145.
Hoff-Olsen, P. et al. 1999. Extraction of DNA from decomposed human tissue. An evaluation of five extraction methods for short tandem repeat typing. *Forensic Sci Int* 105 (3):171.
Ma, H. W., J. Cheng, and B. Caddy. 1994. Extraction of high quality genomic DNA from microsamples of human blood. *J Forensic Sci Soc* 34 (4):231.
Montpetit, S. A., I. T. Fitch, and P. T. O'Donnell. 2005. A simple automated instrument for DNA extraction in forensic casework. *J Forensic Sci* 50 (3):555.
Nagy, M. et al. 2005. Optimization and validation of a fully automated silica-coated magnetic beads purification technology in forensics. *Forensic Sci Int* 152 (1):13.
Scherczinger, C. A. et al. 1997. DNA extraction from liquid blood using QIAamp. *J Forensic Sci* 42 (5):893.
Sinclair, K., and V. M. McKechnie. 2000. DNA extraction from stamps and envelope flaps using QIAamp and QIAshredder. *J Forensic Sci* 45 (1):229.
Wolfe, K. A. et al. 2002. Toward a microchip-based solid-phase extraction method for isolation of nucleic acids. *Electrophoresis* 23 (5):727.

## Study Questions

1. In the organic extraction method:
   (a) DNA is in the solvent phase (phenol, etc.).
   (b) DNA is in the buffer phase (water, etc.).
   (c) Protein is in the buffer phase (water, etc.).
   (d) Both (a) and (c).

2. In differential extraction:
   (a) DTT forms the disulfide bond to lyse the sperm membrane.

(b) DTT forms the disulfide bond to lyse the epithelium membrane.
(c) DTT breaks the disulfide bond to lyse the sperm membrane.
(d) DTT breaks the disulfide bond to lyse the epithelium membrane.

3. In the Chelex method:
   (a) DNA is extracted by Chelex.
   (b) $Mg^{2+}$ is needed for extraction.
   (c) Chelex binds to the DNA degradation enzyme and inhibits its activity.
   (d) Chelex removes $Mg^{2+}$.

4. The extraction negative control:
   (a) Monitors contamination from evidence collection to final analysis.
   (b) Monitors contamination occurring during extraction only.
   (c) Monitors contamination from extraction to amplification.
   (d) Monitors contamination from extraction to final analysis.

5. When a blood stain is extracted using Chelex method:
   (a) It is important to remove the supernatant from the initial soak.
   (b) It is important to extract the supernatant from the initial soak.

6. Sperm cells and epithelial cells are observed under a microscope. The extracted sperm cell fraction has no DNA; the epithelial cell fraction contains DNA. Which of the following has gone wrong?
   (a) The PBS buffer
   (b) The proteinase K
   (c) The DTT
   (d) All of the above

7. During Chelex extraction, the extract was heated to 100°C, cooled down, and stored. Why is Chelex-extracted DNA not reannealed to double stranded?

# DNA Quantitation

# 13

Determining the amount of DNA in a sample is essential for polymerase chain reaction (PCR)-based DNA testing. Because PCR-based DNA testing is very sensitive, a narrow concentration range is required for amplification to be successful. If too much DNA template is used in PCR, the resulting artifacts may interfere with data analysis and interpretation. On the other hand, very low amounts of DNA template may result in a partial DNA profile or failure to attain a profile. Also, forensic evidence samples are often mixtures that may include DNA from nonhuman sources. For instance, DNA from microbial organisms may be present in a sample exposed to such an environment. Thus, it is necessary to use human-specific DNA quantitation methods to selectively determine the amount of human DNA present.

In the U.S., quality assurance guidelines require the use of a quantitation method that estimates the amount of human nuclear DNA in nonreference evidence samples. Additionally, samples may contain PCR inhibitors that may lead to a failure of DNA amplification. Thus, a test that can measure the quality and quantity of the DNA template of a sample is desirable.

In this section, the slot blot, interchelating dye, and quantitative PCR methods are introduced. The quantitative PCR method is the most sensitive of the three methods. Both the slot blot and the quantitative PCR methods can detect primate (including human) DNA, while only the quantitative PCR method can detect the PCR inhibitors.

## 13.1 Slot Blot Assay

The slot blot assay can be used to detect human genomic DNA in a sample. The genomic DNA is denatured and a small volume of a sample is spotted, using a slot blot device, onto a nitrocellulose membrane. The single stranded DNA is then immobilized onto a nylon membrane. The targeted sequence is revealed by hybridization with a labeled 40-nucleotide probe complementary to a primate-specific α–satellite DNA sequence at the D17Z1 locus (Figure 13.1).

Three detecting schemes of the slot blot assay have been developed. Initially, the D17Z1 probe was labeled with radioisotopes that could be visualized

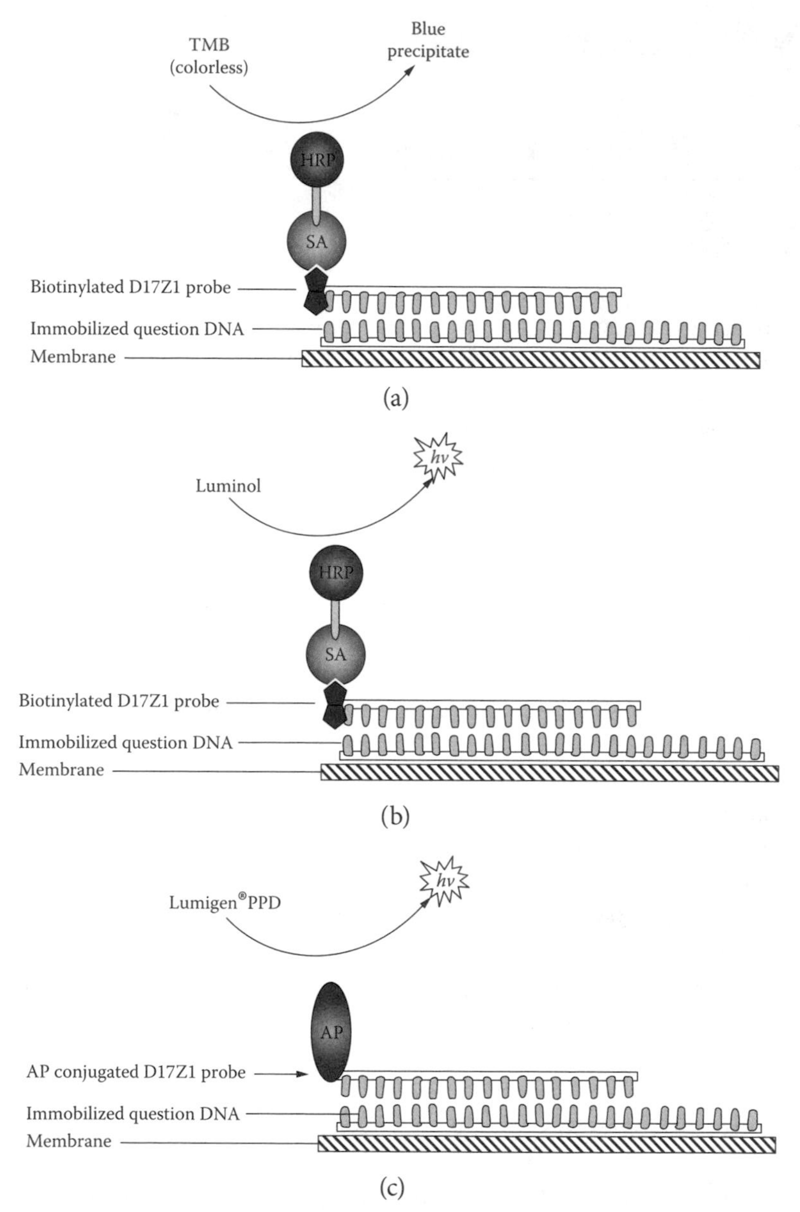


**Figure 13.1** Slot blot assay. Questioned DNA is immobilized onto a solid phase membrane, then hybridized with a biotinylated D17Z1 probe. The detection of the hybridization is carried out by (a) streptavidin (SA) and horseradish peroxidase (HRP) conjugate and a colorimetric reaction is catalyzed by HRP using tetramethylbenzidine (TMB) as a substrate; (b) SA and HRP conjugate; a chemiluminescent reaction is catalyzed by HRP using Luminol as a substrate; (c) immobilized DNA is hybridized with an alkaline phosphatase (AP)-labeled D17Z1 probe. The detection of the hybridization is carried out by a chemiluminescent reaction catalyzed by AP using Lumigen® PPD as a substrate.

by exposing the slot blot membrane to x-ray film. The hazardous radioisotope detection method was then replaced by alkaline phosphatase-labeled and biotinylated probes. The alkaline phosphatase-labeled probe can couple with chemiluminescent detection (Lumi-Phos Plus kit, Lumigen, Inc.). The biotinylated probe (QuantiBlot Human DNA Quantitation Kit, Applied Biosystems) can couple with either colorimetric or chemiluminescent detection.

In colorimetric detection, the biotin moiety of the probe is bound to streptavidin. The streptavidin is conjugated to horseradish peroxidase which catalyzes the oxidation reaction of the tetramethylbenzidine (TMB), a substrate, forming a blue precipitate. With the chemiluminescent detection method, the horseradish peroxidase catalyzes the oxidation reaction of a substrate such as Luminol, emitting photons that can be detected by exposure to x-ray film. The sensitivity of chemiluminescent detection is slightly higher than that of colorimetric detection. The detection mechanisms will be discussed in Chapter 16.

The detected signal intensity is in proportion to the concentration of DNA of the sample in question. Quantitative measurements can be made by comparing an unknown sample to a set of standards with known DNA concentrations (Figure 13.2).

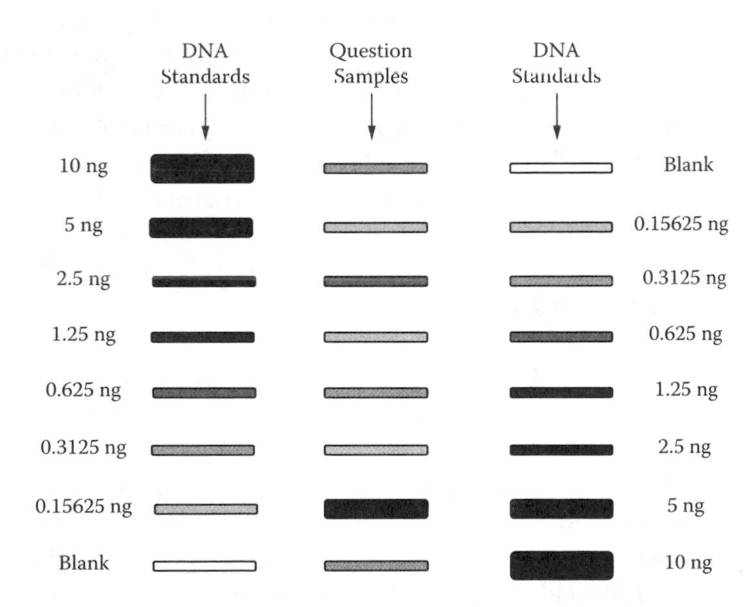


**Figure 13.2** Human DNA quantitation result using slot blot assay. Standards with known amounts of human DNA are applied and comparison is made of unknown samples and a set of standards. The quantities in the unknown samples are estimated by visual comparison to the standards.

The assay typically quantifies DNA over the range of 150 pg to 10 ng, although lower quantities can be detected with longer film exposure times. The D17Z1 sequences of humans and other primates share homology. The probe cannot distinguish human and other primate genomic DNA. This is not a great concern because cases involving nonhuman primates are rare. Nevertheless, the probe does not cross-react with all other species. This assay is being replaced by recently developed quantitative PCR assays.

## 13.2 Fluorescent Interchelating Dye Assay

Small quantities of DNA can be also quantified by using a fluorescent interchelating dye method. The Quant-iT™ PicoGreen® dsDNA reagent Molecular Probes® is a fluorescent interchelating dye that stains double-stranded DNA (dsDNA) for quantitation in a sample (Figure 13.3). The detection limit of this method is approximately 250 pg. Because this dye is not specific to human DNA, this technique is especially useful for the quantitation of known reference samples. For instance, DNA database samples from known sources can be tested using this method. The assay has also been adapted for automation and is thus a high throughput method.

DNA samples are simply added to a solution containing the fluorescent interchelating dye. The fluorescence is measured using a standard spectrofluorometer with excitation and emission wavelengths. A standard curve is first created using samples containing known amounts of DNA. The assay is then performed for the unknown samples and the quantities of DNA in the samples are determined by comparing the results to the standard curve.

## 13.3 Quantitative PCR Assay

Based on the principle of PCR amplification, the amount of PCR product amplified correlates with the initial DNA concentration. Thus, the DNA concentration of a sample can be determined. There are two categories of quantitative PCR methods. ***End-point PCR*** methods measure the amount of amplified product synthesized during PCR at the end of the reaction. Usually, the fluorescence is emitted by the dyes that intercalate into the double stranded DNA. The yield of amplified DNA is detected from the amount of fluorescence emitted by dyes such as SYBR (Figure 13.4).

***Real-time PCR*** methods can quantify the amplified DNA during the exponential phase of PCR. The quantitation result is not affected to a

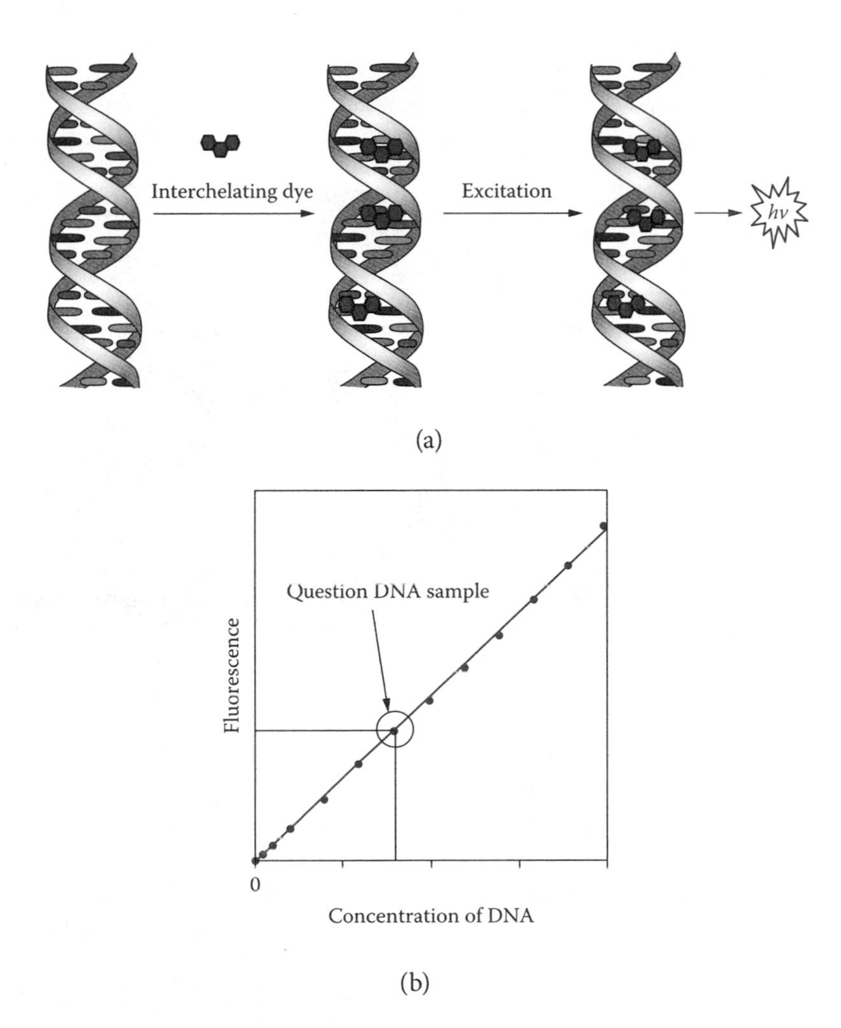


**Figure 13.3** DNA quantitation using interchelating dye. (a) Fluorescent dye is interchelated into DNA. The fluorescence can be measured upon excitation. (b) A standard curve can be constructed using known amounts of DNA standards. The amount of questioned DNA can be determined by comparing the standard curve.

significant extent by slight variations in PCR conditions. Thus, the precision of the quantitation of target sequences is improved with this method.

### 13.3.1 Real-Time Quantitative PCR

Real-time quantitative PCR was developed in the early 1990s, and it analyzes the cycle-to-cycle change in a fluorescence signal resulting from amplification of a target sequence during PCR. A fluorescent reporter is used to

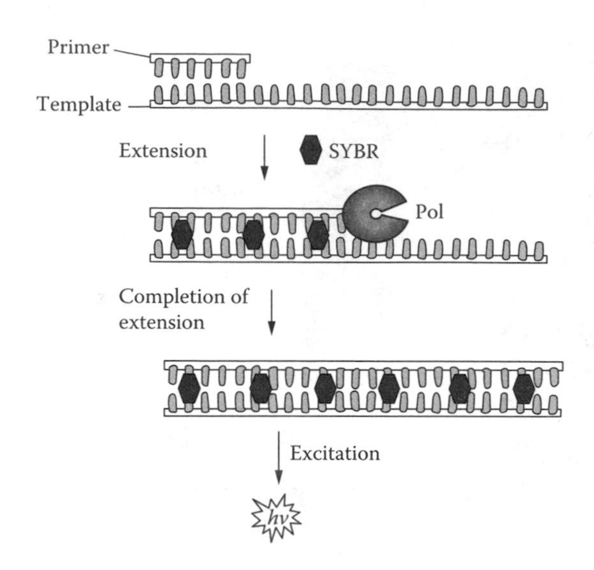


**Figure 13.4** End-point PCR using SYBR Green™ detection. During extension in which DNA synthesis occurs, the green dye binds to the double stranded amplicons. Upon excitation, the emission intensity of the dye can be measured. Pol = *Taq* polymerase.

monitor the PCR. The fluorescence of the reporter molecule increases as products accumulate with each round of amplification. Real-time quantitative PCR is commonly used because of the following advantages:

- Better objectivity than the QuantiBlot method
- Increased sensitivity with a large dynamic range (30 pg to 100 ng or wider, depending upon assay)
- More accurate measurement of small quantities of DNA in samples
- Fewer laboratory manipulations; amenable to automation
- Ability to detect PCR inhibitors

Although capacity is limited, the technique is amenable to multiplexing to detect more than one type of DNA target sequence in a single reaction. Real-time quantitative PCR uses commercially available fluorescence-detecting thermocyclers to amplify specific DNA sequences and measure their concentrations simultaneously. The reporter can be a nonspecific intercalating double stranded DNA-binding dye or a sequence-specific fluorescent-labeled oligonucleotide probe. The target sequences are amplified and detected by the same instrument and the reporter fluorescence is monitored externally. Thus the reaction tubes do not need to be opened. This minimizes aerosol contamination and reduces the risk of false positive results. A widely used real-time PCR probe technique is the TaqMan method (Applied Biosystems).

**Figure 13.5** TaqMan probe. Each probe is labeled with a reporter dye (R) on the 5′ end and a fluorescence quencher (Q) on the 3′ end. MGB = minor groove binder.

#### *13.3.1.1 TaqMan method*

This method employs the 5′ exonuclease activity of *Taq* polymerase to cleave the probe during the extension phase of PCR (also known as the 5′ exonuclease assay). The probe is designed to anneal to the target sequence between the upstream and downstream primers and is added to the PCR mixture together with primers (Figure 13.5). The probe *Tm* (melting temperature) should be higher than the amplification primer *Tm*. Thus, probe binding prior to primer binding during the annealing phase is essential. A minor groove binder (MGB) is often linked at the 3′ end of the probe, which increases the *Tm* values allowing for the use of shorter probes.

The oligonucleotide probe is labeled with both a ***reporter*** fluorescent dye and a quencher dye. While the probe is intact, the proximity of the ***quencher*** greatly reduces the fluorescence emitted by the reporter dye via fluorescent resonance energy transfers or FRET, in which the energy transfer can be affected by altering the relative spatial arrangement of photon donor and acceptor molecules. When the probe molecule is cleaved, the quenching is disrupted and the probe fluoresces. A quencher is usually attached either at the 3′ end or any thymine position. The probe is also labeled with a reporter dye, usually 6-carboxyfluorescein (6-FAM), at the 5′ end. During the extension phase of the PCR cycle, the 5′ exonuclease activity of *Taq* polymerase cleaves the reporter dye from the probe. Because the reporter dye is no longer in close proximity to the quencher, it begins to fluoresce (Figure 13.6). Thus, the intensity of fluorescence is in direct proportion to the amount of target DNA synthesized during the PCR.

The instrument plots the rate of accumulation of amplified DNA over the course of an entire PCR. The greater the initial concentration of target sequences in the reaction mixture, the fewer cycles required to achieve a particular yield of amplified product. The initial concentration of target sequences can therefore be expressed as the ***cycle threshold*** ($C_T$), defined as the number of PCR cycles required to reach a threshold of amplification.

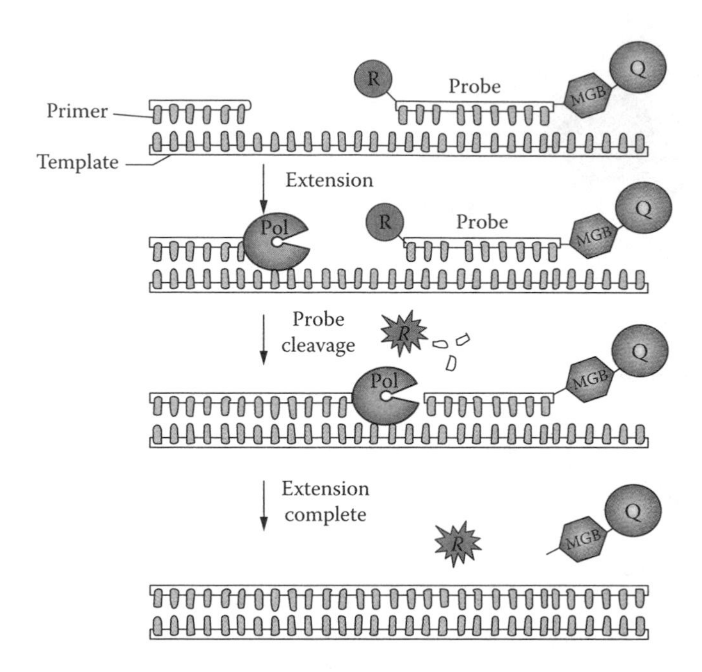


**Figure 13.6** Real-time PCR with TaqMan detection. The TaqMan probe is shown. During extension, the 5′ nuclease activity of *Taq* polymerase (Pol) cleaves the probe. Reporter dye is released during each cycle of PCR.

A plot of $C_T$ against the $\log_{10}$ of the initial concentration of a set of DNA standard yields a straight line as a standard curve. The target sequences in an unknown sample can be quantified by comparing to this standard curve (Figure 13.7).

Commercial real-time PCR kits for human DNA and Y-chromosome DNA quantitation are available. Additionally, the real-time PCR method for mtDNA quantitation has been documented.

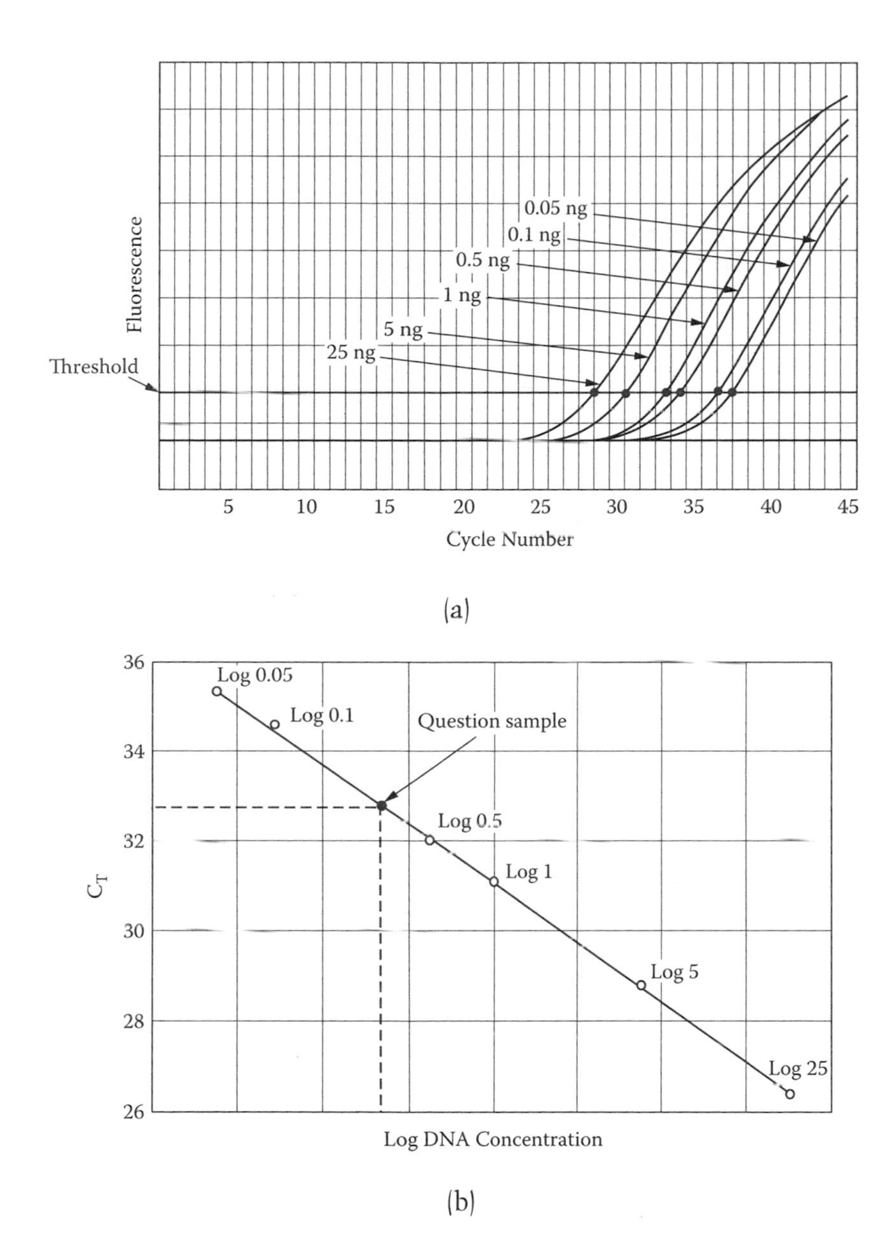


**Figure 13.7** Real-time quantitative PCR. (a) Amplification curves for a dilution series of standards with known quantities of DNA. $C_T$ is the cycle threshold at which the amplification curve crosses the threshold as indicated by the red line. (b) A standard curve based on data obtained from the amplification curves. The quantity of DNA in a questioned sample can be determined from the standard curve.

## Bibliography

Ahn, S. J., J. Costa, and J. R. Emanuel. 1996. PicoGreen quantitation of DNA: Effective evaluation of samples pre- or post-PCR. *Nucleic Acids Res* 24 (13):2623.

Budowle, B. et al. 1995. Simple protocols for typing forensic biological evidence: Chemiluminescent detection for human DNA quantitation and restriction fragment length polymorphism (RFLP) analyses and manual typing of polymerase chain reaction (PCR) amplified polymorphisms. *Electrophoresis* 16 (9):1559.

Budowle, B. et al. 2001. Using a CCD camera imaging system as a recording device to quantify human DNA by slot blot hybridization. *Biotechniques* 30 (3):680.

Budowle, B. et al. 2000. *DNA Typing Protocols: Molecular Biology and Forensic Analysis*. Natick, MA: Eaton Publishing.

Butler, J. M. 2005. *Forensic DNA Typing: Biology, Technology, and Genetics of STR Markers*, 2nd ed. Burlington, MA: Elsevier.

Duewer, D. L. et al. 2001. NIST mixed stain studies #1 and #2: Interlaboratory comparison of DNA quantification practice and short tandem repeat multiplex performance with multiple-source samples. *J Forensic Sci* 46 (5):1199.

Fox, J. C., C. A. Cave, and J. W. Schumm. 2003. Development, characterization, and validation of a sensitive primate-specific quantification assay for forensic analysis. *Biotechniques* 34 (2):314.

Hayn, S. et al. 2004. Evaluation of an automated liquid hybridization method for DNA quantitation. *J Forensic Sci* 49 (1):87.

Kihlgren, A., A. Beckman, and S. Holgersson. 1998. Using D3S1358 for quantification of DNA amenable to PCR and for genotype screening. *Progr Forensic Genet* 7:3.

Kline, M. C. et al. 2003. NIST Mixed Stain Study 3: DNA quantitation accuracy and its influence on short tandem repeat multiplex signal intensity. *Anal Chem* 75 (10):2463.

Mandrekar, M. N. et al. 2001. Development of a human DNA quantitation system. *Croat Med J* 42 (3):336.

Morrison, T. B., J. J. Weis, and C. T. Wittwer. 1998. Quantification of low-copy transcripts by continuous SYBR Green I monitoring during amplification. *Biotechniques* 24 (6):954.

Nicklas, J. A., and E. Buel. 2003. Quantification of DNA in forensic samples. *Anal Bioanal Chem* 376 (8):1160.

Sifis, M. E., K. Both, and L. A. Burgoyne. 2002. A more sensitive method for the quantitation of genomic DNA by Alu amplification. *J Forensic Sci* 47 (3):589.

Walsh, P. S., J. Varlaro, and R. Reynolds. 1992. A rapid chemiluminescent method for quantitation of human DNA. *Nucleic Acids Res* 20 (19):5061.

Waye, J. S. et al. 1989. A simple and sensitive method for quantifying human genomic DNA in forensic specimen extracts. *Biotechniques* 7 (8):852.

## Quantitative PCR Methods

Alonso, A., and P. Martin. 2005. A real-time PCR protocol to determine the number of amelogenin (X-Y) gene copies from forensic DNA samples. *Methods Mol Biol* 297:31.

Alonso, A. et al. 2004. Real-time PCR designs to estimate nuclear and mitochondrial DNA copy number in forensic and ancient DNA studies. *Forensic Sci Int* 139 (2–3):141.

Alonso, A. et al. 2003. Specific quantification of human genomes from low copy number DNA samples in forensic and ancient DNA studies. *Croat Med J* 44 (3):273.

Andreasson, H., and M. Allen. 2003. Rapid quantification and sex determination of forensic evidence materials. *J Forensic Sci* 48 (6):1280.

Andreasson, H., U. Gyllensten, and M. Allen. 2002. Real-time DNA quantification of nuclear and mitochondrial DNA in forensic analysis. *Biotechniques* 33 (2):402.

Andreasson, H. et al. 2006. Nuclear and mitochondrial DNA quantification of various forensic materials. *Forensic Sci Int* 164 (1):56.

Green, R.L. et al. 2005. Developmental validation of the quantifiler real-time PCR kits for the quantification of human nuclear DNA samples. *J Forensic Sci* 50 (4):809.

Nicklas, J. A., and E. Buel. 2003. Development of an Alu-based, real-time PCR method for quantitation of human DNA in forensic samples. *J Forensic Sci* 48 (5):936.

Niederstatter, H. et al. 2006. Characterization of mtDNA SNP typing and mixture ratio assessment with simultaneous real-time PCR quantification of both allelic states. *Int J Legal Med* 120 (1):18.

Ong, Y. L., and A. Irvine. 2002. Quantitative real-time PCR: A critique of method and practical considerations. *Hematology* 7 (1):59.

Richard, M. L., R. H. Frappier, and J. C. Newman. 2003. Developmental validation of a real-time quantitative PCR assay for automated quantification of human DNA. *J Forensic Sci* 48 (5):1041.

Swango, K. L. et al. 2006. A quantitative PCR assay for the assessment of DNA degradation in forensic samples. *Forensic Sci Int* 158 (1):14.

Timken, M. D. et al. 2005. A duplex real-time qPCR assay for the quantification of human nuclear and mitochondrial DNA in forensic samples: implications for quantifying DNA in degraded samples. *J Forensic Sci* 50 (5):1044.

Walker, J. A. et al. 2005. Multiplex polymerase chain reaction for simultaneous quantitation of human nuclear, mitochondrial, and male Y-chromosome DNA: Application in human identification. *Anal Biochem* 337 (1):89.

## Study Questions

1. Which of the following is true?
   (a) Most DNA quantitation methods do not require standard curves.
   (b) Most DNA quantitation methods require standard curves.

2. The probe in the real-time PCR method for DNA quantitation contains:
   (a) A reporter domain
   (b) A quencher domain
   (c) Both reporter and quencher domains
   (d) All of the above

3. In the real-time PCR method for DNA quantitation:
   (a) $C_T$ is the time at which the amplification curve of the sample crosses the threshold.
   (b) $T_m$ is the time at which the amplification curve of the sample crosses the threshold.

4. Which of the following would be true if the water bath temperature were set higher than usual during a slot blot DNA quantitation?
   (a) DNA bands would not be observed.
   (b) DNA bands would be enhanced.

5. Absence of a signal or low sensitivity of a slot blot DNA quantitation could be due to:
   (a) Inactivity of substrate for reporter enzyme
   (b) Degradation of DNA probe
   (c) Inactivity of oxidant
   (d) All of the above

6. Detecting signals of nonhuman DNA samples, such as bacterial DNA, during slot blot DNA quantitation may be due to:
   (a) Excessive salt concentration in wash solution
   (b) Excessively high temperature of water bath

7. The D17Z1 probe for DNA quantitation:
   (a) Is human-specific
   (b) Is mammal-specific
   (c) Is primate-specific
   (d) May detect bacterial DNA

# DNA Amplification by Polymerase Chain Reaction

# 14

## 14.1 Basic Principles of Polymerase Chain Reaction

The ***polymerase chain reaction*** (***PCR***) allows the exponential amplification of specific sequences of DNA to yield sufficient amplified products (***amplicons***) for various downstream applications. The technique is highly sensitive and can amplify very small quantities of DNA. It is very useful for the analysis of limited quantities of evidence in which use of the restriction fragment length polymorphism (RFLP) method is not possible. PCR-based assays are rapid and robust. Thus, PCR forms the basis of many forensic DNA assays such as DNA quantitation, short tandem repeat (STR) profiling, and mitochondrial DNA (mtDNA) sequencing.

The concept of synthesizing DNA by a cycling process was first proposed in the early 1970s. In the mid 1980s, PCR technology was finally developed by Kary Mullis and his coworkers (Cetus Corporation) to amplify the β-globin gene for the diagnosis of sickle cell anemia. In the late 1980s, a thermostable polymerase from *Thermus aquaticus* was employed for PCR. This step greatly increased efficiency and allowed the process to be automated. The result was a powerful impact on molecular biology. In 1993, Mullis was awarded the Nobel Prize for the invention of PCR technology.

In the early 1990s, the technique for the simultaneous amplification and detection of the accumulation of amplicons at each PCR cycle was developed and the concept of ***real time PCR*** was born. This process allows the monitoring of amplicon production at each cycle of the PCR process. The fundamental processes were studied by characterizing the amplification kinetics of PCR using a graph that plotted the amount of amplicon yield at each cycle versus the cycle number (Figure 14.1).

An S-shaped amplification curve is obtained and divided into an exponential phase, a linear phase, and a plateau. During the exponential phase, the amplicon accumulates exponentially. It was revealed that the amplicon accumulation during PCR was correlated to the starting copy number of DNA template. Thus, the amount of amplicon produced during the exponential amplification phase can be used to determine the amount of starting material. This relationship can be further examined using a plot of cycle number versus a log scale of the serial dilution of the starting concentration of DNA template, which results in a linear relationship (see chapter 13). It

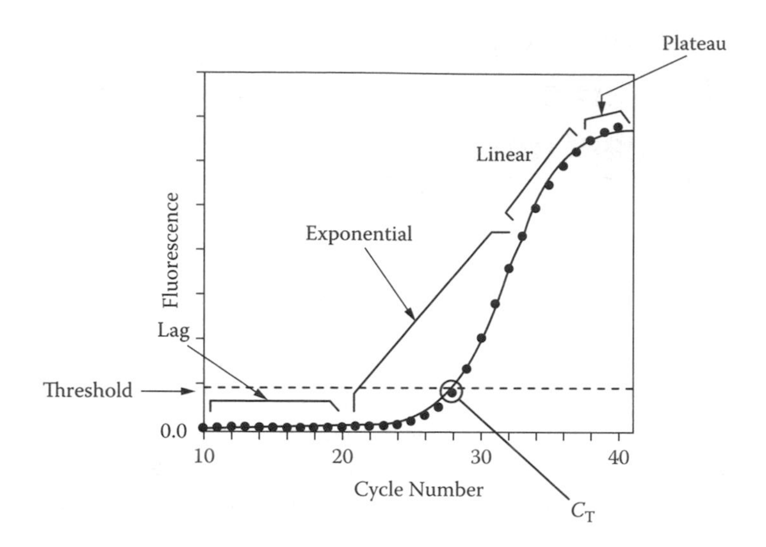


**Figure 14.1** PCR amplification curve. A fluorescence signal results from accumulation of amplicon. Location of signal threshold is indicated as dashed line. The $C_T$ value for the curve is the cycle at which the amplification curve crosses the threshold.

demonstrates that fewer cycles are needed if larger quantities of starting DNA template are present. The slope of this linear curve (Figure 13.7b) is known as the ***amplification efficiency***. The exponential phase continues until one of the components in the reaction becomes limited. At this point, the amplification efficiency decreases and the amplicon no longer accumulates exponentially, and PCR enters the linear phase of the curve.

## 14.2 Essential PCR Components

A PCR reaction requires thermostable DNA polymerases, primers and other components as described below.

### 14.2.1 Thermostable DNA Polymerases

A wide variety of DNA polymerases are available. They vary in fidelity, efficiency, and ability to synthesize longer DNA fragments. Nonetheless, *Taq* polymerase is the most commonly used enzyme for routine applications (0.5 to 5 units per reaction). Currently, AmpliTaq Gold™ DNA polymerase (Applied Biosystems) is the most common DNA polymerase for forensic applications.

PCR reactions are usually set up at room temperature so that nonspecific annealing between primers and template DNA can occur, resulting in the formation of nonspecific amplicons. Additionally, annealing between the primers can occur to form primer dimers. Nonspecific annealing interferes with PCR amplification by reducing the amplification efficiency of the specific sequences of interest.

Such interference can be minimized by a ***hot start PCR*** approach. The AmpliTaq Gold™ DNA polymerase, a modified enzyme, remains in an inactive form until activated with a pH below 7 prior to the PCR cycling in which the inhibitory motif is inactivated. The pH of the buffer system used in the PCR reaction is temperature-sensitive; increasing the temperature decreases the pH of the solution. Thus, the activation of the enzyme can actually be carried out by a heating step at 95°C prior to the start of the cycling. During the heating process, the DNA strands also denature, which can prevent the formation of nonspecific PCR products.

### 14.2.2 PCR Primers

PCR primers are the oligonucleotides that are complementary to the sequences that flank the target region of the template. A pair (forward and reverse) of primers (typically 0.1 to 1 μM) is required. Properly designed primers are critical to the success of a PCR reaction. Computer software is available to assist and optimize the designing of primers.

A primer must be specific to the target sequence or nonspecific products that might interfere with the proper interpretation of a DNA profile might be produced. The primers within a pair should have similar *Tm* values (melting temperatures). The estimated *Tm* values of a primer pair should not differ by more than 5°C. The *Tm* of an oligonucleotide primer can be predicted and calculated using the following equation:

$$Tm = 81.5°C + 16.6 (\log_{10} [K^+]) + 0.41 (\% [G + C]) - (\frac{675}{n})$$

where $[K^+]$ is the concentration of potassium ion; [G + C] is the GC content (%) of the oligonucleotide; and n is the number of bases in the oligonucleotide. The equation shows that the *Tm* can be affected by primer base composition (GC content) and primer length. The GC content of an oligonucleotide primer should be 40% to 60%, and the length of an oligonucleotide primer should be 15 to 25 base pairs although longer primers can be used.

A primer should not contain self-complementary sequences that may form hairpin structures interfering with the annealing of primers and the template. Additionally, the primers in a pair should not share similar sequences to avoid the annealing that will amplify the primers, creating products known as ***primer dimers*** that compete with the target DNA template for PCR components.

***Multiplex PCR*** allows for the simultaneous amplification of more than one region of a template in a single reaction. For a multiplex reaction, the annealing temperatures should be similar among multiplex PCR primer pairs. Again, the regions of complementary sequences should be avoided to prevent the formation of primer dimers. The primers should lead to similar amplification efficiencies among the loci tested and should be designed to yield proper sizes of amplicons to be resolvable in downstream separation and detection procedures such as electrophoresis.

### 14.2.3 Other Components

Essential components include template DNA with target sequences either in linear form (nuclear genomic DNA) or circular form (mitochondrial DNA). Both single- and double-stranded DNA are applicable for PCR. Typically, 1 to 2.5 ng of template DNA is utilized for PCR of forensic applications.

Deoxynucleoside triphosphates (dNTPs) are the substrates for DNA synthesis. A PCR assay usually contains equal molar amounts (typically, 200 μM) of dATP, dCTP, dGTP and dTTP.

Divalent cations, such as $Mg^{2+}$ are required for the enzymatic activity of DNA polymerases. A PCR usually contains 1.5 to 2.5 mM $Mg^{2+}$. Monovalent cations, such as $K^+$ (50 mM) are usually recommended and a buffer is often utilized to maintain pH between 8.3 and 8.8 at room temperature.

## 14.3 Parameters

A wide range of PCR cycling protocols have been used for various applications. Figure 14.2 shows representative PCR cycling conditions commonly used by forensic DNA laboratories.

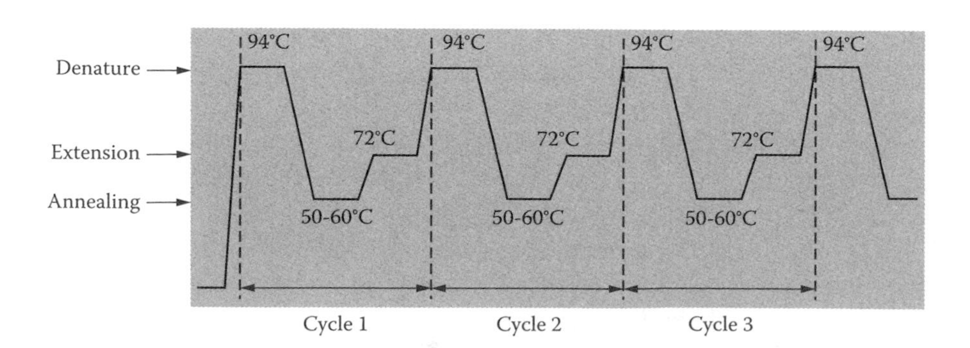


**Figure 14.2** Temperature parameters during thermal cycling of PCR process. The first three cycles are shown.

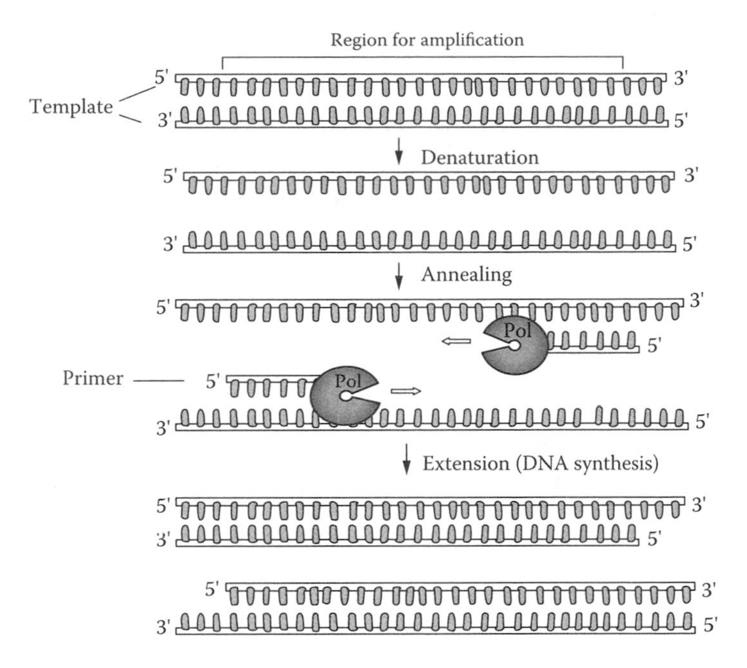


**Figure 14.3** First cycle of PCR. The amplified region is determined by the positions of primers. The direction (5′ to 3′) of DNA synthesis is indicated by arrows. Pol = *Taq* polymerase.



## 14.3.1 Cycles

PCR utilizes a number of cycles for the replication of a specific region of a DNA template. During each cycle, a copy of the target DNA sequence is synthesized. A PCR cycle consists of three elements: denaturation, annealing, and extension. At the beginning of each cycle, the two complementary DNA template strands are separated at high temperatures (94 to 95°C) in a process called ***denaturation***. The temperature is then decreased to allow ***annealing*** between the oligonucleotide primers and the template. The temperature for annealing is usually 3 to 5°C lower than the *Tm* of the oligonucleotide primer.

The annealing temperature is critical. If it is too high, a very low quantity of amplicon is yielded because of failure of annealing between the primer and the template. If the annealing temperature is too low, nonspecific amplification can occur. Next, optimal temperature for the DNA polymerase is reached, thus allowing for DNA replication (***extension***). By the end of each cycle, the copy number of amplicon is nearly doubled (Figure 14.3).

The number of cycles needed for PCR depends primarily on the number of copies of starting DNA template. The relationship can be expressed as the following equation:

$$N_x = N_0 (1 + E)^x$$

where X is the number of PCR cycles; $N_x$ is the copy number of the amplicon after x cycles of PCR; $N_0$ is the initial copy number of the template; and E is the amplification efficiency of the *Taq* polymerase. For example, in a 28-cycle PCR amplification, the DNA template can theoretically be amplified by a factor of approximately $10^8$. If the cycle number is increased to 34, a factor of $10^{10}$ can theoretically be reached.

PCR amplification can be carried out using an instrument known as a ***thermal cycler***. Precise and accurate temperatures at denaturation, annealing, and extension are critical to success. Various available thermal cyclers differ in the number of samples they can process, the sizes of the sample tubes, and temperature control features.

### 14.3.2 Controls

Controls monitor the effectiveness of PCR amplification. A ***positive control*** shows that PCR components such as reagents and thermal cycling parameters are working properly for amplifying a specific region of DNA. A standard DNA template should be used as a positive control and amplified with the same PCR components used on the rest of the samples. The negative control and reagent blank are discussed in Section 14.4.4.

## 14.4 Factors Affecting PCR

### 14.4.1 Template Degradation

It is important to prevent degradation of the DNA during the collection and processing of evidence. Degradation causes DNA to break into smaller fragments. If the damage occurs at a region to be amplified, the result can be failure in PCR amplification. In a degraded sample, the longer the amplicon length, the higher the risk of failure.

### 14.4.2 Low Copy Number of Template

Low copy number (LCN) of DNA is often encountered in forensic samples. When amplifying very low levels (approximately 100 pg of DNA) of a template, a phenomenon in which one of the two alleles fails to be detected from a heterozygote and can falsely be identified as a homozygosity is often observed. This phenomenon is also known as the ***stochastic effect*** in which the two alleles in a heterozygous individual are unequally detected at a low level of starting DNA template. Approaches such as increasing the cycle number have been introduced to address the LCN problem.

### 14.4.3 Inhibitors

The presence of inhibitors can cause PCR amplification failure. A number of PCR inhibitors commonly encountered in evidence samples include heme molecules from blood, indigo dyes from fabrics, and melanin from hair samples. Thus, it is important to remove PCR inhibitors during DNA extraction. Additional procedures can be used to remove PCR inhibitors such as the use of filtration devices.

### 14.4.4 Contamination

PCR is a highly sensitive method and thus procedures that minimize the risk of contamination are necessary. To prevent contamination, pre- and post-PCR samples should be processed in separate areas. Additionally, reagents, supplies and equipment used for pre- and post-PCR steps should be separated as well. Protective gear should include laboratory coats and disposable gloves. Facial masks and hair caps may be used if necessary. Aerosol-resistant pipette tips and DNA-free solutions and test tubes should also be used.

The levels of contamination must be monitored. The controls used to monitor PCR contamination include ***reagent blanks*** (containing all extraction reagents but no sample) that monitor contamination from extraction to final analysis. Contamination detected in a reagent blank but not in a negative control indicates that the reagents used for extraction are contaminated. ***Negative controls*** (containing all PCR reagents and no DNA template) monitor contamination from amplification to final analysis. Contamination observed in a negative control but not in a reagent blank indicates that the contamination occurred during the amplification step. A collection of DNA profiles of each member of a laboratory should be readily available for comparisons. Sources of laboratory contamination can be identified by comparing results with an analyst's DNA profile.

## 14.5 Reverse Transcriptase PCR for RNA-Based Assays

### 14.5.1 Essential Features of RNA

Like DNA, ***ribonucleic acid (RNA)*** is a linear molecule containing four types of nucleotides linked by phosphodiester bonds (Figure 14.4 and Figure 14.5). However, certain properties of RNA differ from those of DNA. Unlike DNA containing deoxyribose (see Figure 14.6), the sugar residue of RNA is ribose which has a hydroxyl (OH) group at the 2′ carbon position. RNA contains uracil (U) in place of thymine in DNA (Figure 14.7). RNA is typically found in cells as a single stranded molecule, while DNA is double stranded.

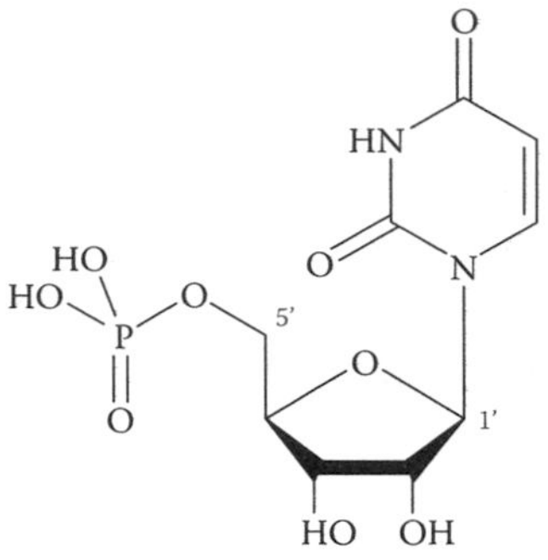

**Figure 14.4** The structure of ribonucleotide. Each ribonucleotide in a RNA polymer consists of three components: ribose, a nitrogenous base (uracil is shown as an example), and a phosphate group. The 1′ carbon of the ribose is attached to the nitrogen of the nitrogenous base. A phosphate group is attached to the 5′ carbon of the ribose.

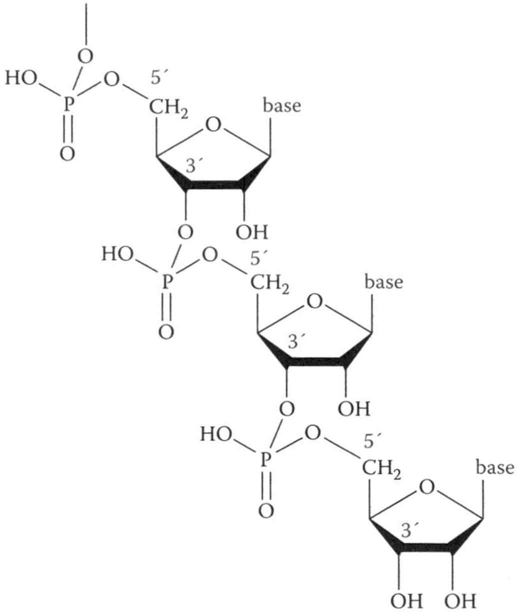

**Figure 14.5** Chemical structure of RNA (polyribonucleotide) chain.

Polyribonucleotides can form complementary helices with DNA strands by base pairing (except uracil pairs with adenine); see Figure 14.8. The RNA synthesized from a DNA template is called messenger RNA (mRNA) and usually contains a polyadenine tail at the 3′ end of the molecule (Figure 14.9).

HO 5′ O OH 4′ 1′ 3′ 2′ OH OH

HO 5′ O OH 4′ 1′ 3′ 2′ OH

(a) (b)

**Figure 14.6** Comparison of (a) ribose and (b) deoxyribose. The hydroxyl group attached to the 2′ carbon of ribose is replaced by a hydrogen group in deoxyribose.

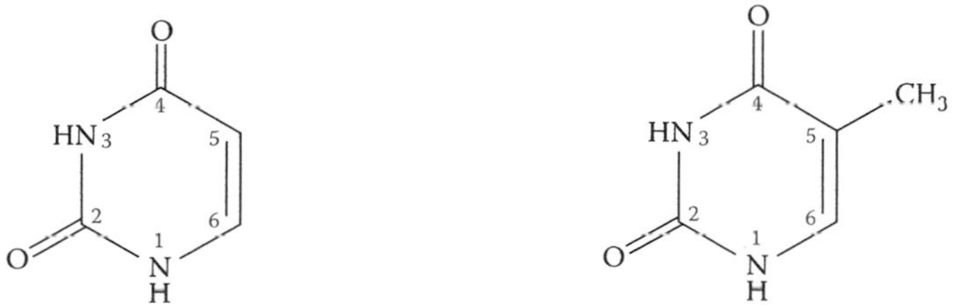

O 4 HN 3 5 2 6 O 1 N H

O 4 $CH_3$ HN 3 5 2 6 O 1 N H

(a) (b)

**Figure 14.7** Comparison of (a) uracil and (b) thymine. Thymine has a methyl group attached at the 5–carbon position.

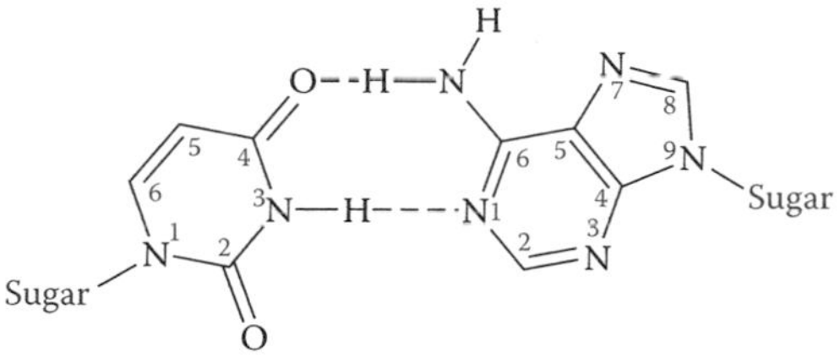

H O--H—N N 7 8 5 4 6 5 9 N 6 3 N—H---N 1 4 Sugar 1 2 3 N 2 N Sugar O

**Figure 14.8** Base pairing of uracil (left) and adenine (right).

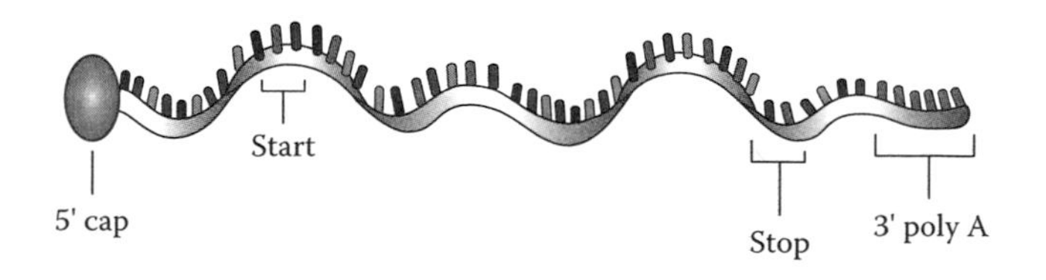


**Figure 14.9** Structure of mRNA. Eukaryotic mRNA is modified at the 5′ end (cap) and 3′ end (polyadenine tail) needed for protein synthesis. Start and stop codons are required for the initiation and termination of protein synthesis.

### 14.5.2 Central Dogma

The pathway for the flow of genetic information is called the ***central dogma*** — a term coined by Francis Crick in 1956. According to the central dogma, DNA serves as the template for its self-synthesis in a process called ***DNA replication***. With RNA synthesis or ***transcription***, the process is carried out using the DNA as a template. Conversely, RNA chains can be used as templates for the synthesis of a DNA strand of complementary sequence in which the end product is referred to as ***complementary DNA (cDNA)***. Protein synthesis, also known as ***translation***, is directed by an RNA template but RNA sequences are never determined by protein templates (Figure 14.10).

The flow of genetic information from RNA to DNA is referred to as ***reverse transcription***. It was discovered independently by David Baltimore and Howard Temin in 1970 and they shared the Nobel Prize for their work. Reverse transcription is carried out by a ***reverse transcriptase*** that forms the basis of reverse transcriptase PCR as described below.

### 14.5.3 Reverse Transcriptase PCR

Reverse transcriptase PCR (RT-PCR) is used to amplify cDNA copies of RNA. RT-PCR is highly sensitive and can be used to detect very small quantities of mRNA. Thus, it can be utilized to measure levels of gene expression even when the RNA of interest is expressed at very low levels. RT-PCR usually consists of two steps.

The first step is synthesizing single stranded cDNA from a template mRNA. An oligodeoxynucleotide primer is utilized to hybridize specifically to a chosen region in a particular target mRNA, followed by an extension phase carried out by reverse transcriptase to synthesize a copy of the cDNA strand. In the second step, the cDNA is amplified by PCR with a pair of oligonucleotide primers corresponding to a specific sequence in the cDNA (Figure 14.11). Two types of PCR methods can be utilized for this purpose. The

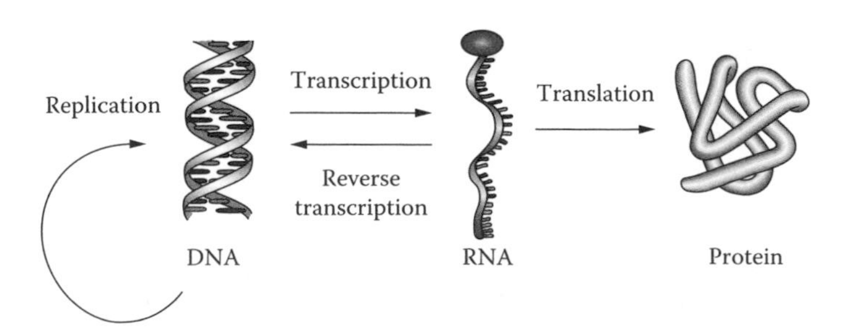


**Figure 14.10** Pathway for flow of genetic information.

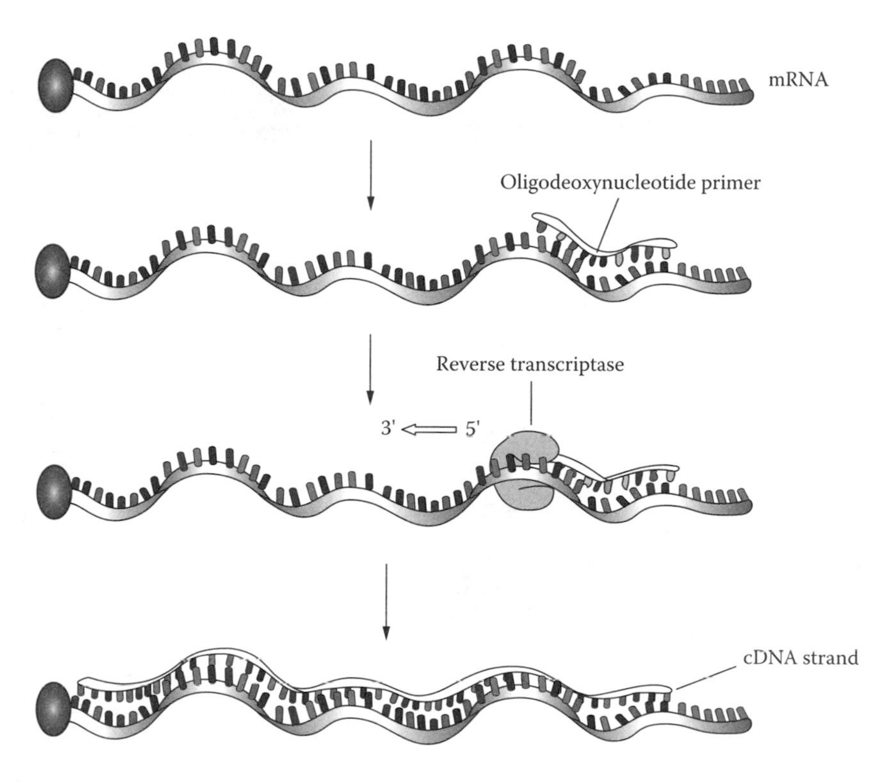


**Figure 14.11** cDNA synthesis. The synthesis of the first strand from mRNA can be carried out using an oligodeoxynucleotide primer and reverse transcriptase.

end-point PCR method measures the amount of amplified product synthesized during PCR at the end of the PCR amplification. The detection of the amplified product indicates the presence of the mRNA of interest. With the real-time PCR method, the amplified product is quantified during the exponential phase of PCR.

### 14.5.4 Forensic Applications of RNA-Based Assays

RNA-based assays have been developed to identify biological fluids. The assays are based on the expression of certain genes in certain cell or tissue types. Candidate tissue-specific genes that may be useful for identification of biological fluid have been identified. Thus, the techniques used in the identification of biological fluids are based on the detection of specific types of mRNA expressed exclusively in certain cells. These assays utilize RT-PCR to detect gene expression levels of mRNAs. The tissue-specific genes utilized for biological fluid identification are summarized in Table 14.1. Compared to conventional assays used for biological fluid identification, the RNA-based

**Table 14.1 Application of RNA-Based Assays for Biological Fluid Identification**

| Biological Fluid | Gene Symbol | Description | Real-Time PCR | End-Point PCR |
|---|---|---|---|---|
| Blood | *HBA1* | Hemoglobin α1 | ☑ | |
| | *PBGD* | Porphobilinogen deaminase | | ☑ |
| | *SPTB* | β–Spectrin | | ☑ |
| Menstrual blood | *MMP7* | Matrix metalloproteinase | | ☑ |
| Saliva | *HTN3* | Histatin 3 | | ☑ |
| | *STATH* | Statherin | | ☑ |
| Semen | *KLK3* | Kallikrein 3 | ☑ | |
| | *PRM1* | Protamine 1 | | ☑ |
| | *PRM2* | Protamine 2 | | ☑ |
| Vaginal Secretions | *HBD1* | β–Defensin 1 | | ☑ |
| | *MUC4* | Mucin 4 | ☑ | ☑ |

*Source:* Adapted from Juusola, J., and J. Ballantyne. 2005. *Forensic Sci Int* 152:1; Nussbaumer, C. et al. 2006. *Forensic Sci Int* 157:181.

assay has higher specificity and is amenable to automation. However, one limitation is that the RNA is unstable due to degradation by endogenous ribonucleases.

## Bibliography


Afonina, I. A. et al. 2002. Minor groove binder-conjugated DNA probes for quantitative DNA detection by hybridization-triggered fluorescence. *Biotechniques* 32 (4):940.

Bassam, B. J. et al. 1996. Nucleic acid sequence detection systems: Revolutionary automation for monitoring and reporting PCR products. *Austral Biotechnol* 6:285.

Birch, D. E. et al. 1996. Simplified hot start PCR. *Nature* 318.

Bloch, W. 1991. A biochemical perspective of the polymerase chain reaction. *Biochemistry* 30 (11):2735.

Budowle, B. et al. 2000. *DNA Typing Protocols: Molecular Biology and Forensic Analysis*. Natick, MA: Eaton Publishing.

Butler, J. M. 2005. *Forensic DNA Typing: Biology, Technology, and Genetics of STR Markers*, 2nd ed. Burlington, MA: Elsevier.

Butler, J. M. et al. A novel multiplex for simultaneous amplification of 20 Y chromosome STR markers. *Forensic Sci Int* 129 (1):10.

Cardullo, R. A. et al. 1988. Detection of nucleic acid hybridization by nonradiative fluorescence resonance energy transfer. *Proc Natl Acad Sci USA* 85 (23):8790.

Chung, D. T. et al. 2004. A study on the effects of degradation and template concentration on the amplification efficiency of the STR Miniplex primer sets. *J Forensic Sci* 49 (4):733.

Comey, C. T. et al. 1993. PCR amplification and typing of the HLA DQ alpha gene in forensic samples. *J Forensic Sci* 38 (2):239.

Dieffenbach, C. W., and G. S. Dveksler. 2003. *PCR Primer: A Laboratory Manual*, 2nd ed. Cold Spring Harbor, NY: Cold Spring Harbor Laboratory Press.

Dieffenbach, C. W., T. M. Lowe, and G. S. Dveksler. 1993. General concepts for PCR primer design. *PCR Methods Appl* 3 (3):S30.

Eckert, K. A., and T. A. Kunkel. 1990. High fidelity DNA synthesis by the *Thermus aquaticus* DNA polymerase. *Nucleic Acids Res* 18 (13):3739.

_____. 1991. DNA polymerase fidelity and the polymerase chain reaction. *PCR Methods Appl* 1 (1):17.

Edwards, M. C., and R. A. Gibbs. 1994. Multiplex PCR: Advantages, development, and applications. *PCR Methods Appl* 3 (4):S65.

Erlich, H. A., D. Gelfand, and J. J. Sninsky. 1991. Recent advances in the polymerase chain reaction. *Science* 252 (5013):1643.

Erlich, H. A. et al. Reliability of the HLA-DQ alpha PCR-based oligonucleotide typing system. *J Forensic Sci* 35 (5):1017.

Gasparini, P. et al. 1989. Amplification of DNA from epithelial cells in urine. *New Engl J Med* 320 (12):809.

Gill, P. et al. 2000. An investigation of the rigor of interpretation rules for STRs derived from less than 100 pg of DNA. *Forensic Sci Int* 112 (1):17.

Gyllensten, U. B., and H. A. Erlich. 1988. Generation of single-stranded DNA by the polymerase chain reaction and its application to direct sequencing of the HLA-DQA locus. *Proc Natl Acad Sci USA* 85 (20):7652.

Haff, L. et al. 1991. A high-performance system for automation of the polymerase chain reaction. *Biotechniques* 10 (1):102-3, 106.

Henegariu, O. et al. 1997. Multiplex PCR: Critical parameters and step-by-step protocol. *Biotechniques* 23 (3):504.

Higuchi, R. et al. 1992. Simultaneous amplification and detection of specific DNA sequences. *Biotechnology* 10 (4):413.

Higuchi, R. et al. 1993. Kinetic PCR analysis: Real-time monitoring of DNA amplification reactions. *Biotechnology* 11 (9):1026.

Hochmeister, M. N. et al. 1995. Confirmation of the identity of human skeletal remains using multiplex PCR amplification and typing kits. *J Forensic Sci* 40 (4):701.

Holland, P. M. et al. 1991. Detection of specific polymerase chain reaction product by utilizing the 5′–3′ exonuclease activity of *Thermus aquaticus* DNA polymerase. *Proc Natl Acad Sci USA* 88 (16):7276.

Isacsson, J. et al. 2000. Rapid and specific detection of PCR products using light-up probes. *Mol Cell Probes* 14 (5):321.

Ju, J. et al. 1995. Fluorescence energy transfer dye-labeled primers for DNA sequencing and analysis. *Proc Natl Acad Sci USA* 92 (10):4347.

Kimpton, C. et al. 1994. Evaluation of an automated DNA profiling system employing multiplex amplification of four tetrameric STR loci. *Int J Legal Med* 106 (6):302.

Krawczak, M. et al. 1989. Polymerase chain reaction: replication errors and reliability of gene diagnosis. *Nucleic Acids Res* 17 (6):2197.

Kunkel, T. A. 1992. DNA replication fidelity. *J Biol Chem* 267 (26):18251.

Kutyavin, I. V. et al. 2000. 3′ minor groove binder DNA probes increase sequence specificity at PCR extension temperatures. *Nucleic Acids Res* 28 (2):655.

Kwok, S., and R. Higuchi. 1989. Avoiding false positives with PCR. *Nature* 339 (6221):237.

Lee, L. G. et al. 1999. Seven-color, homogeneous detection of six PCR products. *Biotechniques* 27 (2):342.

Markoulatos, P., N. Siafakas, and M. Moncany. 2002. Multiplex polymerase chain reaction: A practical approach. *J Clin Lab Anal* 16 (1):47.

Mitsuhashi, M. 1996. Technical report part 2: Basic requirements for designing optimal PCR primers. *J Clin Lab Anal* 10 (5):285.

Moretti, T., B. Koons, and B. Budowle. 1998. Enhancement of PCR amplification yield and specificity using AmpliTaq Gold DNA polymerase. *Biotechniques* 25 (4):716.

Nazarenko, I. A., S. K. Bhatnagar, and R. J. Hohman. 1997. A closed tube format for amplification and detection of DNA based on energy transfer. *Nucleic Acids Res* 25 (12):2516.

Reynolds, R., G. Sensabaugh, and E. Blake. 1991. Analysis of genetic markers in forensic DNA samples using the polymerase chain reaction. *Anal Chem* 63 (1):2.

Rutty, G. N., A. Hopwood, and V. Tucker. 2003. The effectiveness of protective clothing in the reduction of potential DNA contamination of the scene of crime. *Int J Legal Med* 117 (3):170.

Saiki, R. K. et al. 1985. Enzymatic amplification of beta-globin genomic sequences and restriction site analysis for diagnosis of sickle cell anemia. *Science* 230 (4732):1350.

Saiki, R. K. et al. 1988. Primer-directed enzymatic amplification of DNA with a thermostable DNA polymerase. *Science* 239 (4839):487.

Sanchez, J. J. et al. 2003. Multiplex PCR and minisequencing of SNPs: A model with 35 Y chromosome SNPs. *Forensic Sci Int* 137 (1):74.

Schoske, R. et al. 2003. Multiplex PCR design strategy used for the simultaneous amplification of 10 Y chromosome short tandem repeat (STR) loci. *Anal Bioanal Chem* 375 (3):333.

Selvin, P. R. 1995. Fluorescence resonance energy transfer. *Methods Enzymol* 246:300.

Shuber, A. P., V. J. Grondin, and K. W. Klinger. 1995. A simplified procedure for developing multiplex PCRs. *Genome Res* 5 (5):488.

Solinas, A., N. Thelwell, and T. Brown. 2002. Intramolecular TaqMan probes for genetic analysis. *Chem Commun (Camb)* (19):2272.

Tindall, K. R., and T. A. Kunkel. 1988. Fidelity of DNA synthesis by the *Thermus aquaticus* DNA polymerase. *Biochemistry* 27 (16):6008.

Wallin, J. M. et al. 2002. Constructing universal multiplex PCR systems for comparative genotyping. *J Forensic Sci* 47 (1):52.

Walsh, P. S., H. A. Erlich, and R. Higuchi. 1992. Preferential PCR amplification of alleles: Mechanisms and solutions. *PCR Methods Appl* 1 (4):241.

Westwood, S. A., and D. J. Werrett. 1990. An evaluation of the polymerase chain reaction method for forensic applications. *Forensic Sci Int* 45 (3):201.

Whitcombe, D. et al. 1998. A homogeneous fluorescence assay for PCR amplicons: Its application to real-time, single-tube genotyping. *Clin Chem* 44 (5):918.

## RNA-Based Assays

Amin, K. M. et al. 1993. The exon-intron organization of the human erythroid beta-spectrin gene. *Genomics* 18 (1):118.

Bauer, M., A. Kraus, and D. Patzelt. 1999. Detection of epithelial cells in dried blood stains by reverse transcriptase-polymerase chain reaction. *J Forensic Sci* 44 (6):1232.

Bauer, M., and D. Patzelt. 2002. Evaluation of mRNA markers for the identification of menstrual blood. *J Forensic Sci* 47 (6):1278.

____. 2003. Protamine mRNA as molecular marker for spermatozoa in semen stains. *Int J Legal Med* 117 (3):175.

Chu, Z. L. et al. 1994. Erythroid-specific processing of human beta spectrin I pre-mRNA. *Blood* 84 (6):1992.

Gelmini, S. et al. Real-time quantitative reverse transcriptase-polymerase chain reaction (RT-PCR) for the measurement of prostate-specific antigen mRNA in the peripheral blood of patients with prostate carcinoma using the TaqMan detection system. *Clin Chem Lab Med* 39 (5):385.

Gipson, I. K. et al. 1997. Mucin genes expressed by human female reproductive tract epithelia. *Biol Reprod* 56 (4):999.

Gipson, I. K. et al. 1999. MUC4 and MUC5B transcripts are the prevalent mucin messenger ribonucleic acids of the human endocervix. *Biol Reprod* 60 (1):58.

Gubin, A. N., and J. L. Miller. 2001. Human erythroid porphobilinogen deaminase exists in two splice variants. *Blood* 97 (3):815.

Hochmeister, M. N. et al. 1999. Validation studies of an immunochromatographic one-step test for the forensic identification of human blood. *J Forensic Sci* 44 (3):597.

Juusola, J., and J. Ballantyne. 2003. Messenger RNA profiling: Prototype method to supplant conventional methods for body fluid identification. *Forensic Sci Int* 135 (2):85.

____. 2005. Multiplex mRNA profiling for the identification of body fluids. *Forensic Sci Int* 152 (1):1.

Nussbaumer, C., E. Gharehbaghi-Schnell, and I. Korschineck. 2006. Messenger RNA profiling: A novel method for body fluid identification by real-time PCR. *Forensic Sci Int* 157 (2–3):181.

Phang, T. W. et al. 1994. Amplification of cDNA via RT-PCR using RNA extracted from postmortem tissues. *J Forensic Sci* 39 (5):1275.

Sabatini, L. M., Y. Z. He, and E. A. Azen. 1990. Structure and sequence determination of the gene encoding human salivary statherin. *Gene* 89 (2):245.

Salamonsen, L. A., and D. E. Woolley. 1996. Matrix metalloproteinases in normal menstruation. *Hum Reprod* 11 Suppl 2:124.

_____. 1999. Menstruation: Induction by matrix metalloproteinases and inflammatory cells. *J Reprod Immunol* 44 (1–2):1.

Steger, K. et al. 2000. Expression of protamine-1 and -2 mRNA during human spermiogenesis. *Mol Hum Reprod* 6 (3):219.

Valore, E. V. et al. 1998. Human beta-defensin-1: An antimicrobial peptide of urogenital tissues. *J Clin Invest* 101 (8):1633.

## Study Questions

1. The negative control in amplification monitors contamination:
   (a) From evidence collection to final analysis
   (b) From DNA extraction to final analysis
   (c) Occurring only during amplification
   (d) From amplification to final analysis

2. The negative control in amplification contains all PCR components except:
   (a) Template
   (b) Polymerase
   (c) Magnesium
   (d) dNTPs

3. For multiplex STR analysis, fluorescent labeling of PCR products may be accomplished by:
   (a) Fluorescently labeled 5′ end-labeled primers
   (b) Fluorescently labeled dNTPs
   (c) Fluorescent intercalating dye

4. The order of steps in PCR cycling is:
   (a) Annealing, denaturation, and extension
   (b) Denaturation, annealing, and extension
   (c) Extension, denaturation, and annealing
   (d) Extension, annealing, and denaturation

5. The role of magnesium in a PCR reagent is:
   (a) To inhibit DNA degradation
   (b) Required by DNA polymerase
   (c) Required for the annealing step
   (d) Required by the denaturation step

6. AmpliTaq Gold polymerase is:
   (a) Activated by heat
   (b) Inactivated by heat
   (c) Always active and not affected by heat

7. Which of the following may be affected by PCR inhibitors?
   (a) PCR-based assays
   (b) Restriction enzyme-based assays
   (c) Serology-based assays
   (d) All of the above

8. How can you overcome problems with PCR inhibition?
   (a) Add more DNA polymerase.
   (b) Add more DNA template.
   (c) Remove inhibitor using filtration devices.
   (d) All of the above.

9. During a hot start PCR:
   (a) AmpliTaq Gold polymerase is activated.
   (b) AmpliTaq Gold polymerase is inactivated.
   (c) AmpliTaq Gold polymerase is added.

10. Which of the following is true?
    (a) DNA polymerase also has nuclease activity.
    (b) DNA polymerase has no nuclease activity.
    (c) Nuclease has polymerase activity.
    (d) None of the above.

11. Contaminated DNA may be detected if:
    (a) The contamination occurred during DNA extraction.
    (b) The contaminatin occurred during DNA amplification.
    (c) All of the above.

12. During set-up of a PCR, Chelex resin was accidentally loaded from a DNA sample into the PCR reagent. The Chelex resin:
    (a) May affect the reaction.
    (b) Will not affect the reaction.

# 15 DNA Electrophoresis

It is necessary to separate various sizes of DNA fragments so that a DNA fragment in question can be identified and analyzed. This can be achieved by the electrophoresis technique in which fragments are separated based on the migration of charged macromolecules in an electric field.

## 15.1 Basic Principles

DNA is a negatively charged molecule in an aqueous environment, with the phosphate groups of DNA nucleotides carrying negative charges. DNA molecules migrate from the negative electrode toward the positive electrode in an electric field during electrophoresis. The ***electrical potential*** is the force responsible for moving the charged macromolecules during electrophoresis. Thus, proper ionic strength of an electrophoretic buffer is necessary to achieve efficient electrical conductance.

The ***electrophoretic mobility*** of macromolecules is primarily determined by their ***charge-to-mass ratios*** and their shapes. However, because the phosphate group of every DNA nucleotide carries a negative charge, the charge-to-mass ratio of DNA molecules is the same even if the length of the DNA fragment varies. The shapes of linear DNA molecules are similar and forensic DNA testing usually analyzes linear DNA molecules. Therefore, the electrophoretic separation of different sized DNA fragments is based more on the effects of a molecular sieve mechanism provided by the supporting matrix in which the fragments travel (Figure 15.1).

It is obviously easier for smaller molecules to migrate though the pores of a molecular sieve than larger molecules; this is why they migrate faster through a matrix. Hence, the electrophoretic mobility increases as the size decreases. Conversely, larger DNA molecules migrate much more slowly because they experience more friction as they travel through the net of pores in a gel. Therefore, the separation of DNA molecules with different sizes can be accomplished.

### 15.1.1 Estimation of DNA Size

The ***relative mobility*** (***$R_f$***) of a DNA molecule during electrophoresis can be calculated as the distance of band migration divided by the distance of tracking dye migration. The DNA band migration is the distance from

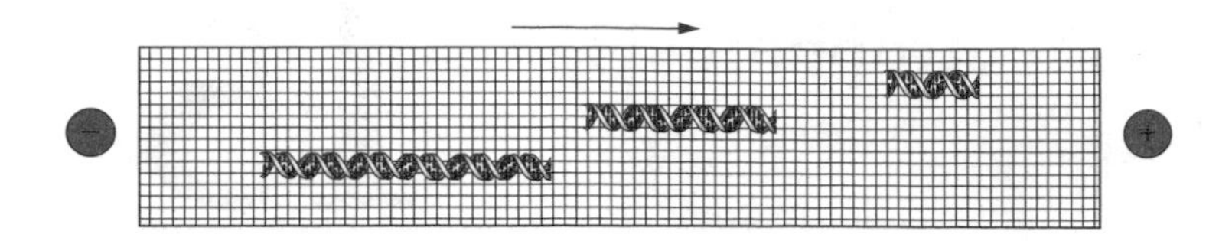

**Figure 15.1** Molecular sieve model for separation of DNA fragments through electrophoresis which involves the migration of DNA molecules through a matrix. Smaller DNA molecules migrate faster than their larger counterparts.

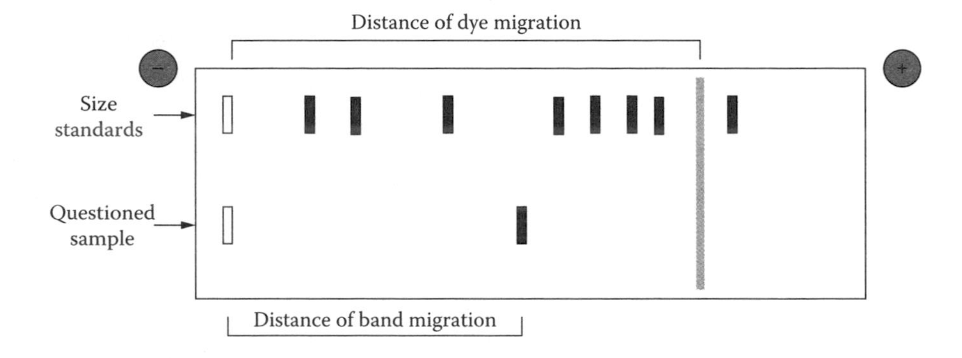


**Figure 15.2** Measurement of electromobility of DNA. Questioned DNA samples are analyzed concurrently with a set of standard DNA fragments of known size. The distances of DNA migration can thus be measured. The samples are usually mixed with a gel loading buffer prior to loading to the gel. The buffer contains dyes that add color to the sample to facilitate the process of loading. Dyes such as bromophenol blue and xylene cyanol FF migrate toward the positive electrode during electrophoresis. They can be used for tracking purposes as well.

sample origin to the center of the band. The tracking dye migration distance extends from the origin to the center of the dye band (Figure 15.2). To estimate the size of DNA, standards containing DNA of known size and questioned samples are run on the same gel at the same time. The standards can be used to estimate the size of an unknown DNA molecule. A plot of $\log_{10}$ base pair of the standards versus $R_f$ for a given gel can be constructed. A linear relationship (over a size range; varies by electrophoresis system) between $\log_{10}$ base pair of the DNA molecule and the $R_f$ can be observed. The $R_f$ of the test sample is interpolated on the plot from which the size of an unknown DNA molecule can be determined (Figure 15.3).

## 15.2 Supporting Matrices

Although the actual separation occurs in an aqueous phase, most variants of electrophoresis use a physical support material also called a ***matrix***. As

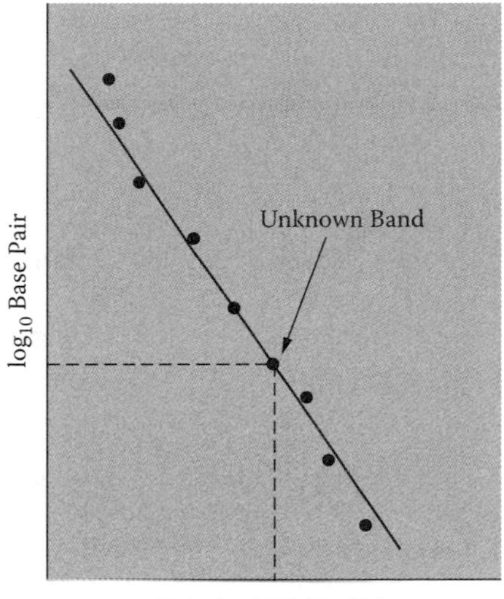


**Figure 15.3** Estimation of size of a DNA fragment. A plot of size standards versus relative migration during electrophoresis is shown. The size of the questioned DNA fragment can be estimated using this plot.

discussed above, the matrix can be used as a molecular sieve for the separation of DNA molecules and it also reduces diffusion and convection during electrophoresis. Agarose and polyacrylamide are commonly used electrophoresis matrices because of good reproducibility, reliability, and versatility.

### 15.2.1 Agarose

Agarose is a linear polymer composed of alternating residues of D- and L-galactose (Figure 15.4). A gelatinized agarose forms a three-dimensional sieve with pores from 50 to 200 nm in diameter (Figure 15.5). DNA fragments ranging from 50 to 20,000 base pairs in size are best resolved in agarose gels.

The electrophoretic mobility of double stranded DNA through agarose gel matrices is inversely proportional to the $\log_{10}$ of the size of DNA fragments ranging from 50 to 20,000 base pairs. The electrophoretic mobility of a linear DNA fragment is also inversely proportional to the concentration of agarose in the gel. A low concentration of agarose in a gel forms larger sized pores, allowing larger DNA fragments to move through.

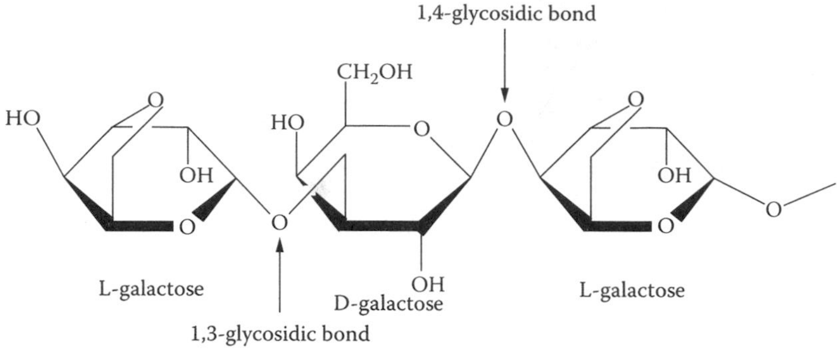

**Figure 15.4** Chemical components of agarose, a linear polysaccharide composed of alternating units of D- and L-galactose. Often, the L-galactose residue has an anhydro bridge between the 3 and 6 positions and is called 3,6-anhydro-L-galactose.

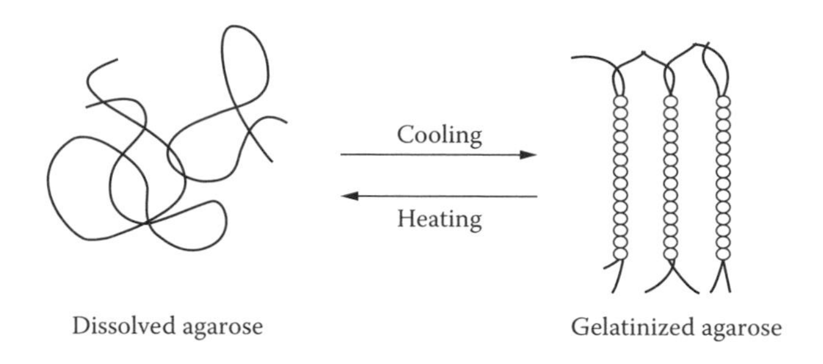


**Figure 15.5** Agarose chains. Gelatinized agarose consists of a three-dimensional network in which chains of agarose form helical fibers that aggregate into supercoiled structures.

## 15.2.2 Polyacrylamide

A polyacrylamide gel matrix is very effective for the separation of smaller fragments of DNA (5 to 500 base pairs). Additionally, a single nucleotide difference in the length of a DNA fragment can be resolved with this type of gel. Polyacrylamide produces much smaller pore sizes than agarose gels and thus has a much higher resolving power than agarose gels for low molecular weight DNA molecules. A polyacrylamide gel matrix is formed by polymerization and cross-linking reactions.

**Polymerization Reaction** — Long linear chains of polyacrylamide are polymerized from acrylamide monomers (Figure 15.6). This polymerization reaction is initiated in the presence of free radicals generated from the reduction of ammonium persulfate (APS) by N,N,N′,N′-tetramethylethylene diamine (TEMED). See Figure 15.7.

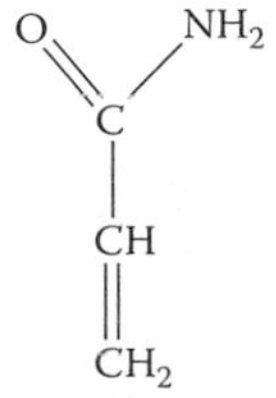

(a) 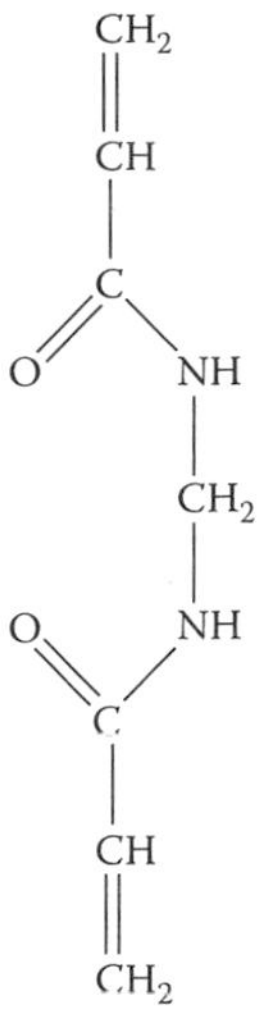 (b)

**Figure 15.6** Chemical structures of (a) acrylamide and (b) N,N′-methylenebisacrylamide (BIS).

**Cross-Linking Reaction** — As shown in Figure 15.7, three-dimensionally cross-linked polyacrylamide chains can be formed with the use of cross-linking agents, N,N′-methylenebisacrylamide (BIS; Figure 15.6). The porosity of the resulting gel is determined by the lengths of these chains and the degree of cross-linking between them. Therefore, the sizes of the pores formed can be adjusted by altering the concentrations of the acrylamide monomer and the cross-linking reagent.

Polyacrylamide can also be used in a capillary electrophoresis matrix. It is very difficult to insert a cross-linked polyacrylamide matrix into a capillary. Therefore, a solution of a linear polymer (noncross-linked) of dimethyl polyacrylamide is used as a matrix for capillary electrophoresis. For instance, POP-4 (4% lineal dimethyl polyacrylamide polymer) is used for forensic STR analysis (see Chapter 18) and POP-6 (6% lineal dimethyl polyacrylamide polymer) is used for mitochondrial DNA sequencing (see Chapter 21). These linear polymers are commercially available (Applied Biosystems).

**Denaturing Polyacrylamide Electrophoresis** — Polyacrylamide-based electrophoresis can be carried out with both double and single stranded DNA. Electrophoresis performed under the conditions of single stranded DNA is called ***denaturing electrophoresis***. Denatured DNA migrates through the gel as linear molecules at a rate independent of base composition and sequence. With short single stranded fragments, molecules differing in size by a single nucleotide can be separated. This extraordinary sensitivity to size is extremely

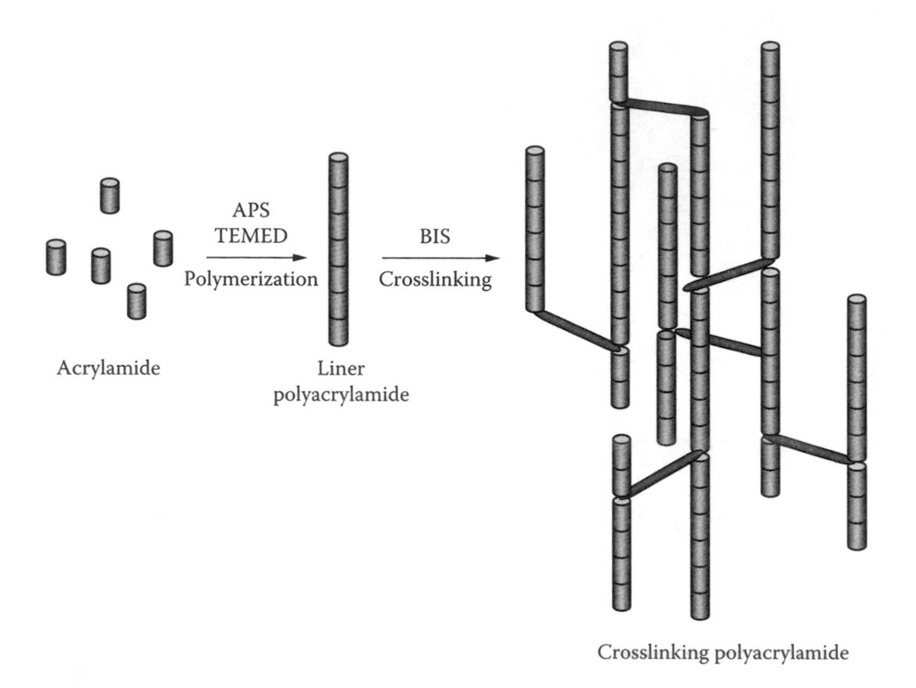


**Figure 15.7** Formation of polyacrylamide gel. Polymerization is followed by cross-linking in which the pore size of the gel is determined by its degree of polymerization and cross-linking.

useful for separation of small DNA fragments used for DNA sequencing and STR analysis in forensic applications. The denaturing condition can be best achieved by adding chemicals such as urea or formamide to the matrix or by increasing the temperature or pH during electrophoresis.

## 15.3 Apparatus and Forensic Applications

Agarose electrophoresis is carried out in a slab gel. Agarose gel can be prepared in a variety of shapes and sizes and can be run in different configurations. The most common gel for forensic use is the horizontal slab. Polyacrylamide can be used either in a slab gel or in a capillary electrophoretic apparatus.

### 15.3.1 Slab Gel Electrophoresis

#### *15.3.1.1 Agarose Gel Electrophoresis*

One forensic application of agarose gel electrophoresis is restriction fragment length polymorphism (RFLP) analysis of variable number tandem repeat (VNTR) loci (see Chapter 17). An agarose gel is used to separate the DNA

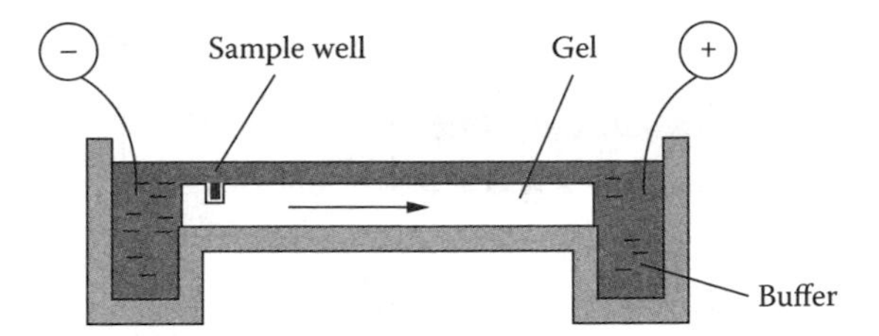


**Figure 15.8** Horizontal agarose slab gel apparatus. The gel can be prepared by heating an agarose suspension to dissolve it. The agarose solution can then be poured into a gel cast and allowed to cool until gelatinized. Electrophoresis buffer is poured over the gel to submerge it. The samples, usually mixed with loading buffer containing dye to allow visualization, are loaded into the wells of the submerged gel using a pipet. The electrodes are indicated.

fragments by size ranging from 500 to 20,000 base pairs of commonly used VNTR loci for forensic testing. This type of electrophoresis is done under non-denatured conditions (double stranded DNA).

Figure 15.8 shows an example of the apparatus used for gel electrophoresis of DNA. A thin slab of agarose gel is prepared by allowing liquid agarose to gelatinize in a cast. The gel contains small wells for loading samples. After the gel is submerged in a buffer tank, the samples are loaded. An electric field is applied and the negatively charged DNA molecules migrate through the agarose. Slab gel electrophoresis is capable of running multiple samples simultaneously. After electrophoresis, the sample is made visible by using a staining reagent (see Chapter 16). The separated DNA samples appear as bands and the sizes of the DNA fragments can be estimated by comparing them with the sizes of standards run concurrently.

#### *15.3.1.2 Polyacrylamide Gel Electrophoresis*

The forensic application of this apparatus is the separation of STR fragments and DNA sequencing reaction products of mitochondrial DNA. The sizes of DNA fragments that can be separated range from 100 to 500 base pairs—much smaller than can be separated with agarose gels. Single nucleotide resolution to distinguish similar sized fragments can be achieved with this technique under denatured conditions with only single stranded DNA.

The polyacrylamide slab gels for this application are usually run in a vertical configuration (Figure 15.9). Detection of DNA bands in polyacrylamide gels will be described in Chapter 16. The sizes of DNA fragments can be calculated by including an internal size standard. As with agarose gel, polyacrylamide gel electrophoresis is capable of running multiple samples

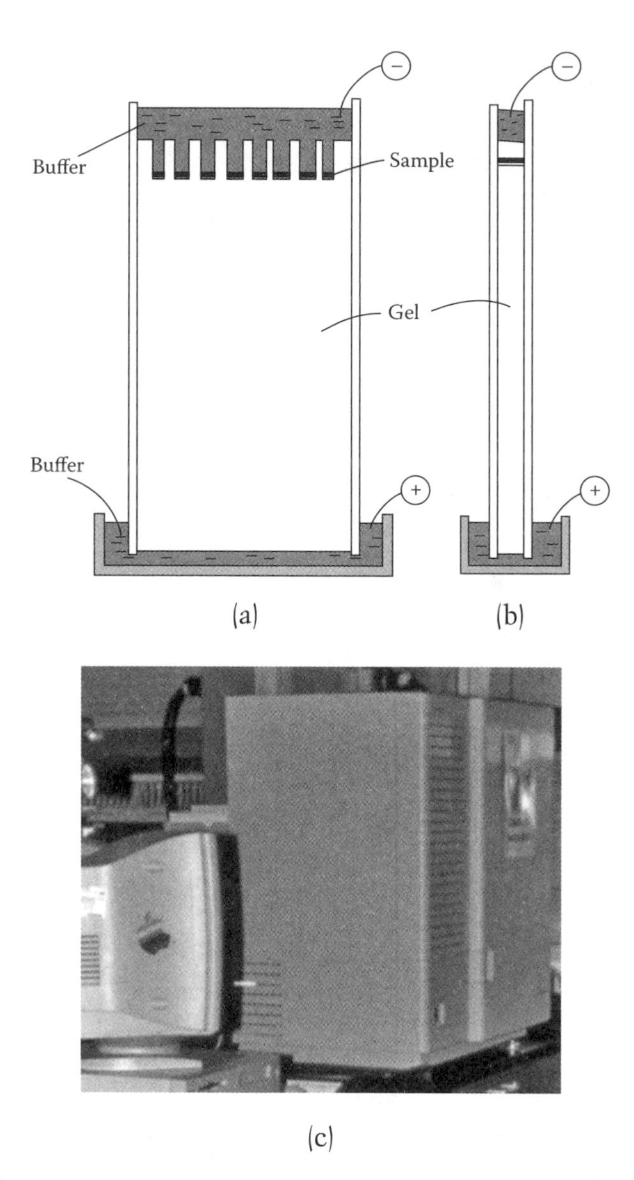


**Figure 15.9** Vertical slab polyacrylamide gel. (a) front view. (b) side view. (c) automated gel electrophoresis instrument (ABI PRISM 377® Genetic Analyzer).

simultaneously and high throughput can be achieved. However, caution should be taken to ensure that cross-contamination does not occur from spilling of samples from adjacent wells. Additionally, polyacrylamide gels are more difficult to prepare than agarose gels.

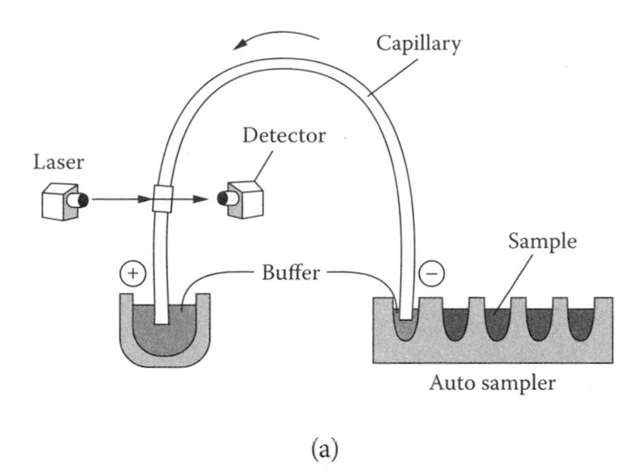


(a)

(b)

**Figure 15.10** Essential components of capillary electrophoresis system (a) including a capillary, buffer reservoirs, two electrodes, a laser excitation source, a fluorescence detector, and an autosampler that holds the sample. Sample injection, electrophoresis, and data collection are automated and controlled by a computer. (b) Photo of capillary electrophoresis instrument (ABI PRISM 310® Genetic Analyzer).

## 15.3.2 Capillary Electrophoresis

Capillary electrophoresis (Figure 15.10) is newer than the slab gel platform method. Linear dimethyl polyacrylamide is used as the matrix. This type of electrophoresis is conducted under a denatured condition for forensic applications such as STR analysis and mitochondrial DNA sequencing analysis. The denatured condition is achieved by including urea in the electrophoresis matrix and formamide during sample preparations. Separation in capillaries is performed at higher voltages; the electric field (V/cm) is hundreds of times higher than that of a slab gel platform. Thus, the separation in capillary electrophoresis is rapid. However, samples can only be analyzed sequentially so the throughput is much more limited than with slab gels. However, this problem can be overcome by employing capillary array systems that can run up to 16 capillaries at one time.

One essential component of the capillary electrophoresis instrument is the capillary, a thin hollow tube made of fused silica with a translucent detection window for the instrument to detect signals from the labeled DNA fragments during electrophoresis. During electrophoresis, the capillary is connected to buffer reservoirs that are connected to electrodes. The injection of samples into the capillary is performed by an autosampler using an ***electrokinetic*** mechanism (injection based on charge of molecule). Only small quantities of sample are required for each injection and any remaining sample can be reused. Capillary electrophoresis instruments are equipped with detection systems utilizing laser excitation sources that excite the labeled DNA fragments. They also include fluorescence detectors that record the signals sent by the fragments (see Chapter 16).

## Bibliography

Biega, L. A., and B. W. Duceman. 1999. Substitution of $H_2O$ for formamide in the sample preparation protocol for STR analysis using the capillary electrophoresis system: The effects on precision, resolution, and capillary life. *J Forensic Sci* 44:3.

Budowle, B. et al. 2000. *DNA Typing Protocols: Molecular Biology and Forensic Analysis*. Natick, MA: Eaton Publishing.

Buel, E., M. B. Schwartz, and M. J. LaFountain. 1998. Capillary electrophoresis STR analysis: Comparison to gel-based systems. *J Forensic Sci* 43 (1):164.

Butler, J. M. 2005. *Forensic DNA Typing: Biology, Technology, and Genetics of STR Markers*, 2nd ed. Burlington, MA: Elsevier.

Butler, J. M. et al. 1994. Rapid analysis of the short tandem repeat HUMTH01 by capillary electrophoresis. *Biotechniques* 17 (6):1062.

Butler, J. M. et al. 2004. Forensic DNA typing by capillary electrophoresis using the ABI Prism 310 and 3100 genetic analyzers for STR analysis. *Electrophoresis* 25 (10–11):1397.

Butler, J. M., C. M. Ruitberg, and P. M. Vallone. 2001. Capillary electrophoresis as a tool for optimization of multiplex PCR reactions. *Fresenius J Anal Chem* 369 (3–4):200.

Frazier, R. R. et al. 1996. Validation of the Applied Biosystems Prism 377 automated sequencer for the forensic short tandem repeat analysis. *Electrophoresis* 17 (10):1550.

Gill, P., P. Koumi, and H. Allen. 2001. Sizing short tandem repeat alleles in capillary array gel electrophoresis instruments. *Electrophoresis* 22 (13):2670.

Isenberg, A. R. et al. 1998. Analysis of two multiplexed short tandem repeat systems using capillary electrophoresis with multiwavelength fluorescence detection. *Electrophoresis* 19 (1):94.

Issaq, H. J., K. C. Chan, and G. M. Muschik. 1997. The effect of column length, applied voltage, gel type, and concentration on the capillary electrophoresis separation of DNA fragments and polymerase chain reaction products. *Electrophoresis* 18 (7):1153.

Koumi, P. et al. 2004. Evaluation and validation of the ABI 3700, ABI 3100, and the MegaBACE 1000 capillary array electrophoresis instruments for use with short tandem repeat microsatellite typing in a forensic environment. *Electrophoresis* 25 (14):2227.

Lazaruk, K. et al. 1998. Genotyping of forensic short tandem repeat (STR) systems based on sizing precision in a capillary electrophoresis instrument. *Electrophoresis* 19 (1):86.

Madabhushi, R. S. 1998. Separation of four-color DNA sequencing extension products in noncovalently coated capillaries using low viscosity polymer solutions. *Electrophoresis* 19 (2):224.

Mansfield, E. S. et al. 1996. Sensitivity, reproducibility, and accuracy in short tandem repeat genotyping using capillary array electrophoresis. *Genome Res* 6 (9):893.

Mansfield, E. S. et al. 1998. Analysis of multiplexed short tandem repeat (STR) systems using capillary array electrophoresis. *Electrophoresis* 19 (1):101.

Rosenblum, B. B. et al. 1997. Improved single-strand DNA sizing accuracy in capillary electrophoresis. *Nucleic Acids Res* 25 (19):3925.

Roy, R. et al. 1996. Producing STR locus patterns from bloodstains and other forensic samples using an infrared fluorescent automated DNA sequencer. *J Forensic Sci* 41 (3):418.

Sgueglia, J. B., S. Geiger, and J. Davis. 2003. Precision studies using the ABI prism 3100 genetic analyzer for forensic DNA analysis. *Anal Bioanal Chem* 376 (8):1247.

Siles, B. A. et al. 1996. The use of a new gel matrix for the separation of DNA fragments: A comparison study between slab gel electrophoresis and capillary electrophoresis. *Appl Theor Electrophor* 6 (1):15.

Tereba, A., K. A. Micka, and J. W. Schumm. 1998. Reuse of denaturing polyacrylamide gels for short tandem repeat analysis. *Biotechniques* 25 (5):892.

Wang, Y. et al. 1995. Rapid sizing of short tandem repeat alleles using capillary array electrophoresis and energy-transfer fluorescent primers. *Anal Chem* 67 (7):1197.

Wenz, H. et al. 1998. High-precision genotyping by denaturing capillary electrophoresis. *Genome Res* 8 (1):69.

## Study Questions

1. During STR analysis, the resolution of the electrophoresis could be improved by a:
   (a) Non-denaturing electrophoresis system
   (b) Denaturing electrophoresis system

2. Preparing samples for capillary electrophoresis with an addition of formamide is intended to produce:
   (a) Non-denaturing DNA strands
   (b) Denaturing DNA strands

3. During capillary electrophoresis:
   (a) DNA migrates away from the anode and moves toward the cathode.
   (b) DNA migrates away from the cathode and moves toward the anode.

4. The laser used for STR fluorescence detection:
   (a) Excites the dye molecules
   (b) Measures the light intensity emitted from the dye molecules

5. During electrokinetic injection:
   (a) More samples may be drawn by increasing the voltage.
   (b) More samples may be drawn by increasing the volume.

6. The electrophoresis separation of STR samples is performed under certain conditions to improve size precision:
   (a) At room temperature
   (b) In the presence of heat

7. Which of the following is true for capillary electrophoresis?
   (a) As the salt level increases in the sample, fewer DNA molecules are injected
   (b) As the salt level increases in the sample, more DNA molecules are injected

8. The absence of electric current during capillary electrophoresis may be due to:
   (a) Low buffer levels in the reservoirs
   (b) Bubbles in the capillary
   (c) All of the above

# Detection Methods 16

A variety of techniques are available for the detection of DNA fragments in forensic DNA analysis. For example, direct detection of DNA fragments in a gel can be achieved via staining. The detection of DNA probes in a hybridization-based assay can be performed with radioisotopes, colorimetric assays, and chemiluminescence labeling. For polymerase chain reaction (PCR)-based assays, DNA primers can be labeled directly with fluorescent dyes.

## 16.1 Direct Detection of DNA in Gels

This section describes two simple and rapid detection methods used for detecting DNA: staining agarose gels with ethidium bromide and staining denatured polyacrylamide gels with silver.

### 16.1.1 Ethidium Bromide Staining

The location of DNA in an agarose gel can be determined directly by staining with low concentrations of fluorescent intercalating dyes such as ethidium bromide (Figure 16.1 and Figure 16.2). Ethidium bromide is a mutagen and carcinogen; safer fluorescent dyes such as SYBR® stains (Molecular Probes) can also be used. These techniques allow amounts of DNA as small as 10 ng per band to be detected in agarose gels. Staining of agarose gels can be achieved by (1) including ethidium bromide in the gel; or (2) staining the gel after electrophoresis in a solution containing ethidium bromide followed by a washing step known as destaining. Ethidium bromide-stained gels are documented with a standard ultraviolet transilluminator at ~300 nm using a camera with an orange filter.

### 16.1.2 Silver Staining

Electrophoretically separated DNA fragments can also be detected with silver nitrate staining. Silver staining of polyacrylamide gels has been used for the amplified fragment length polymorphism (AFLP) method of variable number tandem repeat (VNTR) profiling. The sensitivity of silver staining is approximately 100 times higher than that obtained with ethidium bromide and silver staining is less hazardous than ethidium bromide detection. Also,

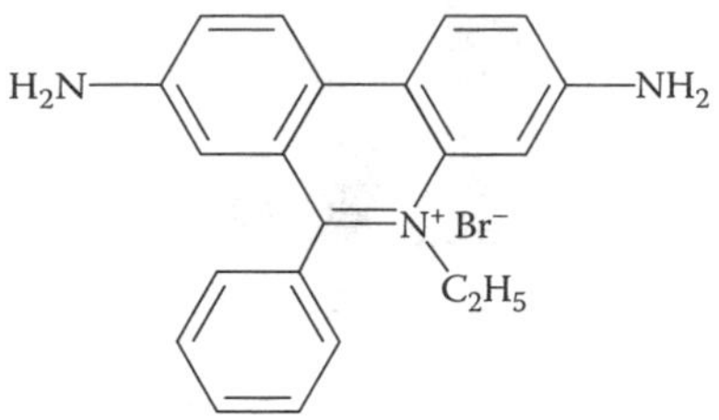

Figure 16.1 Chemical structure of ethidium bromide.

the developing chemicals are readily available at low cost.

Silver staining involves processing a gel followed by exposing it to a series of chemicals. First, the gel is submerged in a silver nitrate solution. The silver ions bind to the DNA and are subsequently reduced using formaldehyde to form a deposit of metallic silver on the DNA in the gel (Figure 16.3). A photograph of the gel with images of the silver-stained DNA strands is kept as a permanent record. Alternatively, the gels may be sealed and preserved for record purposes. One disadvantage of this method is that silver stains RNA and proteins along with DNA. The presence of restriction enzymes and polymerase should therefore be minimized. Additionally, bands from both complementary DNA strands may be detected in a denatured polyacrylamide gel, which leads to a two-band pattern.

## 16.2 Detection of DNA Probes in Hybridization-Based Assay

### 16.2.1 Radioisotope Labeled Probes

Radioisotope probe labeling was used for early versions of VNTR testing and DNA quantitation. Labeling can be accomplished in several ways based

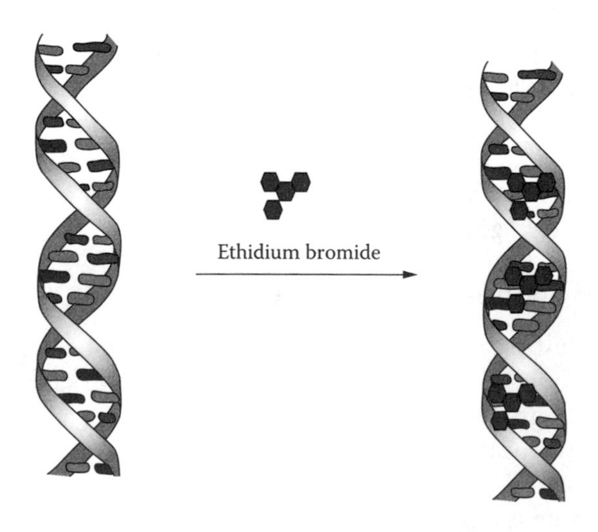


**Figure 16.2** Binding of intercalating agent to DNA. Ethidium bromide can be inserted into the DNA helix.

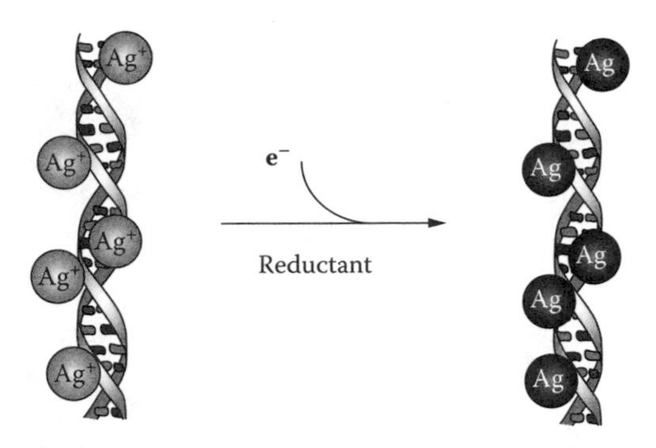


**Figure 16.3** Silver staining of DNA. Silver ($Ag^{+}$) ions bind to DNA and are reduced to metallic silver (Ag) to form dark particles.

on enzymatic reactions. For example, nick translation incorporates labeled deoxyribonucleotides (dNTPs) into double stranded DNA. DNase I is used to introduce single strand nicks within the DNA fragment to be labeled. Next, DNA Polymerase I recognizes the nicks and replaces the pre-existing nucleotides with new strands containing labeled dNTPs, resulting in the generation of $^{32}P$-labeled double stranded DNA molecules. $^{32}P$ is the most common radioisotope used in this technique. Nick translation can utilize any dNTP labeled with $^{32}P$. Prior to hybridization, the probe is denatured by boiling for a few minutes followed by rapid cooling on ice. $^{32}P$ has a very short half-life of approximately 14 days.

These probes can be visualized by exposing the DNA-containing membrane to x-ray film. The radioactive object is commonly placed in an x-ray cassette. The energy released from the decay of the radioisotopes is absorbed by the silver halide grains in the film emulsion and forms a latent image. A chemical development process amplifies the latent image and renders it visible on film (Figure 16.4). Because most $^{32}P$ emissions pass through the thin film emulsion without contributing to the final image, the detection process may require long exposure times. Signal intensity can be enhanced, however, by using intensifying screens at low temperatures. The screens emit photons upon receipt of radioactive β-particles, thus further enhancing signals. More sensitive x-ray film can also be used.

### 16.2.2 Enzyme-Conjugated Probe with Chemiluminescence Reporting System

One disadvantage of using $^{32}P$ is that it is a safety hazard. Additionally, autoradiography is a lengthy process. Therefore, non-radioisotopic detection methods have become popular alternative approaches. The use of alkaline

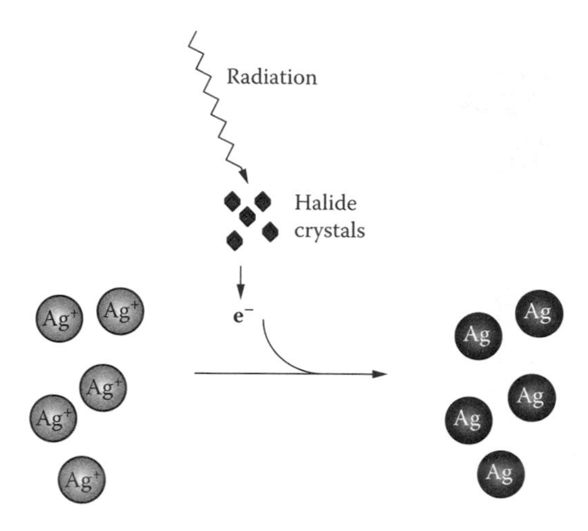


**Figure 16.4** Autoradiography. Exposure to radiation causes halide crystals to release electrons, thus reducing silver ($Ag^+$) ions to metallic silver (Ag).

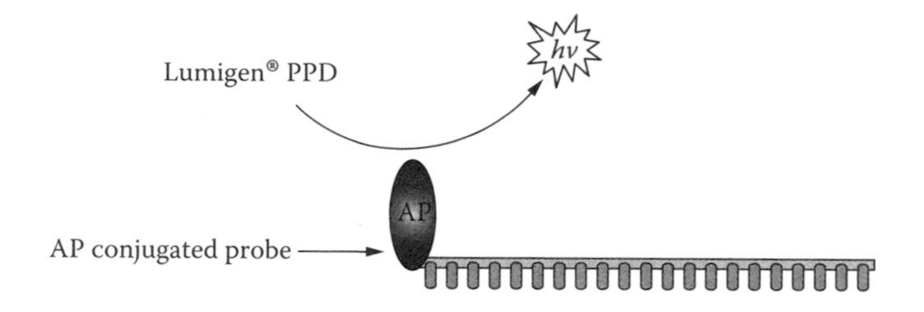


**Figure 16.5** Detection system using alkaline phosphatase conjugated probe. Chemiluminescence is generated by using the Lumigen® PPD as an alkaline phosphatase (AP) substrate.

phosphatase-conjugated probes with chemiluminescent substrates comprises a highly sensitive non-radioisotopic detection system.

***Alkaline Phosphatase*** can cleave the phosphate groups from a variety of substrate molecules. Its enzymatic activity can be measured using dioxetane-based chemiluminescent substrates such as Lumigen® PPD (Figure 16.5). The Lumi-Phos Plus kit of Lumigen Inc. contains this substrate and can serve as a detection system for slot blot assays for DNA quantitation (see chapter 13) and RFLP assays for VNTR profiling (see chapter 17). Alkaline phosphatase catalyzes the cleavage of the phosphate ester of Lumigen® PPD, resulting in the release of a photon (Figure 16.6). The Lumigen® PPD substrate yields a long lasting light emission that can be detected by exposure to x-ray film. This system provides a highly sensitive chemiluminescent detection method

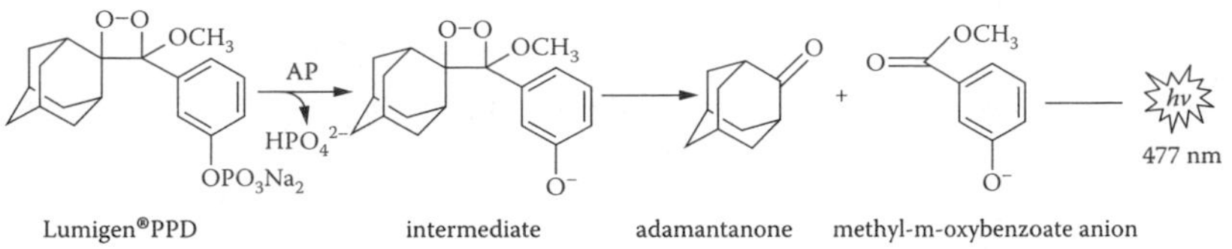

**Figure 16.6** Lumi-Phos Plus contains Lumigen® PPD (4-methoxy-4-[3-phosphate phenyl]spiro[1,2-dioxetane-3,2′-adamantane], disodium salt). Alkaline phosphatase catalyzes the removal of the phosphate group of Lumigen® PPD and generates a chemiluminescent intermediate that is subsequently broken down to the excited state of methyl-m-oxybenzoate. Decay of the excited state end product releases a photon.

**Table 16.1 Forensic Applications of Enzyme Reporting Systems for Detecting DNA**

| Reporter Enzyme | Labeling | Detection Mechanism | Substrate | Forensic Application |
|---|---|---|---|---|
| Alkaline phosphatase (AP) | AP-conjugated probe | Chemiluminescent | Lumigen®PPD | RFLP; blot assay for DNA quantitation |
| Horseradish peroxidase (HRP) | Biotinylated probe, recognized by streptavidin conjugated with HRP | Colorimetric/chemiluminescent | TMB/Luminol | Blot assay for DNA quantitation |
| | Biotinylated primer, recognized by streptavidin conjugated with HRP | Colorimetric | TMB | Reverse blot assay for DQA1 typing and for mtDNA typing |

for alkaline phosphatase-conjugated DNA probes in solution or on a solid matrix such as a membrane (Table 16.1).

## 16.2.3 Biotinylation of DNA with Colorimetric Reporting System

### *16.2.3.1 Biotin*

Biotin, also known as vitamin H, is a water-soluble molecule found in egg yolk (Figure 16.7). It can be incorporated onto oligonucleotide probes without interfering with the ability of probes to hybridize because of its small size (molecular weight: 244.31). Signals from a biotinylated probe can be detected

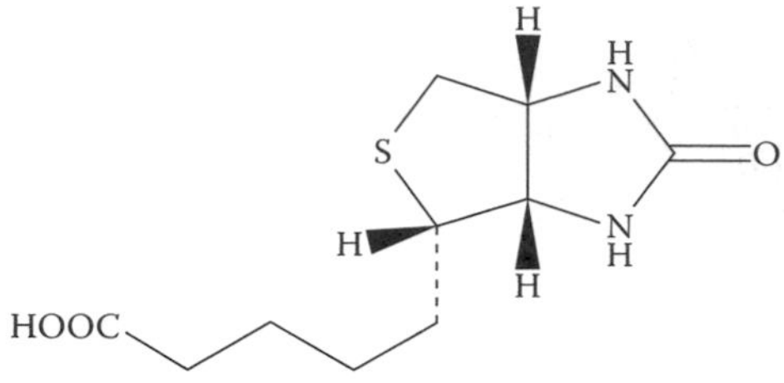

**Figure 16.7** Chemical structure of biotin, also known as vitamin H. Molecular formula: $C_{10}H_{16}N_2O_3S$. Molecular weight: 244.31.

with an enzyme-conjugated avidin system. Two steps are required to detect biotin-labeled probes. First, an avidin conjugate consisting of a reporter enzyme is added. Then, the reporter enzyme is assayed with substrates.

#### *16.2.3.2 Enzyme-Conjugated Avidin*

Avidin is a glycoprotein found in egg white; it binds to biotin with extremely high affinity. Thus, a biotin–avidin complex is very stable. However, avidin detection has a high background due to nonspecific binding. The nonspecific binding can be reduced by replacing avidin with its streptavidin counterpart from *Streptomyces avidinii*. To detect binding, an enzyme-conjugated streptavidin such as horseradish peroxidase (HRP)-conjugated streptavidin can be used. HRP isolated from horseradish roots contains heme residues that catalyze the oxidation reactions of substrates (see chapter 6).

#### *16.2.3.3 Reporter Enzyme Assay*

HRP can be assayed with colorimetric, chemiluminescent, or fluorogenic substrates. One common colorimetric substrate for forensic DNA testing is 3,3′,5,5′-tetramethylbenzidine (TMB) which is oxidized by the peroxidase to form an insoluble precipitate of intense blue at an acidic pH (Figure 16.8). Because the colored precipitate is difficult to remove from membranes, TMB is not suitable if reprobing for RFLP analysis is required. This technique has been used for forensic DNA testing such as slot blot assays for DNA quantitation (Table 16.1). A chemiluminescent substrate such as a Luminol-based reagent can also be employed with HRP. The peroxidase catalyzes the oxidation of Luminol to form a chemiluminescent product (Figure 16.8).

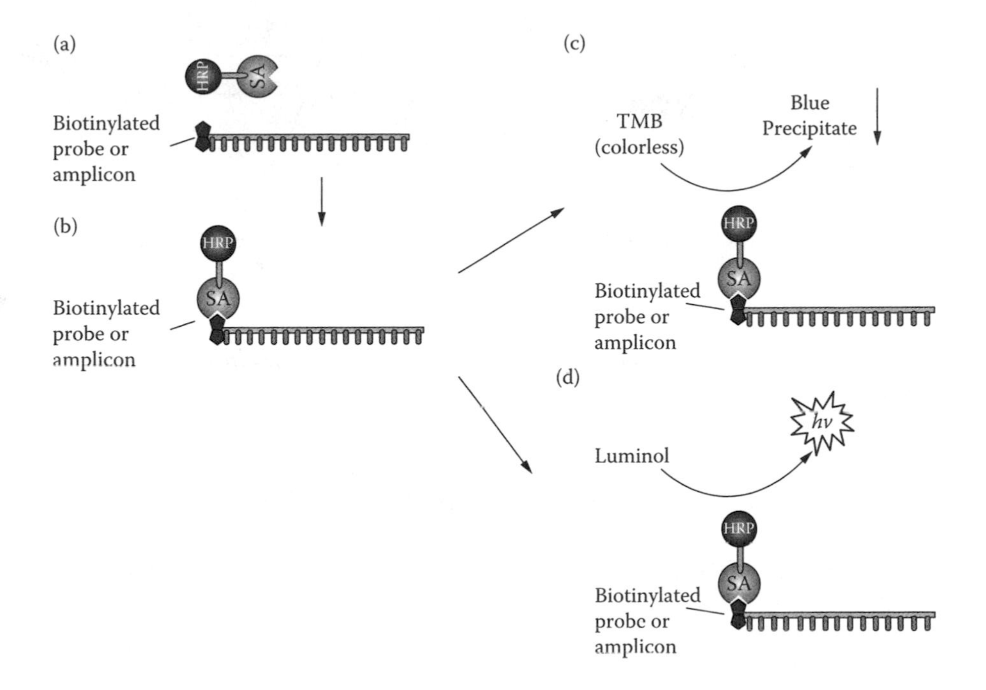


**Figure 16.8** Detection system using biotinylated DNA probes with colorimetric and chemiluminescent reactions. (a) Biotinylated probe is incubated with a streptavidin (SA) and horseradish peroxidase (HRP) conjugate complex. (b) Biotin is recognized by the complex. Reporter enzyme assays can be carried out using either a colorimetric reaction with a TMB substrate (c) or a chemiluminescent reaction using Luminol analogs (d).

## 16.3 Detection Methods for PCR-Based Assays

### 16.3.1 Fluorescence Labeling

#### *16.3.1.1 Fluorescent Dyes*

Advantages of fluorescence detection methods include a higher sensitivity and broader dynamic range than comparable colorimetric detection methods. Furthermore, they have the capacity for simultaneous multiparameter analysis of complex samples such as multiplex PCR products with different fluorescent labels (Figure 16.9). Common fluorescent dyes used in DNA labeling emit fluorescence in the range of 400 to 600 nm (Table 16.2; Figure 16.10).

#### *16.3.1.2 Labeling Methods*

Fluorescent dye labeling can be incorporated into a DNA fragment using a 5′-end fluorescently labeled oligonucleotide primer (Figure 16.11). The dye-

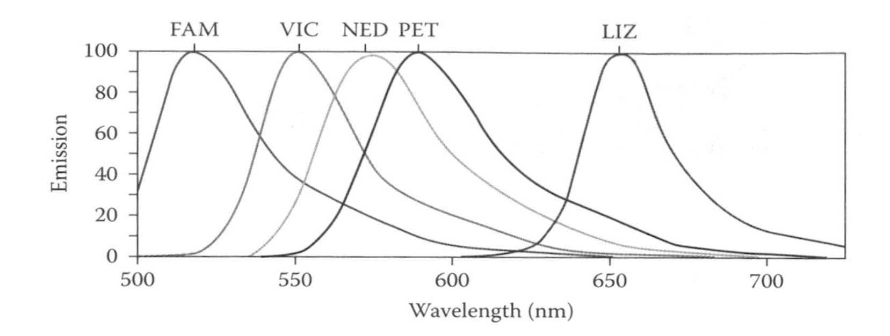


**Figure 16.9** Emission spectra of common fluorescent dyes used for forensic DNA analysis.

**Table 16.2 Properties of Fluorescent Dyes and Their Applications in Commercial STR Kits**

| Fluorescent Dye | Chemical Name | Excitation Maxima (nm) | Emission Maxima (nm) | Indentifiler (Applied Biosystems) | PowerPlex (Promega) |
|---|---|---|---|---|---|
| FAM | Carboxyfluorescein | 494 | 518 | ☑ | |
| Fluorescein | 3′,6′-dihydroxyspiro [isobenzofuran-1(3H),9'-(9H) xanthen]-3-one | 492 | 520 | | ☑ |
| SYBR® Green 1 | | 494 | 521 | | |
| SYBR® Gold | | 495 | 537 | | |
| TET | Carboxy-2′4,7,7′ -tetrachlorofluorescein | 521 | 544 | | |
| JOE | Carboxy-4′,5′-dichloro-2′,7′ -dimethoxyfluorescein | 520 | 548 | | ☑ |
| VIC | | 538 | 554 | ☑ | |
| HEX | Carboxy-2′,4,4′,5′,7,7′ -hexachlorofluorescein | 535 | 556 | | |
| NED | | 546 | 575 | ☑ | |
| Rhodamine Red | | 580 | 590 | | |
| PET | | 558 | 595 | ☑ | |
| Texas Red | | 583 | 603 | | |
| TMR (TAMRA) | Carboxytetrame-thylrhodamine | 565 | 605 | | ☑ |
| ROX (CXR) | Carboxy-X-rhodamine | 585 | 605 | | ☑ |
| LIZ | | 638 | 655 | ☑ | |

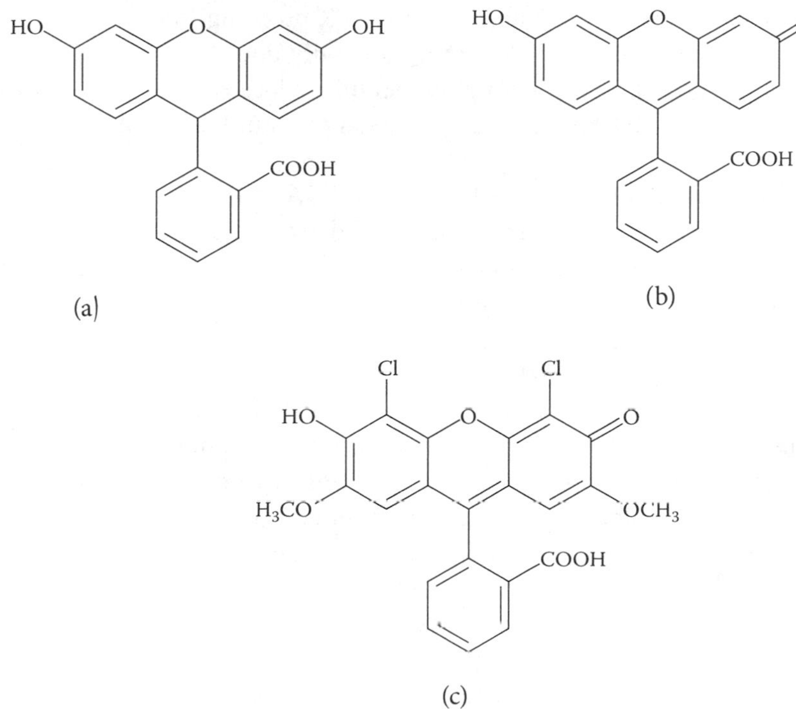

(a)

(b)

(c)

**Figure 16.10** Chemical structures of representative fluorescent dyes. (a) fluorescein, (b) FAM, (c) JOE.

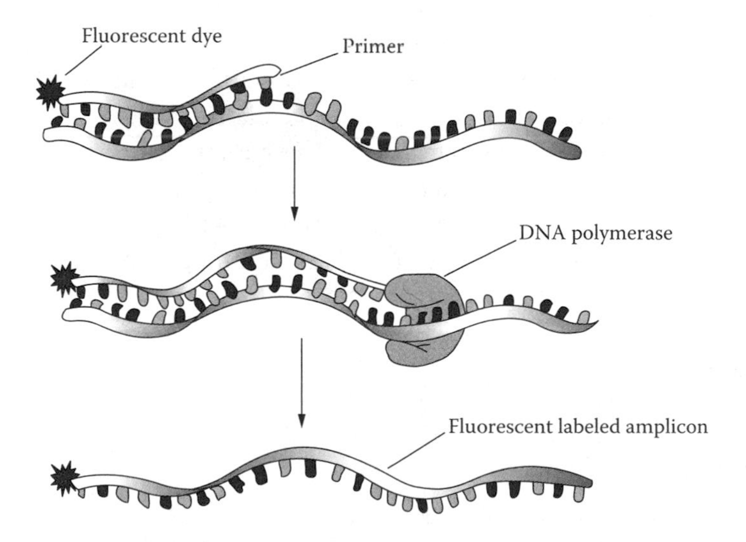


**Figure 16.11** A fluorescent dye labeled primer can be used for the amplification of DNA. The dye is conjugated at the 5′ end of the primer. The amplified product is fluorescently labeled.

labeled primer method is usually used for STR profiling (see chapter 18) in which only one primer is labeled; therefore, only one strand is visible. The two-band pattern observed with silver staining does not appear with this method. Additionally, dye-labeled primers allow multiplex PCR amplifications in the same tube.

Alternatively, fluorescent dye labeling of DNA fragments can be carried out by incorporating fluorescently labeled dideoxynucleotides (ddNTPs) in the PCR product. This labeling method is usually used in DNA sequencing such as mtDNA sequence profiling (see chapter 21).

#### 16.3.1.3 Fluorophore Detection

***Fluorophore*** is a component of a florescent dye molecule which causes the molecule to be fluorescent. First, a laser strikes a fluorophore attached to the end of a DNA fragment. A photon from a laser source excites an electron of the fluorophore of a labeled DNA fragment from its ground energy state to an excited transition state. The excited electron then descends to its ground state and releases a photon. The emitted photon has a higher wavelength than the excitation photon. The absorption and emission wavelengths vary depending on chemical structures and conditions.

Lasers are commonly used as ***excitation sources*** because laser light emissions have high intensity and are monochromatic (single wavelength). The argon ion gas laser is frequently used in applications such as fluorescence-labeled STR and mtDNA sequence analysis because typical fluorescent dyes match the excitation properties of the argon laser.

***Optical filters*** are used to detect light emitted only at a particular wavelength. Signals from multiple fluorophores in the same sample can be recorded separately using optical filters and a mathematical ***matrix*** (fluorophore separation algorithm). The function of a matrix is to subtract the color overlaps of the various fluorescent dyes. The matrix can be established by analysis of standard samples and also by using data analysis software.

A ***fluorescence detector*** is a photosensitive device that measures the light intensity emitted from a fluorophore with a photomultiplier tube or charge-coupled device (CCD). The signal from the fluorophore is collected and converted to an electronic signal expressed in an arbitrary unit such as a ***relative fluorescence unit*** (RFU).

## Bibliography

Bassam, B. J., G. Caetano-Anolles, and P. M. Gresshoff. 1991. Fast and sensitive silver staining of DNA in polyacrylamide gels. *Anal Biochem* 196 (1):80.

Bronstein, I., and P. McGrath. 1989. Chemiluminescence lights up. *Nature* 338 (6216):599.

Bronstein, I. et al. 1990. Rapid and sensitive detection of DNA in Southern blots with chemiluminescence. *Biotechniques* 8 (3):310.

Brunk, C. F., and L. Simpson. 1977. Comparison of various ultraviolet sources for fluorescent detection of ethidium bromide-DNA complexes in polyacrylamide gels. *Anal Biochem* 82 (2):455.

Budowle, B. et al. 1995. Simple protocols for typing forensic biological evidence: Chemiluminescent detection for human DNA quantitation and restriction fragment length polymorphism (RFLP) analyses and manual typing of polymerase chain reaction (PCR) amplified polymorphisms. *Electrophoresis* 16 (9):1559.

Budowle, B. et al. 2000. *DNA Typing Protocols: Molecular Biology and Forensic Analysis*. Natick, MA: Eaton Publishing.

Buel, E., and M. Schwartz. 1995. The use of DAPI as a replacement for ethidium bromide in forensic DNA analysis. *J Forensic Sci* 40 (2):275.

Butler, J. M. 2005. *Forensic DNA Typing: Biology, Technology, and Genetics of STR Markers*, 2nd ed. Burlington, MA: Elsevier.

Decorte, R., and J. J. Cassiman. 1996. Evaluation of the ALF DNA sequencer for high-speed sizing of short tandem repeat alleles. *Electrophoresis* 17 (10):1542.

Feinberg, A. P., and B. Vogelstein. 1983. A technique for radiolabeling DNA restriction endonuclease fragments to high specific activity. *Anal Biochem* 132 (1):6.

French, B. T., H. M. Maul, and G. G. Maul. 1986. Screening cDNA expression libraries with monoclonal and polyclonal antibodies using an amplified biotin-avidin-peroxidase technique. *Anal Biochem* 156 (2):417.

Gillespie, P. G., and A. J. Hudspeth. 1991. Chemiluminescence detection of proteins from single cells. *Proc Natl Acad Sci USA* 88 (6):2563.

Giusti, A. M., and B. Budowle. 1995. A chemiluminescence-based detection system for human DNA quantitation and restriction fragment length polymorphism (RFLP) analysis. *Appl Theor Electrophor* 5 (2):89.

Green, N. M. 1975. Avidin. *Adv Protein Chem* 29:85.

Green, N. M., and E. J. Toms. 1973. The properties of subunits of avidin coupled to sepharose. *Biochem J* 133 (4):687.

Johnson, E. D., and T. M. Kotowski. 1996. Chemiluminescent detection of RFLP patterns in forensic DNA analysis. *J Forensic Sci* 41 (4):569.

Laskey, R. A., and A. D. Mills. 1977. Enhanced autoradiographic detection of $^{32}P$ and $^{125}I$ using intensifying screens and hypersensitized film. *FEBS Lett* 82 (2):314.

LePecq, J. B., and C. Paoletti. 1967. A fluorescent complex between ethidium bromide and nucleic acids: Physical–chemical characterization. *J Mol Biol* 27 (1):87.

Mansfield, E. S., and M. N. Kronick. 1993. Alternative labeling techniques for automated fluorescence based analysis of PCR products. *Biotechniques* 15 (2):274.

Mayrand, P. E. et al. 1992. The use of fluorescence detection and internal lane standards to size PCR products automatically. *Appl Theor Electrophor* 3 (1):1.

McKimm-Breschkin, J. L. 1990. The use of tetramethylbenzidine for solid phase immunoassays. *J Immunol Methods* 135 (1-2):277.

Merril, C. R. 1990. Silver staining of proteins and DNA. *Nature* 343 (6260):779.

Mitchell, L. G., A. Bodenteich, and C. R. Merril. 1994. Use of silver staining to detect nucleic acids. *Methods Mol Biol* 31:197.

Morin, P. A., and D. G. Smith. 1995. Nonradioactive detection of hypervariable simple sequence repeats in short polyacrylamide gels. *Biotechniques* 19 (2):223.

Renz, M., and C. Kurz. 1984. A colorimetric method for DNA hybridization. *Nucleic Acids Res* 12 (8):3435.

Roberts, I. et al. 1991. A comparison of the sensitivity and specificity of enzyme immunoassays and time-resolved fluoroimmunoassay. *J Immunol Methods* 143 (1):49.

Southern, E. M. 1975. Detection of specific sequences among DNA fragments separated by gel electrophoresis. *J Mol Biol* 98 (3):503.

Stone, T., and I. Durrant. 1991. Enhanced chemiluminescence for the detection of membrane-bound nucleic acid sequences: Advantages of the Amersham system. *Genet Anal Tech Appl* 8 (8):230.

Tizard, R. et al. 1990. Imaging of DNA sequences with chemiluminescence. *Proc Natl Acad Sci USA* 87 (12):4514.

Tuma, R. S. et al. 1999. Characterization of SYBR Gold nucleic acid gel stain: A dye optimized for use with 300-nm ultraviolet transilluminators. *Anal Biochem* 268 (2):278.

Waring, M. J. 1965. Complex formation between ethidium bromide and nucleic acids. *J Mol Biol* 13 (1):269.

Watkins, T. I., and G. Woolfe. 1952. Effect of changing the quaternizing group on the trypanocidal activity of dimidium bromide. *Nature* 169 (4299):506.

## Study Questions

1. Which of the following is the substrate of horseradish peroxidase?
   (a) Luminol and its derivatives
   (b) TMB
   (c) All of the above

2. Which of the following is the substrate of alkaline phosphatase?
   (a) Luminol and its derivatives
   (b) TMB
   (c) All of the above

3. Which of the following is usually labeled in an alkaline phosphatase-based detection method?
   (a) A probe
   (b) A primer
   (c) Both of the above

4. Which of the following is usually labeled with biotin for a hybridization-based assay?
   (a) A probe
   (b) A primer
   (c) All of the above

5. Which of the following is usually labeled with a fluorescent dye for a PCR-based assay?
   (a) Dideoxynucleotides
   (b) A primer
   (c) Both of the above

# Forensic DNA Profiling

# V

# Variable Number Tandem Repeat Profiling

# 17

The human genome is abundant in tandem repeats. ***Minisatellites*** were first defined as a class of tandem repeats in the 1980s. Some of these repeats share a GC-rich core sequence. Subsequently, tandem repeats with higher AT contents of core sequence have been documented. The minisatellites are also called ***variable number tandem repeats*** (***VNTRs***) as shown in Figure 17.1. The repeat unit length of a VNTR can range from several to hundreds of base pairs (bp). The tandem repeat arrays can be kilobases (Kb, corresponding to $10^3$ bp) long, and the numbers of tandem repeat units in some VNTR loci are highly variable, leading to variable lengths of DNA fragments. A genotype is defined by a particular number of tandem repeat units at a given locus.

## 17.1 VNTR Loci for Forensic Testing

Table 17.1 lists the common VNTR loci. The chromosomal location of each locus should not be linked. However, loci located far apart on the same chromosome or on different chromosomes can be used. The loci selected should be compatible with the restriction endonuclease cleavage sites as described below. Many VNTR loci used for forensic applications are highly polymorphic. Hundreds of different genotypes per locus can be observed among the population. The discriminating power of VNTR loci used for forensic testing can be measured by ***population match probability*** (*P*m). The lower the *P*m, the less likely a match will occur between two randomly chosen individuals. A *P*m of up to $10^{-12}$ can be achieved by testing several VNTR loci.

## 17.2 Restriction Fragment Length Polymorphism (RFLP)

VNTR profiling utilizes RFLP—the first method used in forensic DNA testing (Figure 17.2). It employs ***restriction endonucleases*** that recognize and cleave specific sites along the DNA sequence. Cleavage of a DNA sample with a particular restriction enzyme results in a reproducible set of restriction fragments of various lengths. Appropriate restriction endonucleases should be selected so that the genomic DNA is cleaved at sites that flank the VNTR

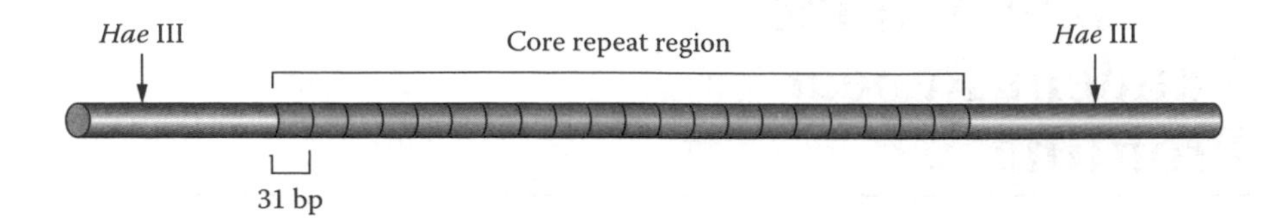


**Figure 17.1** VNTR locus D2S44 (2q21.3-2q22). Each repeat unit consists of 31 bp. *Hae*III = restriction site.

**Table 17.1 Common VNTR Loci**

| Locus | Chromosome Location | Repeat Unit Length (bp) | *HaeIII* Fragment Size (kb) | Probe |
|---|---|---|---|---|
| D1S7 | 1 | 9 | 0.5–12 | MS1 |
| D2S44 | 2 | 31 | 0.7–8.5 | yNH24 |
| D4S139 | 4 | 31 | 2–12 | pH30 |
| D10S28 | 10 | 33 | 0.4–10 | pTBQ7 |
| D14S13 | 14 | 15 | 0.7–12 | pCMM101 |
| D16S85 | 16 | 17 | 0.2–5 | 3'HVR |
| D17S26 | 17 | 18 | 0.7–11 | pEFD52 |
| D17S79 | 17 | 38 | 0.5–3 | V1 |

*Source:* Adapted from Budowle, B. et al. 2000. *DNA Typing Protocols: Molecular Biology and Forensic Analysis.* Natick, MA: Eaton Publishing; Office of Justice Programs. 2000. Future of forensic DNA testing: Predictions of the Research and Development Working Group. National Institute of Justice, U.S. Department of Justice.

core repeat region. The resulting fragments are then separated according to their sizes by gel electrophoresis through a standard agarose gel.

The DNA is then processed using the ***Southern transfer and hybridization*** technique. The DNA is denatured and transferred from the gel to a supporting matrix such as a nylon or nitrocellulose membrane. The DNA immobilized on the membrane is then hybridized with a labeled ***probe***. Only bands complementary to the probe are recognized by detection systems such as autoradiography. Using the RFLP technique, the length variations among restriction sites can be detected. Most forensic applications focus on the length variations of VNTR regions located between two restriction sites.

In summary, the RFLP method includes genomic DNA preparation, restriction endonuclease digestion of the genomic DNA into fragments, agarose gel electrophoretic separation of the DNA fragments according to size, transfer of DNA fragments, hybridization with locus-specific probes using

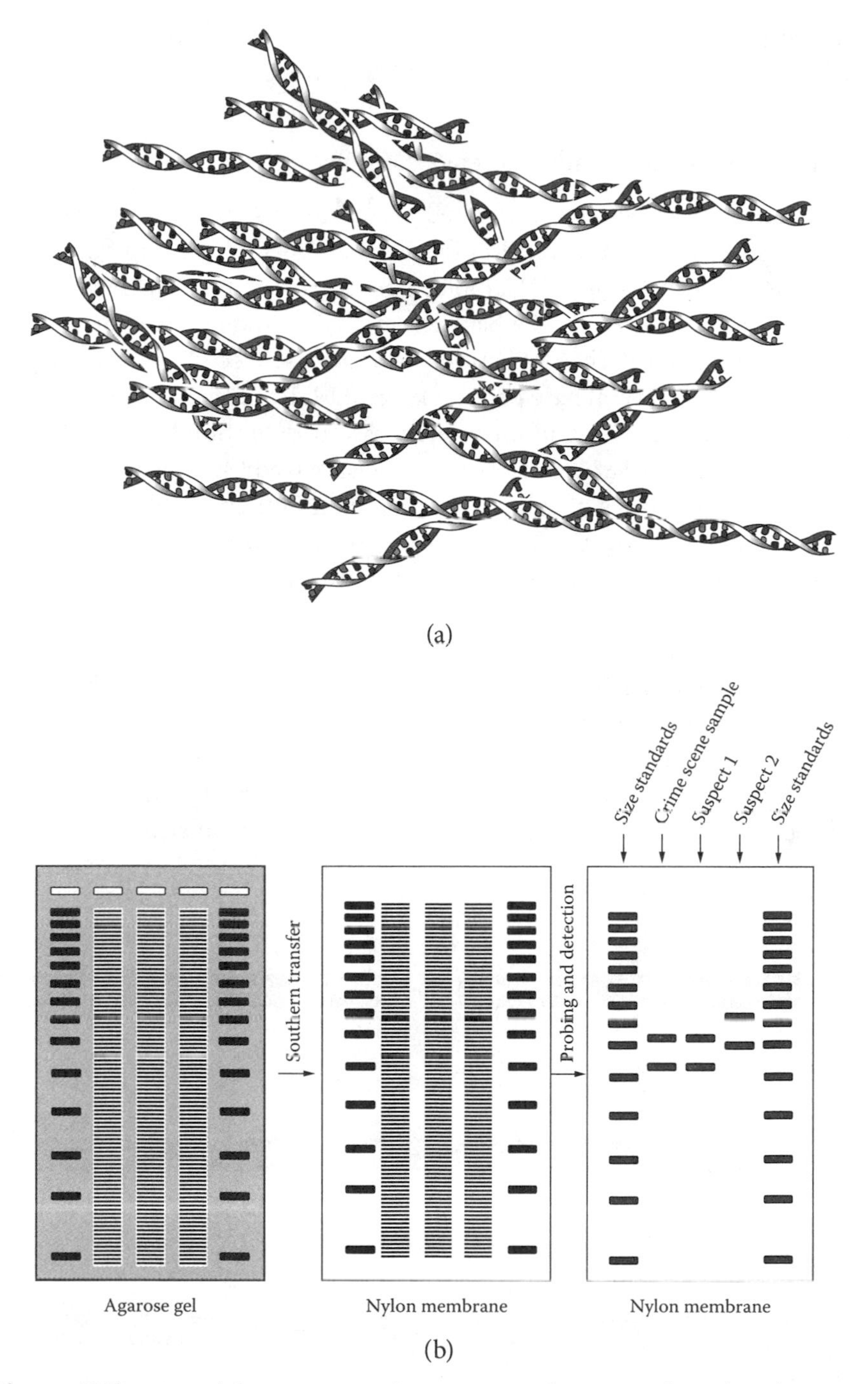


**Figure 17.2** RFLP. (a) Restriction fragments with various lengths of genomic DNA. (b) Restriction fragments are separated by gel elelctrophoresis. DNA is transferred to a solid phase and probed. The signal is detected and the DNA fragment of interest can be observed. Band patterns of heterozygous loci of an individual are shown.

Southern transfer and a hybridization method, and detection of locus-specific bands by autoradiography or chemiluminescence.

### 17.2.1 Restriction Endonuclease Digestion

Restriction endonucleases are isolated from various prokaryotic bacteria. These enzymes protect the bacteria from phage infections using host endonucleases to destroy foreign DNA molecules. To protect their own DNA from digestion by endonucleases, the bacteria methylate the cytosine or adenine residues of their DNA. The restriction endonucleases recognize and cleave a short motif that is several base pairs in length. Many of them have their own specific palindromic recognition sequences. To date, hundreds of restriction endonucleases have been described. They are traditionally classified into three types on the basis of subunit composition and enzymatic properties. Type II enzymes cleave DNA at defined positions near to or within their recognition sequences and they are commonly used in molecular research laboratories. The type II restriction endonucleases are also used for forensic DNA testing (Figure 17.3).

The preferred restriction endonucleases for forensic use are those that cleave at the flanking regions of VNTR repeat units but not within the repeat units. Moreover, the DNA fragments from the restriction digestion can be separated by agarose electrophoresis. The enzymatic activities are not affected by the methylation of human DNA. The VNTR loci must be compatible

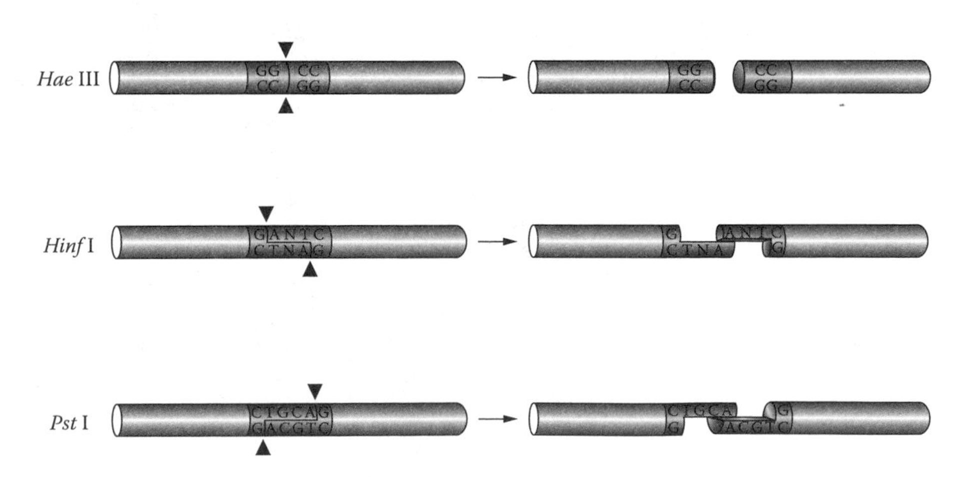


**Figure 17.3** Restriction sites for *Hae*III, *Hinf*I, and *Pst*I. *Hae*III digestion produces a blunt end. *Hinf*I and *Pst*I digestions produce sticky ends. N = any nucleotide.

with the restriction endonucleases selected for RFLP analysis. The core repeat sequences of the VNTR should not contain the restriction site for the enzyme used in the assay.

Most forensic laboratories use a single restriction endonuclease for a panel of VNTR loci because the DNA in evidence samples is often insufficient for performing multiple tests with different restriction endonucleases. For instance, *Hinf*I-based RFLP was commonly used in European forensic laboratories, and *Hae*III-based RFLP was used in North American and some European forensic laboratories. Other restriction endonuclease-based RFLPs such as *Pst*I are available. Several VNTR loci are suitable for RFLP analysis with these restriction endonucleases. The use of common restriction endonucleases allows the comparison of data of various laboratories.

*Hae*III isolated from *Haemophilus aegypticus* presents several advantages for forensic RFLP analysis. It recognizes a four-base sequence, 5′-GGCC-3′, and cleaves the DNA between the internal G and C residues of the recognition site (GG/CC). *Hinf* I recognizes a five-base restriction site and *Pst*I recognizes a six-base restriction site. Hypothetically, the number of four-base restriction sites is likely to occur more often than five- and six-base restriction sites in the human genome. Thus, *Hae*III restriction sites will occur more frequently than *Hinf* I and *Pst*I sites. Furthermore, *Hae*III-cleaved DNA fragments are smaller than those of *Hinf* I and *Pst*I, so the *Hae*III-generated VNTR allele sizes are easier to separate using conventional agarose gel electrophoresis. *Hae*III's enzymatic activity is not affected by the methylation of human genomic DNA. Its enzymatic activities also appear unaffected when a reaction proceeds under nonoptimal conditions. Additionally, low star activity is observed.

The restriction endonuclease digested DNA samples are separated in an agarose gel, also called an ***analytic gel***. After electrophoresis, a smear of various sizes of DNA fragments can be observed.

### 17.2.2 Southern Transfer and Hybridization

Also known as ***Southern blotting***, this technique was named after Sir Edwin Southern who developed it in the United Kingdom in the mid-1970s. The method can be used to transfer DNA to a solid matrix so that it can be detected with a probe. This method is still used today in many research laboratories. To transfer the DNA from the agarose gel, the DNA in the gel must be denatured into single-stranded DNA which is then transferred by capillary action to a solid matrix such as a piece of nylon membrane. The single-stranded DNA fragments transferred can be immobilized on a nylon membrane by an ultraviolet cross-linking process (Figure 17.4).

### 17.2.3 Hybridization

Two types of probes exist for VNTR analysis. The ***multi-locus probe*** (***MLP***) technique can detect multiple VNTR loci simultaneously (Figure 17.5). As discussed earlier, some VNTRs form groups in the human genome and share a short GC-rich core sequence of 10 to 15 bp. The MLP consists of this core sequence and produces a complex barcode-like band pattern generated from multiple VNTR loci (Figure 17.6).

MLP was pioneered by Sir Alec Jeffreys in 1984 at the University of Leicester in the United Kingdom and called ***DNA fingerprinting***. Because of its excellent discriminating power, the method was used for parentage testing in immigration disputes with great success. The disadvantages of the MLP approach are that the interpretation of a mixed DNA sample from more than one individual is nearly impossible and MLP does not perform well for degraded or limited quantities of DNA from crime scene samples. For these reasons, MLP analysis has not been widely employed in forensic DNA laboratories.

The solution is the development of probes that recognize a specific region of genomic DNA at a VNTR locus. The technique is called the ***single locus***

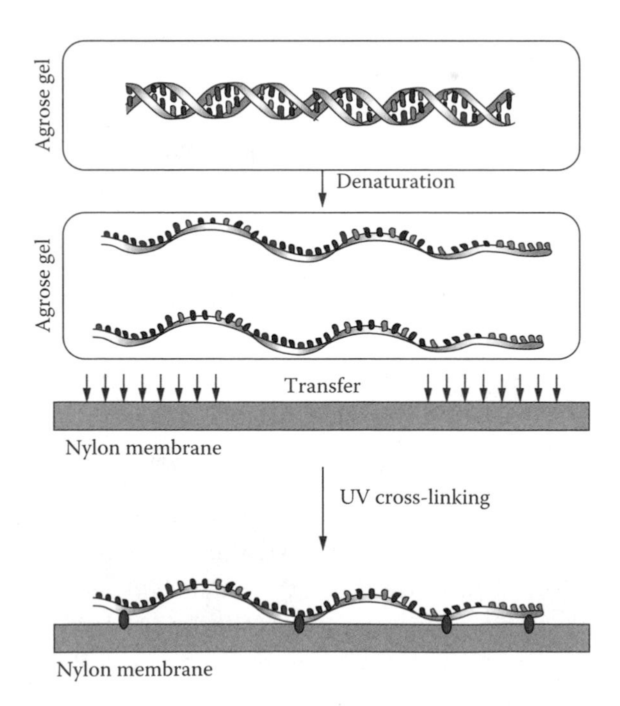


**Figure 17.4** Southern blotting. DNA in agarose gel is denatured into single-stranded DNA and transferred to a solid phase membrane where the single-stranded DNA is immobilized upon ultraviolet cross-linking.

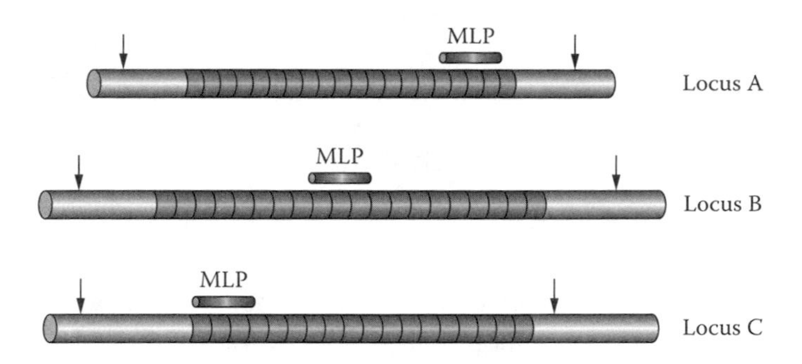


**Figure 17.5** VNTR analysis using MLP. The technique can detect multiple VNTR loci simultaneously. Restriction sites are indicated by arrows.

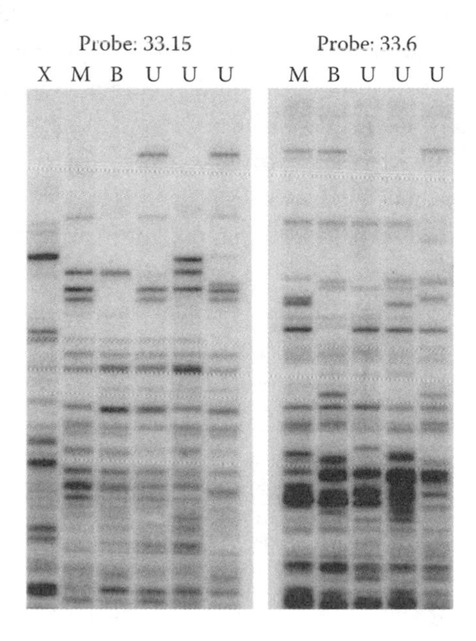


**Figure 17.6** First application of DNA fingerprinting. Two MLPs were used to analyze samples for an immigration case. M = mother. U = three undisputed children. B = male child in dispute. All bands in B can be traced back to M or U. X is an unrelated individual (*Source:* Jeffreys, A. J. 2005. *Nat Med* 11 (10):1035. With permission).

***probe* (*SLP*)** as depicted in Figure 17.7. SLP generates a simple pattern called a ***DNA profile***, consisting of one band for a homozygote and two bands for a heterozygote, respectively. In order to improve the discriminating power of the test, SLP analyses of different VNTR loci can be performed by using different probes sequentially with a single locus at one time.

**Figure 17.7** VNTR analysis using SLP. The technique can detect a single VNTR locus. Restriction sites are indicated by arrows.

SLP is more sensitive than MLP. Furthermore, it can analyze mixed DNA samples from two or more contributors. The sizes of fragments can be estimated and converted to a numerical form applicable for databasing. Therefore, the samples do not have to be analyzed in the same agarose gel for comparisons as required with MLP.

This technique led to the solving of a double murder case in Leicestershire in the 1980s. The case was the first to apply DNA evidence to a criminal investigation. DNA profiling identified the true perpetrator and also excluded an innocent suspect (Figure 17.8). This case demonstrated the great potential of DNA profiling in forensic investigation. SLP then became a common method in most forensic laboratories.

To detect VNTR loci, a labeled SLP probe containing the sequence of the VNTR region is used. The probe is then hybridized to the complementary target sequence of DNA immobilized on the membrane. Any unbound probes are washed away so that they do not interfere with the signal. Two types of detection systems are used for VNTR analysis. Radioisotope labeling, such as with a $^{32}$P-labeled probe, can be used. The hybridized probe can be detected by exposing the membrane to a sheet of x-ray film to generate an autoradiograph. Alternatively, an enzyme-conjugated probe can also be used. Alkaline phosphatase is an example of an enzyme used in this type of probe. Its enzymatic activity can be detected with chemiluminescent substrates. A chemiluminescent signal can be detected by exposure to x-ray film as well.

In forensic DNA analysis, several loci are commonly analyzed sequentially using the same membrane. Once the analysis of the first probe is completed, it is removed by a procedure called stripping. The nylon membrane is then hybridized to the next probe and the process is repeated for each probe to be tested.

Typically, a size standard is run on each gel. Band sizes can thus be estimated by comparison to size standards. However, the VNTR alleles that differ by only one or two repeat units are usually not distinguishable. For this reason, bands of similar size are grouped into ***bins*** and genotypes can be determined using the ***fixed bin*** method. A sample with a known VNTR genotype is also run on each gel as a positive control. In the United States, a genomic DNA sample from cell line K562 is used.

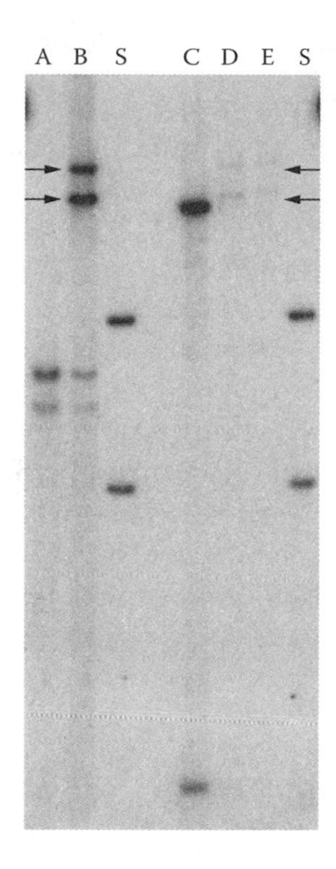


**Figure 17.8** First application of DNA profiling in a criminal investigation. SLP was used to analyze samples. A = hair roots from the first victim. B = mixture of semen and vaginal fluid from first victim. C = blood from second victim. D = vaginal swab from second victim. E = semen stain on clothing from second victim. S = blood from suspect. Alleles (arrows) are matched with the profiles of the two cases but not with the suspect profile (*Source:* Jeffreys, A. J. 2005. *Nat Med* 11 (10):1035. With permission).

DNA samples to be compared are run in parallel lanes on the same gel. The band locations are compared from lane to lane to identify similar patterns. If the VNTR fragments are at corresponding positions, they are declared a match. If not, the result is a nonmatch and the two DNA samples can be deemed to have come from different origins. The following possible conclusions can be made:

- Exclusion (profiles are different)
- Inclusion (profiles match)
- No conclusion can be made from the existing data

Chapter 23 evaluates and discusses the strength of the results.

### 17.2.4 Factors Affecting RFLP Results

Sample conditions, genetic mutations, and artifacts appearing during testing can affect VNTR profiling results and consequently impact data interpretation.

#### *17.2.4.1 DNA Degradation*

RFLP analysis requires the genomic DNA to be intact. DNA degradation results in damage such as creating nicks and breaks in the strand. The more severe the degradation, the smaller the average size of the DNA. When the average size becomes too small, the allele may not be detected. Many VNTR tandem arrays can span several Kb in length. In theory, large sized alleles are more likely to be affected by degradation than smaller sized alleles at a different locus. The smaller alleles may be not affected at all.

A two-banded heterozygous profile might be observed as a one-banded homozygous profile if the larger band is missed due to degradation. This artifact could lead to a false determination of exclusion. However, DNA degradation can be detected prior to conducting RFLP by the use of agarose gel electrophoresis also known as a ***yield gel***. The average size of the DNA can be estimated by comparison to a size standard run on the same gel.

#### *17.2.4.2 Restriction Digestion-Related Artifacts*

**17.2.4.2.1 Partial Restriction Digestion** Complete restriction digestion should be achieved for RFLP analysis. If partial digestion occurs, the partially cleaved DNA strands are longer than the cleaved fragments (Figure 17.9). Thus, partial digestion results in a mixture of fragments with correct sizes and slightly larger fragments. Under this condition, a larger sized uncleaved band, usually lower in intensity than the true bands, could be observed. The multibanded pattern due to partial digestion could be observed at other loci analyzed in the same nylon membrane.

Detection of more than two bands in an RFLP profile may lead to a false interpretation and be incorrectly concluded to be a mixture. However, partial digestion can be detected. A small portion of a sample can be analyzed in an agarose or ***test gel***. Electrophoresis can be carried out and comparisons can be made between uncleaved and completely digested standard samples of DNA. Usually, a smear of various sizes of DNA fragments can be observed if restriction digestion is completed. Conversely, high molecular weight genomic DNA can still be observed from partially digested DNA samples. If partial digestion occurs, procedures such as additional purification of the DNA sample can be carried out. The amounts of DNA, restriction enzymes, and other components should be carefully calculated. Adjustments of the incubation conditions for digestion can also aid in achieving complete digestion.

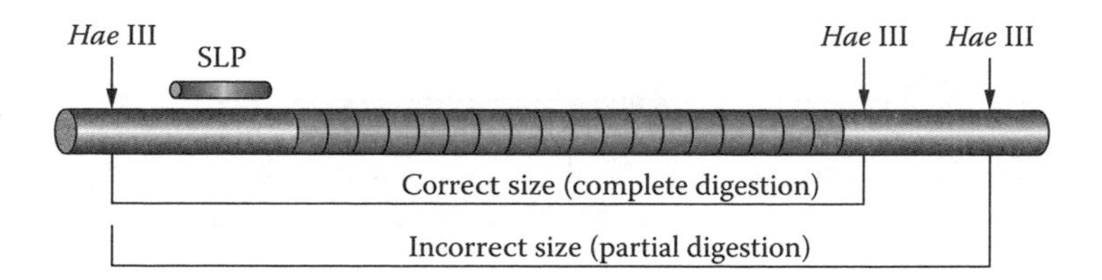


**Figure 17.9** Effects of partial restriction digestion on RFLP profile. Only the restriction fragments detectable by the probe are shown. *Hae*III restriction sites are indicated by arrows.

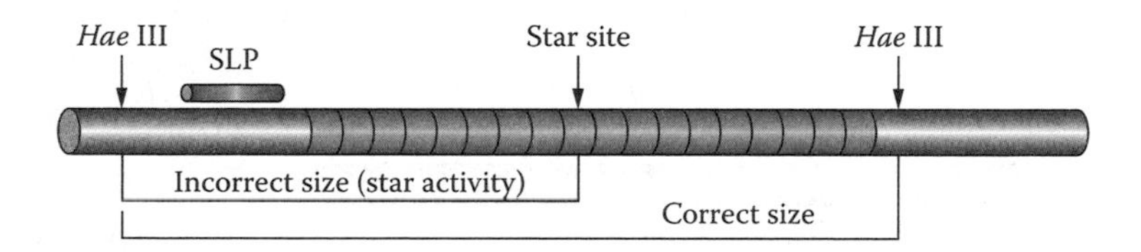


**Figure 17.10** Star effects on RFLP profiles. Only the restriction fragments detectable by the probe are shown. *Hae*III restriction sites and star sites are indicated by arrows.

**17.2.4.2.2 Star Activity** ***Star activity*** refers to a deviation of the specificity of a cleavage site of a restriction enzyme. For instance, *Hae*III cleaves at the GGCC DNA sequence. Under some conditions, the enzymatic specificity is reduced and cleaves at a sequence slightly different from GGCC. If the start site is presented at an internal location of a VNTR locus, the enzyme would cleave the GGCC sequences and additionally cleave at the internal star site (Figure 17.10). This would result in an additional band smaller than the true alleles, although the intensity of this band usually is not the same as other bands. Thus, a multiband pattern is observed. However, the star site presented externally to a VNTR locus cannot be detected by the probe and does not affect the profiling results.

**17.2.4.2.3 Point Mutations** A ***point mutation*** is caused by the substitution, deletion, or insertion of a single nucleotide. Point mutations at a restriction site abolish the site and the result is a band with a slightly larger size than the true allele (Figure 17.11). The point mutation could be present internally in a VNTR sequence. If such a point mutation creates a restriction enzyme site, the enzyme will cleave at the regular site and at the mutation site and yield two smaller bands. If the created restriction site is located internal to the probe binding region, both bands will be detected for that allele. These rare mutations obey Mendelian inheritance.

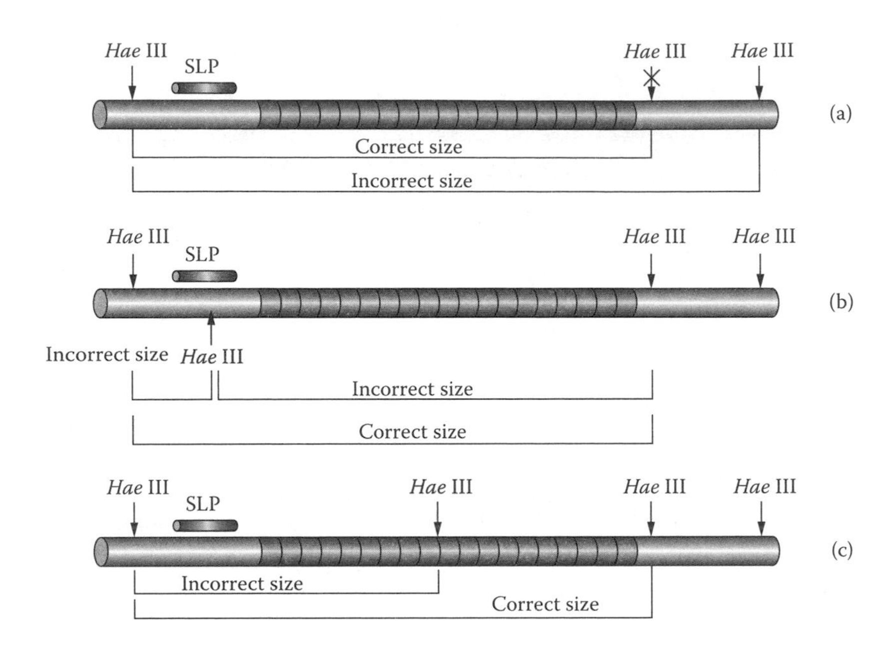


**Figure 17.11** Effects of point mutations on RFLP profile. Only the restriction fragments detectable by the probe are shown. *Hae*III restriction sites are indicated by arrows. (a) Point mutation (in red) abolishes the *Hae*III restriction site. (b) A point mutation (in red) creates an internal *Hae*III restriction site residing within the probe binding region. (c) A point mutation (in red) creates an internal *Hae*III restriction site residing outside the probe binding region.

### 17.2.4.3 *Electrophoresis and Blotting Artifacts*

**17.2.4.3.1 Partial Stripping** If more than one VNTR locus is analyzed sequentially using the same membrane, the probe must be removed by the stripping process before the application of the next probe. Any probe remaining on the membrane due to partial stripping may generate additional bands when the next probe is analyzed. However, the bands due to partial stripping are usually faint and have the same electrophoretic mobility as the previous autoradiograph.

**17.2.4.3.2 Separation Resolution Limits and Band Shifting** Agarose gel electrophoresis cannot resolve restriction fragments that differ by one or a few repeat units, especially for high molecular weight fragments. These bands may not be separated and will appear as a single band. This may lead to a false interpretation as a homozygous profile. Additionally, minor variations in the electrophoretic mobility of DNA fragments, known as band shifting, can cause two samples from the same individual to appear different.

**17.2.4.3.3 Bands Running off Bottom of Gel** The commonly used VNTR loci generate bands from hundreds of bp to 20 kb in length. The small sized bands have higher electrophoretic mobility and may run off the front edge of a gel during electrophoresis and fail to be detected. This phenomenon may also lead to a false interpretation as a homozygous profile.

## 17.3 Amplified Fragment Length Polymorphism (AFLP)

Some VNTR loci have relatively short alleles (fewer than 1 kb). These loci are applicable for PCR amplification. This technique is called ***amplified fragment length polymorphism (AFLP)***. One locus, D1S80, was used by forensic DNA laboratories for AFLP analysis. Fragments in the range of 14 to 42 repeat units (16 bp per repeat) were amplified using the AFLP method (Figure 17.12). The amplified DNA fragments were commonly separated according to size using polyacrylamide gel electrophoresis and detected using a silver stain (Figure 17.13).

D1S80 loci are detected as discrete alleles and thus can be compared directly to an allelic ladder (a collection of common alleles used as a standard) on the same gel. This technique represented an improvement over the RFLP system. RFLP allele sizing cannot be performed with precision and the resolution limits of agarose gel electrophoresis are much lower compared to the polyacrylamide gels.

The AFLP technique requires less DNA than the RFLP method and performs better for degraded samples. The AFLP method at the D1S80 locus can be analyzed in a multiplex fashion with an amelogenin locus. The amelogenin gene is used for forensic gender typing applications. Typing the amelogenin gene enables the determination of sex of the contributor of a biological sample.

Due to the wide variation in allele sizes at the D1S80 locus, preferential amplification may be observed. Under certain conditions, the larger alleles may not be consistently amplified as the small alleles, which may cause lower intensity of the larger allele. Additionally, only one locus was

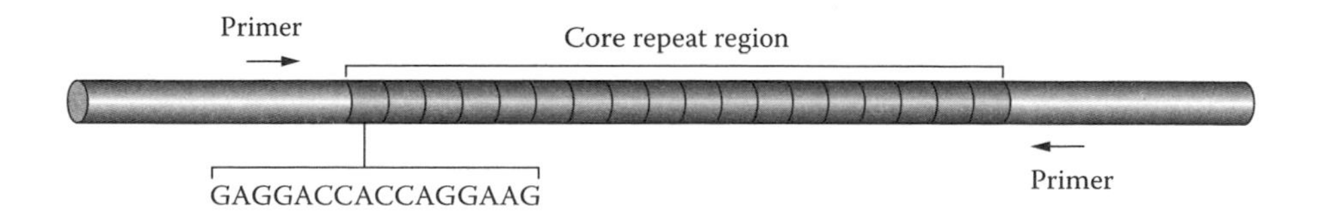


**Figure 17.12** VNTR locus D1S80 (chromosome 1p). Each repeat unit is 16 base pairs long. PCR primers are indicated to amplify the core repeat region.

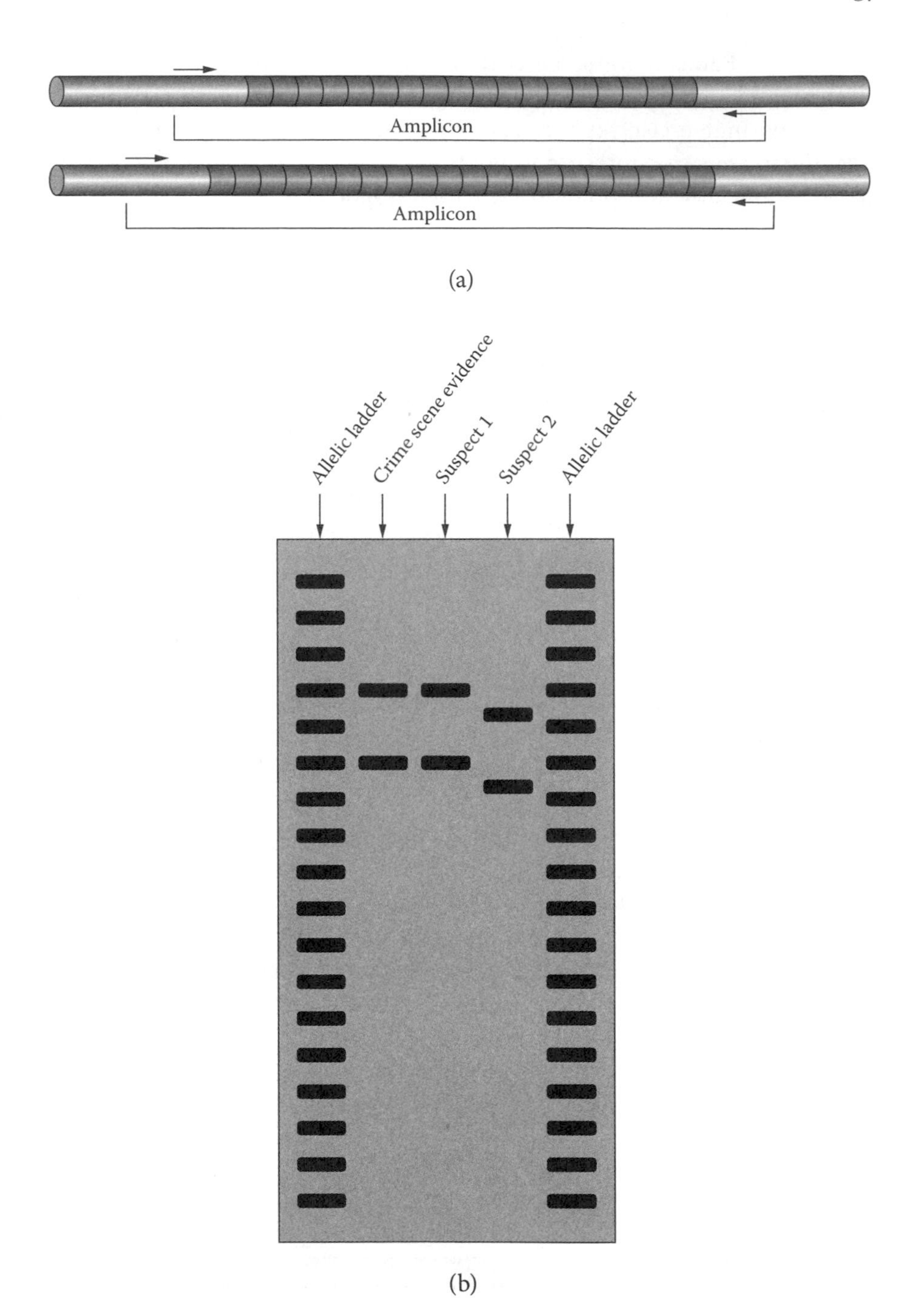


**Figure 17.13** AFLP analysis. (a) Heterozygous D1S80 loci amplified by PCR. PCR primers are indicated as arrows. (b) Silver stained polyacrylamide gel showing D1S80 amplicons along with allelic ladders.

analyzed in this system and the D1S80 locus contains two alleles that are common in some populations. Thus, the discriminating power is reduced compared to RFLP. D1S80 was gradually replaced by multiplex STR systems in the late 1990s.

# Bibliography

## RFLP

Balazs, I. et al. 1989. Human population genetic studies of five hypervariable DNA loci. *Am J Hum Genet* 44 (2):182.

Benzinger, E. A. et al. 1995. Time course and inhibitors of Hae III digestion in the forensic laboratory. *Appl Theor Electrophor* 4 (4):179.

Benzinger, E. A. et al. 1997. Products of partial digestion with Hae III. Part 1. Characterization, casework experience and confirmation of the theory of three-, four- and five-banded RFLP pattern origins using partial digestion. *J Forensic Sci* 42 (5):850.

Benzinger, E. A. et al. 1998. An illustrated guide to RFLP troubleshooting. *J Forensic Sci* 43 (3):665.

Bragg, T. et al. 1988. Isolation and mapping of a polymorphic DNA sequence (cTBQ7) on chromosome 10 [D10S28]. *Nucleic Acids Res* 16 (23):11395.

Budowle, B. et al. 1990. Hae III: A suitable restriction endonuclease for restriction fragment length polymorphism analysis of biological evidence samples. *J Forensic Sci* 35 (3):530.

Budowle, B. et al. 1991. Fixed-bin analysis for statistical evaluation of continuous distributions of allelic data from VNTR loci, for use in forensic comparisons. *Am J Hum Genet* 48 (5):841.

Budowle, B. et al. 1994. The assessment of frequency estimates of Hae III generated VNTR profiles in various reference databases. *J Forensic Sci* 39 (2):319.

Budowle, B. et al. 1995. Simple protocols for typing forensic biological evidence: Chemiluminescent detection for human DNA quantitation and restriction fragment length polymorphism (RFLP) analyses and manual typing of poly merase chain reaction (PCR) amplified polymorphisms. *Electrophoresis* 16 (9):1559.

Budowle, B. et al. 2000. *DNA Typing Protocols: Molecular Biology and Forensic Analysis*. Natick, MA: Eaton Publishing.

Budowle, B. et al. 2000. Restriction fragment-length polymorphism typing of variable number tandem repeat loci, in *DNA Typing Protocols: Molecular Biology and Forensic Analysis*. Natick, MA: Eaton Publishing.

Duewer, D. L. et al. 1997. Interlaboratory comparison of autoradiographic DNA profiling measurements. 4. Protocol effects. *Anal Chem* 69 (10):1882.

Duewer, D. L., K. L. Richie, and D. J. Reeder. 2000. RFLP band size standards: NIST standard reference material 2390. *J Forensic Sci* 45 (5):1093.

Elder, J. K., and E. M. Southern. 1983. Measurement of DNA length by gel electrophoresis II: Comparison of methods for relating mobility to fragment length. *Anal Biochem* 128 (1):227.

Eriksen, B., and O. Svensmark. 1992. Comparison of two molecular weight markers used in DNA profiling. *Int J Legal Med* 105 (3):145.

Gary, K. T., D. L. Duewer, and D. J. Reeder. 1999. Graphical tools for RFLP measurement quality assurance: Laboratory performance charts. *J Forensic Sci* 44 (5):978.

Gill, P., and D. J. Werrett. 1987. Exclusion of a man charged with murder by DNA fingerprinting. *Forensic Sci Int* 35 (2-3):145.

Giusti, A. M., and B. Budowle. 1992. Effect of storage conditions on restriction fragment length polymorphism (RFLP) analysis of deoxyribonucleic acid (DNA) bound to positively charged nylon membranes. *J Forensic Sci* 37 (2):5973.

Hartmann, J. M. et al. 1997. The effect of sampling error and measurement error and its correlation on the estimation of multi-locus fixed-bin VNTR RFLP genotype probabilities. *J Forensic Sci* 42 (2):241.

Hau, P., and N. Watson. 2000. Sequencing and four-state minisatellite variant repeat mapping of the D1S7 locus (MS1) by fluorescence detection. *Electrophoresis* 21 (8):1478.

Jeffreys, A. J., M. Turner, and P. Debenham. 1991. The efficiency of multilocus DNA fingerprint probes for individualization and establishment of family relationships, determined from extensive casework. *Am J Hum Genet* 48 (5):824.

Jeffreys, A. J., V. Wilson, and S. L. Thein. 1985. Individual-specific 'fingerprints' of human DNA. *Nature* 316 (6023):76.

_____. 1992. Hypervariable 'minisatellite' regions in human DNA. 1985. *Biotechnology* 24:467.

Kanter, E. et al. 1986. Analysis of restriction fragment length polymorphisms in deoxyribonucleic acid (DNA) recovered from dried bloodstains. *J Forensic Sci* 31 (2):403.

Kasai, K., Y. Nakamura, and R. White. 1990. Amplification of a variable number of tandem repeats (VNTR) locus (pMCT118) by the polymerase chain reaction (PCR) and its application to forensic science. *J Forensic Sci* 35 (5):1196.

Laber, T. L. et al. 1992. Evaluation of four deoxyribonucleic acid (DNA) extraction protocols for DNA yield and variation in restriction fragment length polymorphism (RFLP) sizes under varying gel conditions. *J Forensic Sci* 37 (2):404.

Laber, T. L. et al. 1994. Validation studies on the forensic analysis of restriction fragment length polymorphism (RFLP) on LE agarose gels without ethidium bromide: Effects of contaminants, sunlight, and the electrophoresis of varying quantities of deoxyribonucleic acid (DNA). *J Forensic Sci* 39 (3):707.

Laber, T. L. et al. 1995. The evaluation and implementation of match criteria for forensic analysis of DNA. *J Forensic Sci* 40 (6):1058.

Lander, E. S. 1989. DNA fingerprinting on trial. *Nature* 339 (6225):501.

Lewis, M. E. et al. 1990. Restriction fragment length polymorphism DNA analysis by the FBI Laboratory protocol using a simple, convenient hardware system. *J Forensic Sci* 35 (5):1186.

McNally, L. et al. 1989. The effects of environment and substrata on deoxyribonucleic acid (DNA): The use of casework samples from New York City. *J Forensic Sci* 34 (5):1070.

Milner, E. et al. 1989. Isolation and mapping of a polymorphic DNA sequence pH30 on chromosome 4[HGM provisional no. D4S139]. *Nucleic Acids Res* 17 (10):4002.

Nakamura, Y. et al. 1987. Isolation and mapping of a polymorphic DNA sequence pYNH24 on chromosome 2 (D2S44). *Nucleic Acids Res* 15 (23):10073.

Rankin, D. R. et al. 1996. Restriction fragment length polymorphism (RFLP) analysis on DNA from human compact bone. *J Forensic Sci* 41 (1):40.

Reed, K. C., and D. A. Mann. 1985. Rapid transfer of DNA from agarose gels to nylon membranes. *Nucleic Acids Res* 13 (20):7207.

Richie, K. L. et al. 1996. Long PCR for VNTR analysis. *J Forensic Sci* 44 (6):1176.

Rossi, U. 1989. International recommendations (as of March, 1988) on the application of methods involving DNA polymorphisms in forensic haematology. *Haematologica* 74 (2):219.

Rudin, N., and K. Inman. 2002. *Introduction to Forensic DNA Analysis*, 2nd ed. Boca Raton, FL: CRC Press.

Sharma, B. R. et al. 1995. A comparative study of genetic variation at five VNTR loci in three ethnic groups of Houston, Texas. *J Forensic Sci* 40 (6):933.

Stolorow, A. M. et al. 1996. Interlaboratory comparison of autoradiographic DNA profiling measurements. 3. Repeatability and reproducibility of restriction fragment length polymorphism band sizing, particularly bands of molecular size > 10K base pairs. *Anal Chem* 68 (11):1941.

Tahir, M. A. et al. 1995. Restriction fragment length polymorphism (RFLP) typing of DNA extracted from nasal secretions. *J Forensic Sci* 40 (3):459.

Tahir, M. A. et al. 1996. Deoxyribonucleic acid profiling by restriction fragment length polymorphism analysis: Compilation of validation studies. *Sci Justice* 36 (3):173.

Tully, G., K. M. Sullivan, and P. Gill. 1993. Analysis of 6 VNTR loci by multiplex PCR and automated fluorescent detection. *Hum Genet* 92 (6):554.

Wyman, A. R. and R. White. 1980. A highly polymorphic locus in human DNA. *Proc Natl Acad Sci USA* 77 (11):6754.

## AFLP

Baechtel, F. S., K. W. Presley, and J. B. Smerick. 1995. D1S80 typing of DNA from simulated forensic specimens. *J Forensic Sci* 40 (4):536.

Baechtel, F. S. et al. 1993. Multigenerational amplification of a reference ladder for alleles at locus D1S80. *J Forensic Sci* 38 (5):1176.

Budowle, B. et al. 1991. Analysis of the VNTR locus D1S80 by the PCR followed by high-resolution PAGE. *Am J Hum Genet* 48 (1):137.

Cosso, S., and R. Reynolds. 1995. Validation of the AmpliFLP D1S80 PCR amplification kit for forensic casework analysis according to TWGDAM guidelines. *J Forensic Sci* 40 (3):424.

Gross, A. M., G. Carmody, and R. A. Guerrieri. 1997. Validation studies for the genetic typing of the D1S80 locus for implementation into forensic casework. *J Forensic Sci* 42 (6):1140.

Kasai, K., Y. Nakamura, and R. White. 1990. Amplification of a variable number of tandem repeats (VNTR) locus (pMCT118) by the polymerase chain reaction (PCR) and its application to forensic science. *J Forensic Sci* 35 (5):1196.

Kloosterman, A. D., B. Budowle, and P. Daselaar. 1993. PCR amplification and detection of the human D1S80 VNTR locus: Amplification conditions, population genetics and application in forensic analysis. *Int J Legal Med* 105 (5):257.
Latorra, D., and M. S. Schanfield. 1996. Analysis of human specificity in AFLP systems APOB, PAH, and D1S80. *Forensic Sci Int* 83 (1):15.
Nakamura, Y. et al. 1988. Isolation and mapping of a polymorphic DNA sequence (pMCT118) on chromosome 1p [D1S80]. *Nucleic Acids Res* 16 (19):9364.
Sajantila, A. et al. 1992. Amplification of reproducible allele markers for amplified fragment length polymorphism analysis. *Biotechniques* 12 (1):16.
Skinker, D. M. et al. 1997. DNA typing of azoospermic semen at the D1S80 locus. *J Forensic Sci* 42 (4):718.
Skowasch, K., P. Wiegand, and B. Brinkmann. 1992. pMCT 118 (D1S80): A new allelic ladder and an improved electrophoretic separation lead to the demonstration of 28 alleles. *Int J Legal Med* 105 (3):165.
Watanabe, G. et al. 1997. Nucleotide substitution in the 5' flanking region of D1S80 locus. *Forensic Sci Int* 89 (1-2):75.

## Study Questions

1. Which of the following DNA samples can be used in the RFLP method?
   (a) DNA extracted by the Chelex method
   (b) DNA extracted by the organic solvent method
   (c) DNA extracted by the silica-based method

2. Failure to obtain an RFLP profile can be caused by:
   (a) DNA degradation
   (b) Star activity of restriction enzyme
   (c) Partial restriction digestion

3. Mixed samples from more than one donor can be analyzed via the:
   (a) MLP approach
   (b) SLP approach

4. Which of the following is more sensitive?
   (a) RFLP
   (b) AFLP

5. In RFLP, an extra band pattern can be caused by:
   (a) Mutations
   (b) Star activity
   (c) Partial restriction digestion

# Autosomal Short Tandem Repeat (STR) Profiling

# 18

An STR is a region of human DNA containing an array of tandem repeats. Arrays range from only a few to about a hundred repeated units. A repeat unit can be 2 to 6 base pairs (bp) long. STRs are also called ***microsatellites*** or ***simple sequence repeats***. The number of repeat units of STR loci can vary greatly among the population. The most commonly used STR loci are 100 to 500 bp in length—shorter than the smallest VNTRs (up to 1000 bp). STR loci have many advantages due to the small size of the alleles:

- STR loci are applicable for PCR amplification.
- STR profiling performs better than VNTR profiling for degraded DNA samples.
- Preferential amplification is reduced at STR loci compared to VNTR loci using AFLP.
- Better electrophoretic resolution of DNA fragments is achieved than VNTR profiling.
- STR loci are applicable for multiplexing amplification.
- STR profiling, As with VNTR profiling, is capable of handling interpretation of mixed DNA profiles from multiple contributors.

Thus STR loci are better candidates for forensic DNA testing than VNTR loci. This section will discuss autosomal STR profiling. Male-specific Y chromosome STR will be discussed in Chapter 19.

## 18.1 Characteristics of STR Loci

More than $10^5$ STRs exist in the human genome. Many STRs have been characterized and used in various types of studies such as genetic mapping and linkage analysis. Some STRs have been characterized specifically for forensic DNA profiling.

### 18.1.1 Core Repeat and Flanking Regions

Each STR locus contains a core repeat region in which the number of tandem repeat units varies among individuals (Figure 18.1). The number of tandem

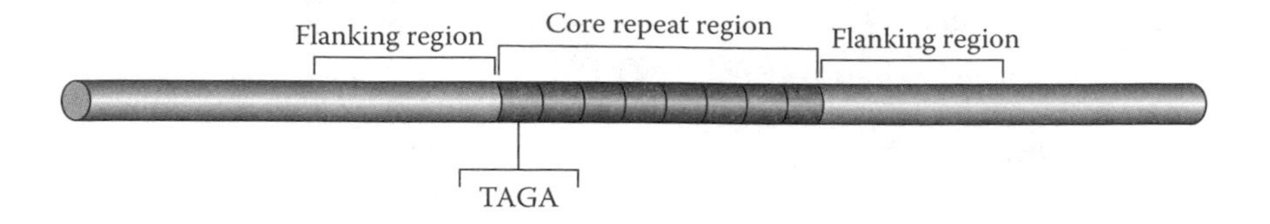


**Figure 18.1** Core repeat and flanking regions of CSF1PO STR locus. It consists of eight repeating units of tetrameric nucleotides (TAGA), thus, designated as allele 8.

repeat units determines genotypes for human identification. The flanking regions surrounding the core repeats are also needed for STR analysis. PCR primers complementary to these flanking regions are used to allow the core repeat regions to be amplified.

### 18.1.2 Repeat Unit Length

Repeat unit length is the number of nucleotides in a single unit of tandem repeat. Dimeric, trimeric, tetrameric, pentameric, and hexameric repeat units appear in the human genome. For example, the human genome has at least $10^4$ tetrameric repeats representing approximately 9% of total STRs. The tetrameric repeats are the most commonly used STR loci for forensic DNA profiling.

Only a few thousand pentameric repeats and a few hundred hexameric repeats exist in the human genome. The pentameric and hexameric repeats are very polymorphic. Only a few pentameric and hexameric repeats are used for forensic applications because they are less abundant in the human genome.

Dimeric and trimeric repeats are very abundant but they are not usually used for forensic applications. High frequencies of stutter peaks that interfere with genotype interpretation are observed when dimeric and trimeric repeats are amplified.

### 18.1.3 Core Repeat Sequences

STR loci compatible for forensic use can be divided into several classes based on their repeat sequences. ***Simple repeats*** consist of tandem repeats with identical repeat unit sequences. Allele designation is based on the number of repeat units in the core repeat region. For example, a D5S818 allele consisting of ten repeating units of the tetrameric nucleotide AGAT is designated as allele 10. ***Compound repeats*** consist of more than one type of simple repeat. ***Complex repeats*** contain several clusters of different tandem repeats with intervening sequences. The designation of complex repeats is based on the sizes of the

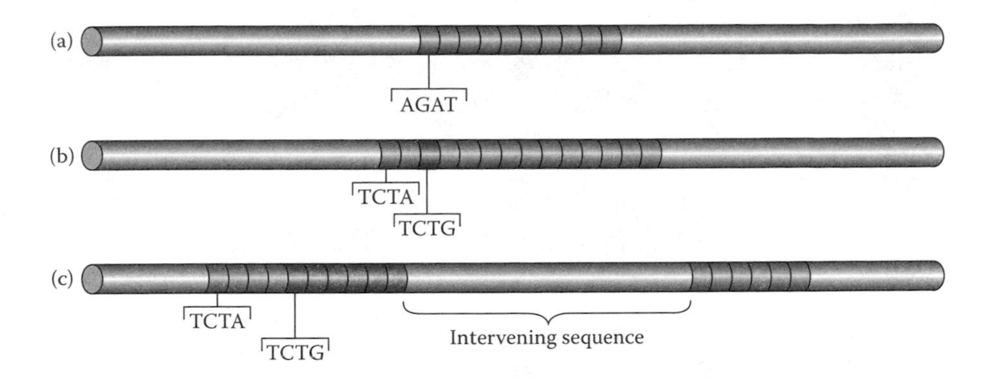


**Figure 18.2** Examples of core repeat sequences. (a) Simple repeat in which D5S818 $[AGAT]_{10}$ is designated as allele 10 consisting of ten repeating units of the tetrameric nucleotides, AGAT. (b) Compound repeats. Allele 14 of D8S1179 consists two types of repeating units, $[TCTA]_2$, $[TCTG]_1$, and $[TCTA]_{11}$. (c) Complex repeats in which allele 24 of D21S11 contains several clusters of different tandem repeats, $[TCTA]_4$, $[TCTG]_6$, $[TCTA]_6$ with intervening sequences (43 base pairs).

alleles. However, size is also dependent upon the primers used for PCR amplification. Figure 18.2 shows representative examples of core repeat sequences.

***Non-consensus alleles*** with incomplete repeat units also appear in the population. These non-consensus alleles, also known as ***microvariants,*** differ from common alleles by one or more nucleotides. They are designated by the number of consensus repeats, followed by a decimal point and the number of nucleotides of the partial repeat, e.g., the TH01 allele 9.3 is 1 nucleotide shorter than allele 10.

Another type of non-consensus allele can result from a limitation of STR analysis. These alleles have the same number of tandem repeats as commonly encountered alleles but contain different sequences. These microvariants cannot be distinguished by STR profiling because their length is identical to the lengths of common alleles.

## 18.2 STR Loci Commonly Used for Forensic DNA Profiling

A number of STR loci have been characterized for forensic DNA profiling (Table 18.1). The discriminating power of an STR locus used for forensic testing can be measured by a parameter known as ***population match probability*** (*P*m). The lower the *P*m, the less likely a match will occur between two randomly chosen individuals. To achieve low *Pm* in forensic STR profiling, a number of unique characteristics of STR loci are desired. First, the STR loci should be highly variable among the population. Second, if more than one locus is selected, the loci should not be linked. The STR loci employed

**Table 18.1 Common STR Loci**

| Locus | Repeat Motif | Repeat Category | Chromosome Location | Physical Position | Structural Gene |
|---|---|---|---|---|---|
| ACTBP2 / SE33 | AAAG | Complex | 6q14 | Chr 6 89.043 Mb | β-actin related pseudogene |
| CSF1PO | TAGA | Simple | 5q33.1 | Chr 5 149.436 Mb | c-fms protooncogene, intron 6 |
| D2S1338 | [TGCC] [TTCC] | Compound | 2q35 | Chr 2 218.705 Mb | Anonymous |
| D3S1358 | [TCTG] [TCTA] | Compound | 3p21.31 | Chr 3 45.557 Mb | Anonymous |
| D5S818 | AGAT | Simple | 5q23.2 | Chr 5 123.139 Mb | Anonymous |
| D7S820 | GATA | Simple | 7q21.11 | Chr 7 83.433 Mb | Anonymous |
| D8S1179 | [TCTA] [TCTG] | Compound | 8q24.13 | Chr 8 125.976 Mb | Anonymous |
| D13S317 | TATC | Simple | 13q31.1 | Chr13 81.620 Mb | Anonymous |
| D16S539 | GATA | Simple | 16q24.1 | Chr 16 84.944 Mb | Anonymous |
| D18S51 | AGAA | Simple | 18q21.33 | Chr 18 59.100 Mb | Anonymous |
| D19S433 | AAGG | Simple | 19q12 | Chr 19 35.109 Mb | Anonymous |
| D21S11 | [TCTA] [TCTG] | Complex | 21q21.1 | Chr 21 19.476 Mb | Anonymous |
| FGA | CTTT | Simple | 4q31.3 | Chr 4 155.866 Mb | α-fibrinogen, intron 3 |
| Penta D | AAAGA | Simple | 21q22.3 | Chr 21 43.880 Mb | Anonymous |
| Penta E | AAAGA | Simple | 15q26.2 | Chr 15 95.175 Mb | Anonymous |
| TH01 | TCAT | Simple | 11p15.5 | Chr 2 2.149 Mb | Tyrosine hydroxylase, intron 1 |
| TPOX | GAAT | Simple | 2p25.3 | Chr 2 1.472 Mb | Thyroid peroxidase, intron 10 |
| VWA | [TCTG] [TCTA] | Compound | 12p13.31 | Chr 12 5.963 Mb | von Willebrand factor, intron 40 |

*Chromosomal location is based on the cytogenetic map and physical position is based on the DNA sequence (Mb = megabase).* Chr = chromosome.

*Sources:* Adapted from Butler, J. M. 2006. *J Forensic Sci* 51 (2):253; Jobling, M. A., and P. Gill. 2004. *Nat Rev Genet* 5 (10):739.

usually are located at different chromosomes. Loci located at the same chromosome can also be used, but should be separated enough to ensure they are not linked (Figure 18.3 and Figure 18.4).

STR loci with fewer amplification artifacts such as stutter products, are desired. STR loci with short allele lengths are preferred for multiplexing STR analysis and the testing of degraded DNA samples.

The application of STR for genetic studies was documented in the early 1990s. The development of the STR multiplex system for forensic profiling was pioneered by Forensic Science Services in the United Kingdom. The first STR multiplex system known as the quadruplex, consisted of four STR loci (F13A1, FES, TH01, and VWA) with a *P*m of $10^{-4}$ (Figure 18.5). In 1995, the first national STR database was established in the United Kingdom. It contained the core set covered by the ***second generation multiplex*** (SGM) consisting of six STR loci (D8S1179, D18S51, D21S11, FGA, TH01, and VWA) with a *P*m value of $10^{-7}$. The SGM system also included the amelogenin locus for gender typing to determine the sexes of DNA contributors. Subsequently,

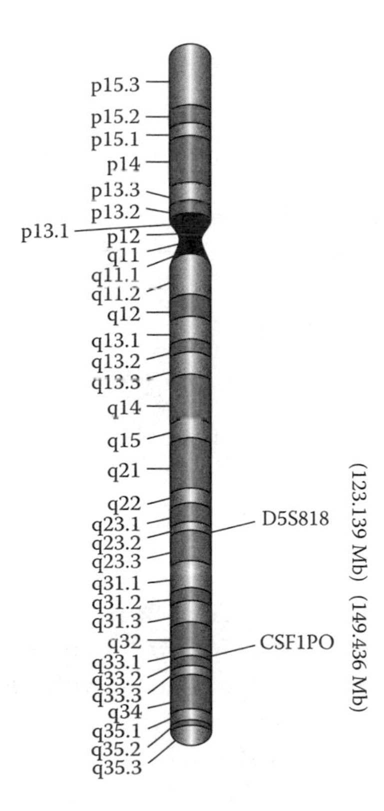


**Figure 18.3** Cytogenetic map showing locations of STR markers on chromosome 5. CSF1PO and D5S818 are separated by 26 Mb (megabases).

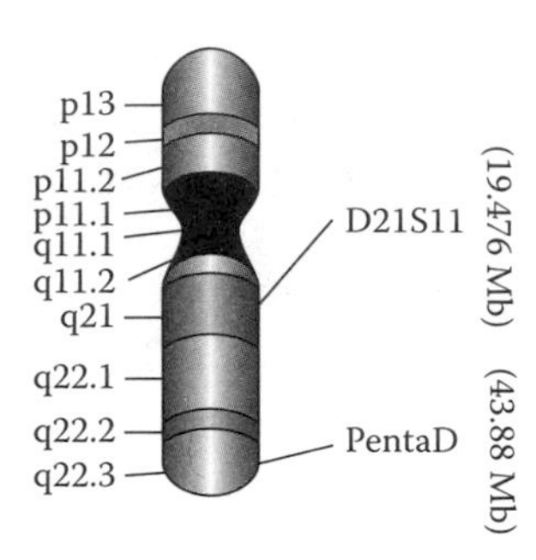


**Figure 18.4** Cytogenetic map showing locations of STR markers on chromosome 21. D21S11 and PentaD are separated by 24 Mb (megabases).

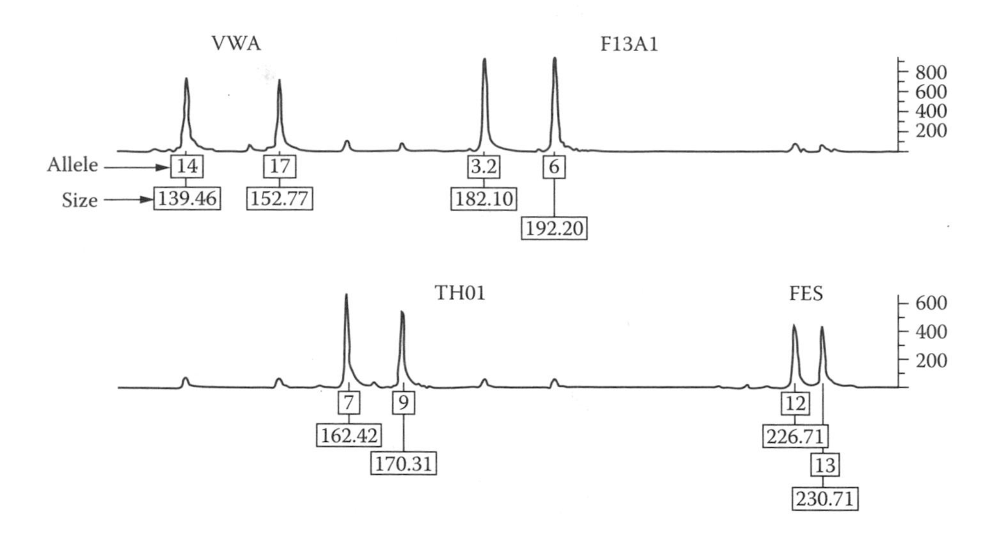


**Figure 18.5** DNA profile obtained using the first STR multiplex system, the quadruplex, showing F13A1, FES, TH01, and VWA loci.

four additional loci were added to the core set with *P*m as low as $10^{-13}$ (SGM Plus). In the United States, the Federal Bureau of Investigation established the ***Combined DNA Index System*** (***CODIS***) in 1998. It contains 13 core STR loci plus the amelogenin gender typing locus with a *P*m of $10^{-15}$. The 13 STR loci are analyzed by a number of commercially available kits (Table 18.2).

To allow for international data exchange, the European DNA Profiling Group (EDNAP) recommended the use of TH01 and VWA loci for all participating European laboratories in 1996. In 1998, the ***European Standard Set*** (***ESS***) of loci was established and included TH01, VWA , FGA, and D21S11

**Table 18.2 Core Loci and Commercial STR Kits**

| | Core Loci | | | | STR Kits | | | |
|---|---|---|---|---|---|---|---|---|
| **Locus** | **CODIS** | **ESS** | **SGM** | **SGM Plus** | **AmpFlSTR Identifiler** | **AmpFlSTR SEfiler (SE)** | **PowerPlex 16** | **PowerPlex ES** |
| ACTBP2/SE33 | | | | | | ☑ | | ☑ |
| CSF1PO | ☑ | | | | ☑ | | ☑ | |
| D2S1338 | | | | ☑ | ☑ | ☑ | | |
| D3S1358 | ☑ | ☑ | | ☑ | ☑ | ☑ | ☑ | ☑ |
| D5S818 | ☑ | | | | ☑ | | ☑ | |
| D7S820 | ☑ | | | | ☑ | | ☑ | |
| D8S1179 | ☑ | ☑ | ☑ | ☑ | ☑ | ☑ | ☑ | ☑ |
| D13S317 | ☑ | | | | ☑ | | ☑ | |
| D16S539 | ☑ | | | ☑ | ☑ | ☑ | ☑ | |
| D18S51 | ☑ | ☑ | ☑ | ☑ | ☑ | ☑ | ☑ | ☑ |
| D19S433 | | | | ☑ | ☑ | ☑ | | |
| D21S11 | ☑ | ☑ | ☑ | ☑ | ☑ | ☑ | ☑ | ☑ |
| FGA | ☑ | ☑ | ☑ | ☑ | ☑ | ☑ | ☑ | ☑ |
| Penta D | | | | | | | ☑ | |
| Penta E | | | | | | | ☑ | |
| TH01 | ☑ | ☑ | ☑ | ☑ | ☑ | ☑ | ☑ | ☑ |
| TPOX | ☑ | | | | ☑ | | ☑ | |
| VWA | ☑ | ☑ | ☑ | ☑ | ☑ | ☑ | ☑ | ☑ |
| Amelogenin | ☑ | ☑ | ☑ | ☑ | ☑ | ☑ | ☑ | ☑ |

for forensic use in Europe. Recently, D3S1358, D8S1179, and D18S51 loci were added to the ESS. Other loci are used as well, such as SE33 (one of the eight in Germany's database).

## 18.3 Forensic STR Analysis

STR loci are amplified using fluorescent dye-labeled primers. A multiplex STR system utilizes multiple fluorescent dyes that can be spectrally resolved. The amplified products are separated and detected via electrophoresis. The various fluorescent dye colors are resolved by the detector and the peaks corresponding to each DNA fragment are identified using specialized computer software. The data collection process generates an ***electropherogram*** that shows a profile of peaks corresponding to each DNA fragment. The area or amplitude of the peak reflects the fluorescent signal intensity. The positions of these peaks represent the electrophoretic mobility of the DNA fragments expressed as data points or sizes (Figure 18.6).

## 18.4 STR Genotyping

As noted earlier, electropherograms are usually plotted as fluorescent signal intensity versus data points or sizes. The data in an electropherogram can then be converted into a genotype. The genotype for a specific STR locus is defined as the number of repeat units of the allele. STR genotype data of different laboratories can be compared easily and is applicable for databasing. The genotyping process requires two steps: (1) the DNA fragments are sized by comparison to an internal size standard and (2) the genotype is determined by using an allelic ladder.

### 18.4.1 Determining Sizes of STR Fragments

The size of an STR fragment is determined by an internal size standard that is mixed in with DNA samples. The standard is labeled with a different colored dye so that it can be spectrally distinguished from DNA fragments of an unknown size (Figure 18.7). The sample is then separated by electrophoresis.

**Figure 18.6** Capillary electrophoresis separation of amplified STR products. Fluorescent dye-labeled amplification products are separated and subsequently detected. Various fluorescent dye colors are resolved by the detector. The peaks corresponding to each DNA fragment are identified.

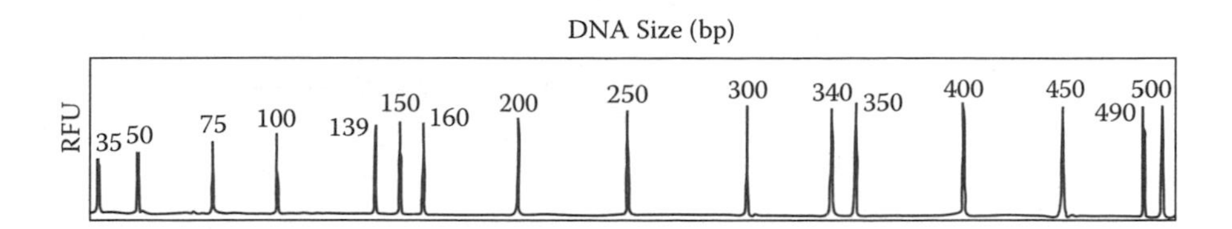


**Figure 18.7** Electropherogram of GeneScan™ 500 size standard (Applied Biosystems). RFU = relative fluorescence unit.

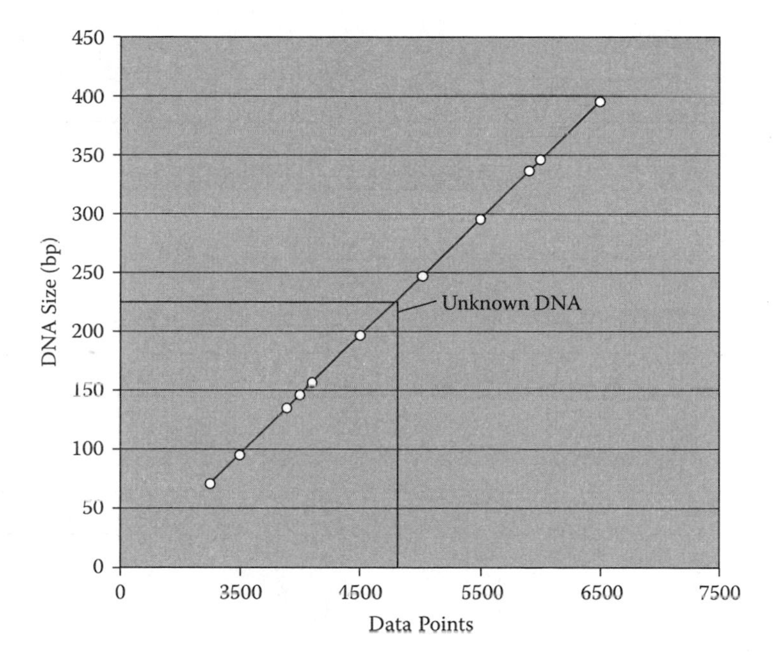


**Figure 18.8** Estimation of size of STR fragment. Size standard curve is plotted. The plot allows the size of an unknown DNA fragment to be estimated.

To determine the sizes of DNA fragment peaks detected in an electropherogram, a standard curve using the internal size standards must be established (Figure 18.8). This means the standard must be detected properly to establish a standard curve. After a size standard curve is generated, DNA fragment size is determined based on the standard curve.

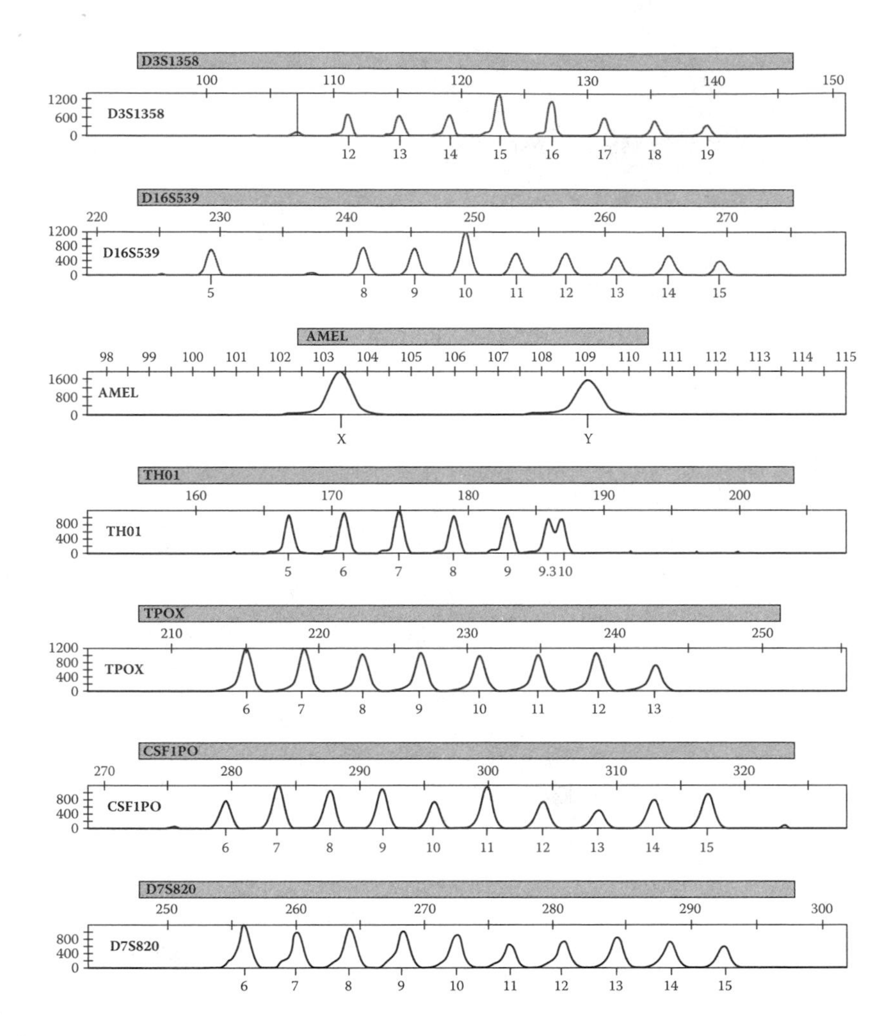


**Figure 18.9** Electropherogram of allelic ladder.

### 18.4.2 Determining Genotypes of STR Fragments

Allelic ladders are important for accurate genotype profiling. An allelic ladder is a collection of synthetic fragments corresponding to common alleles observed in the human population for a given set of STR loci (Figure 18.9). The ladders are compared to data obtained from an electropherogram to determine sample genotype. Thus, each allele in a ladder must be resolved properly in order to determine correct STR alleles for an unknown sample. The sizes of the questioned sample are correlated to sizes for each allele in an allelic ladder to determine the genotype of an unknown sample. The comparison of the unknown and the known allows determination of the allele designation (genotype) of the unknown sample (Figure 18.10). If a rare allele fails to match alleles within an allelic ladder, it is considered an ***off-ladder allele***. If an off-ladder allele is present, the sample should be reanalyzed so that it can be confirmed. The electrophoretic mobility of the rare allele is reproducible.

## 18.5 Factors Affecting Genotyping Results

A number of genetic, amplification, and electrophoresis-related factors may affect the accuracy of genotypic profiles.

### 18.5.1 Mutations

STR loci with low mutation frequencies are desired, in particular, for human identification after mass disasters and also in missing person and paternity cases.

#### *18.5.1.1 Mutations at STR Core Repeat Regions*

Mutations, usually resulting in a gain or loss of a single repeat unit, are observed at some STR loci. If a mutation occurs in the germ cells (cells that form gametes), the mutant allele will be transmitted to and be present in all cell types of the progeny. This type of inheritable mutation in germ cell lineage is called a ***germ line mutation***. The frequency of germ line mutation can be measured by ***mutation rate***, expressed as the number of mutations per generation (germ line transmission). The average mutation rate of commonly used STR loci is about $10^{-4}$ mutations per germ line transmission. However, the mutation rate may vary among different STR loci.

In contrast, ***somatic mutations*** involve mutation only of somatic cells. The germ cells are not affected and thus a mutant allele will not be transmitted to progeny. A somatic mutation occurring at an STR core repeat region can be detected. The ratio of the signal intensities of these alleles varies, depending on the number of mutant cells in the tissue. Somatic mutations usually are tissue type-specific. STR profiles from different tissue types from the same individual can be compared.

#### *18.5.1.2 Chromosomal and Gene Duplications*

Duplication of a single chromosome or part of a chromosome results in three copies of a particular chromosome. This condition, called ***trisomy***, is rare and often associated with genetic diseases such as Down's syndrome (chromosome 21 duplication). Duplications have also been observed in chromosomes 13, 18, and X. Other anomalies include duplication of a single gene or a group of genes instead of an entire chromosome.

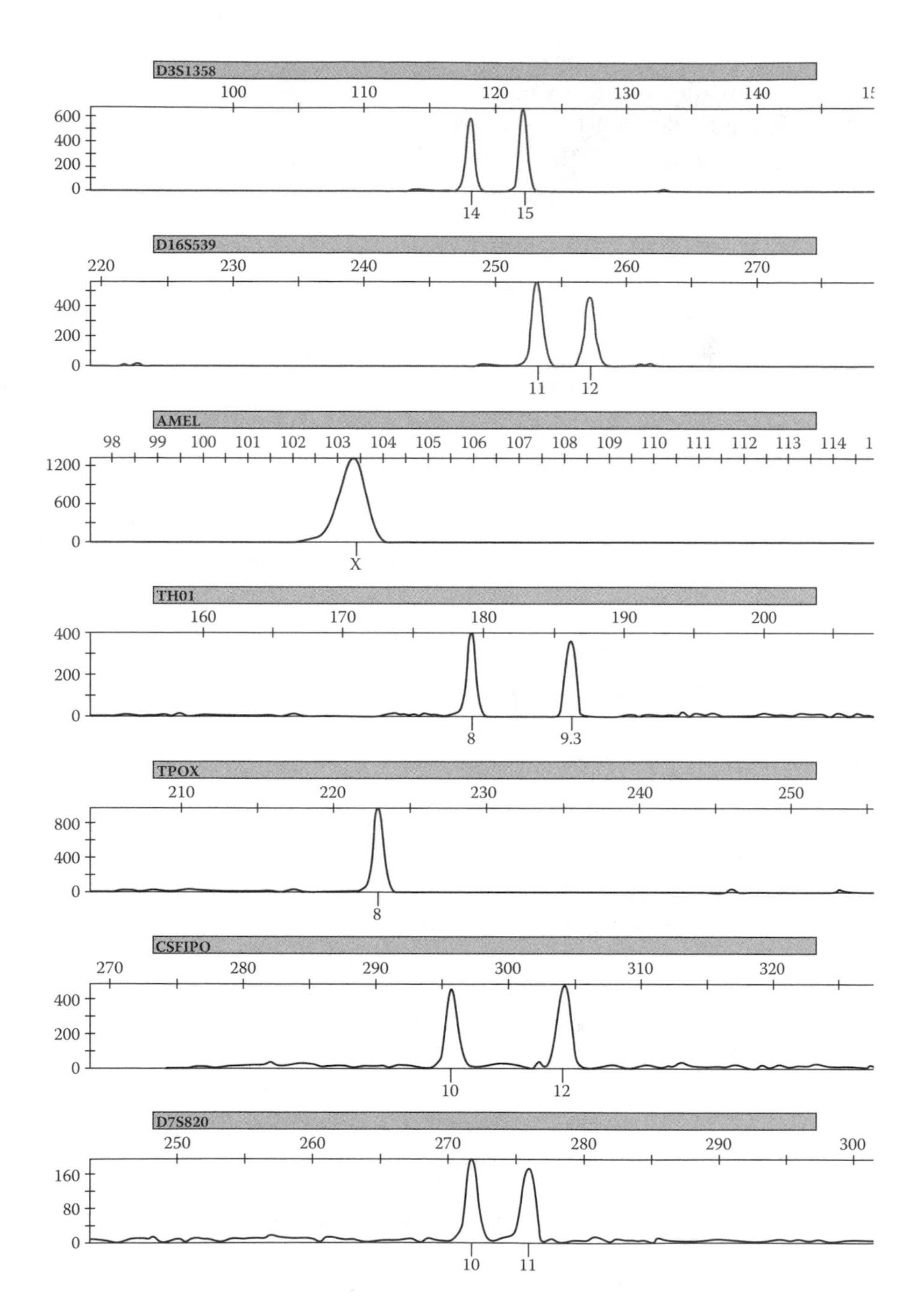


**Figure 18.10** Individual DNA profile.

A duplication bearing a mutation within the STR core repeat region can affect the number of tandem repeat units. A ***triallelic*** or three-banded pattern can be detected at a single locus, but not at other loci of a multiplex STR profile (Figure 18.11). The three alleles usually have equal signal intensity (peak amplitude or peak area). Dozens of triallelic patterns at STR loci commonly used for forensic DNA analysis have been documented. Many occur at the TPOX, FGA, and CSF1PO loci. If the duplicate is not mutated, only two alleles will be observed in a heterozygote. However, the ratio of the peak amplitude of the alleles will be 1:2 (one copy versus two copies) at the locus. The peak amplitude ratio should be equal at other loci.

### *18.5.1.3 Point Mutations*

***Point mutations*** involve change of a base pair through substitution, insertion, or deletion. Insertion or deletion mutations affect the lengths of the STR core repeat region and the amplified flanking region and thus affect the profile. A base pair substitution mutation (except those residing within the primer binding region; see below) will not affect the length of DNA and thus not affect the STR profile.

However, mutations occurring at the primer binding sequence of the flanking region of an STR locus may affect genotype results. If the mutation at the primer binding region abolishes the ability of the primer to anneal, complete failure of the amplification of the allele will result. This phenomenon is known as a ***null allele*** or silent allele (Figure 18.12). Alternative primers can be used to compensate for this mutation. Additionally, a primer that matches the known mutation sequence can be used.

If the mutation does not completely prevent the primer from binding and simply alters the efficiency of the amplification, the resulting signal intensity of the allele will only be decreased. This problem may be solved by modifying the amplification conditions.

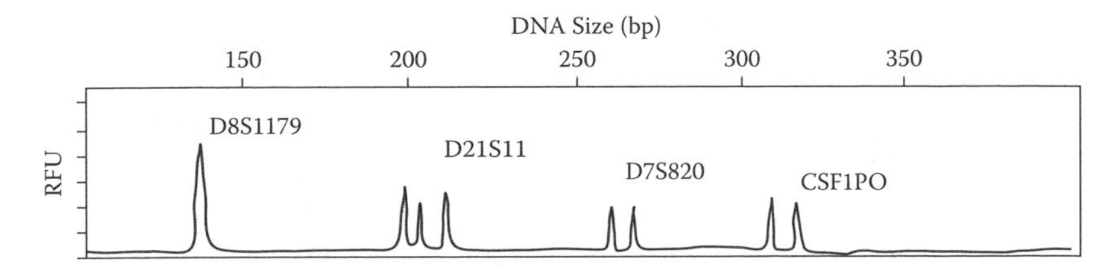


**Figure 18.11** Triallele observed only at D21S11 and not at other STR loci.

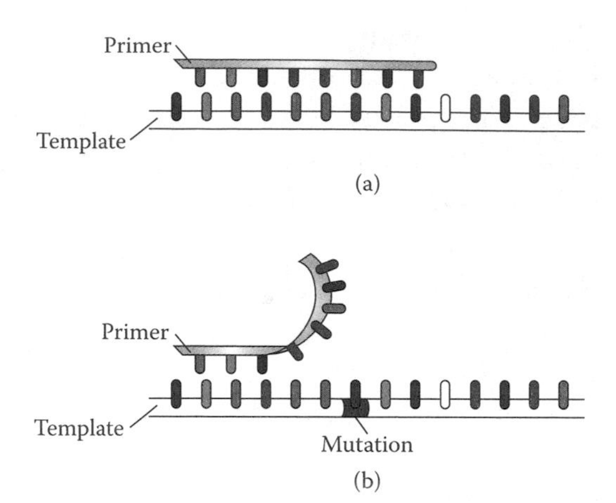


**Figure 18.12** Null allele. An allele present in the sample failed to be amplified by one of the primer sets as a result of a rare mutation at the primer binding sequence of the flanking region. (a) Wild type. (b) Mutation.

## 18.5.2 Amplification Artifacts

### *18.5.2.1 Stuttering*

A stutter is a minor allele peak, also known as a stutter peak, whose repeat units are shorter or longer than the parental allele peak (Figure 18.13). Commonly observed stutters are one repeat unit shorter than the parental allele. It is believed that the stuttering is due to the slippage of polymerase during amplification (Figure 18.14). Stutters with repeat units longer than the parental allele peak can also be observed, but are very rare. Larger alleles appear to yield more stutter than smaller alleles at a given STR locus. The ***stutter ratio*** is defined as the area of the parental peak divided by the area of the stutter peak. The stutter ratio is usually less than 0.15. Less stuttering is observed with pentameric and hexameric repeat unit loci, and complex repeat loci.

### *18.5.2.2 Non-template Adenylation*

During PCR amplification, DNA polymerase often adds an extra nucleotide, usually an adenosine, to the 3′-end of a PCR product. Such a phenomenon is referred to as a ***non-template addition*** resulting in an amplicon that is one base pair longer than the parental allele (designated the +A peak) as shown in Figure 18.15. Commercial multiplex STR kits utilize amplification conditions that favor the adenylation of amplified products. Thus, most amplicons in a sample contain an additional adenosine on the 3′ end (+A peak). Partial

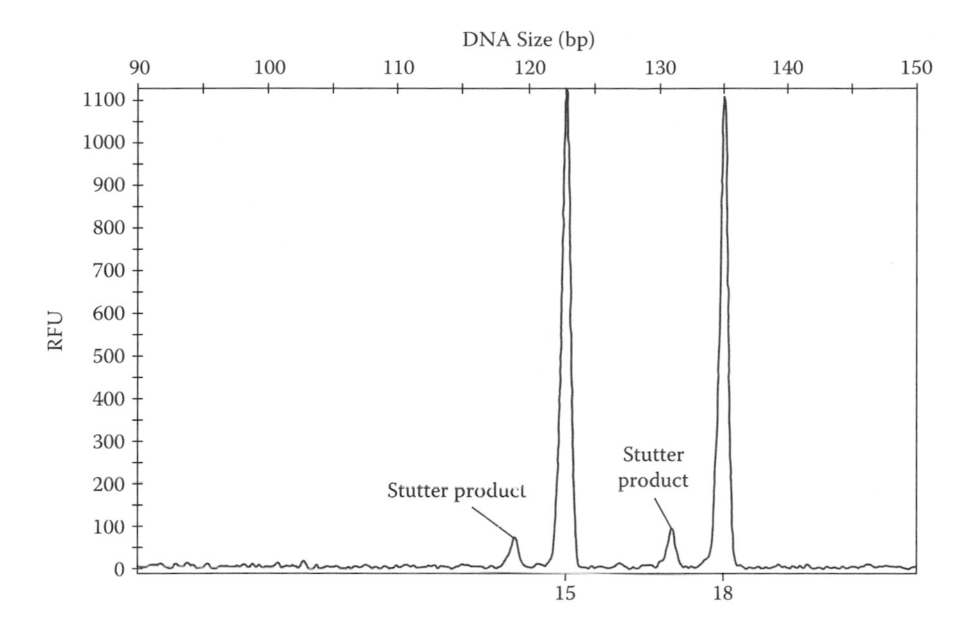


**Figure 18.13** Stutter products.

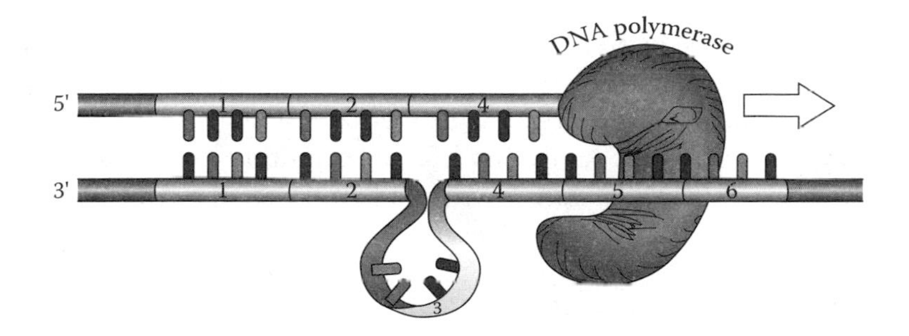


**Figure 18.14** Proposed mechanism for stutter products. During DNA synthesis step of PCR amplification, a DNA polymerase slips and a region of the primer–template complex becomes unpaired, causing the template strand to form a loop. The consequence of this one-repeat loop is a shortened PCR product smaller than the template by a single repeat unit.

non-template addition can occur when too much DNA template is utilized in a PCR reaction; often a mixture of –A and +A peaks can be observed.

### *18.5.2.3 Heterozygote Imbalance*

This imbalance occurs when one of the alleles has greater peak area or amplitude than the other allele within the same locus in which the two alleles of

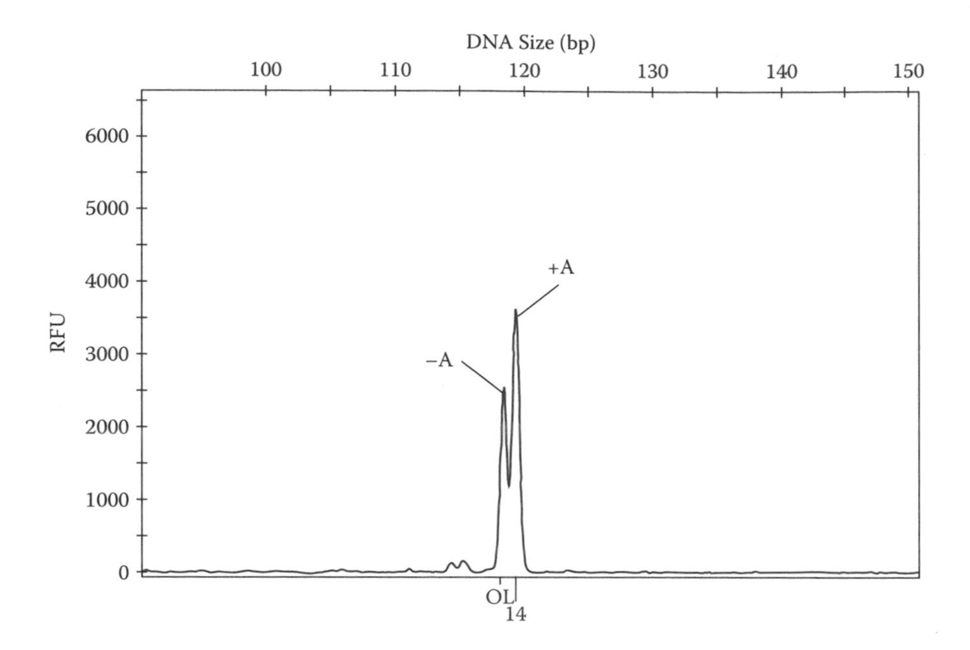


**Figure 18.15** Non-template adenylation. OL = off ladder allele.

a heterozygote are compared (Figure 18.16). It is believed that heterozygote imbalance may arise if (1) the DNA extracted contains unequal copies of DNA template of the two alleles for the heterozygote or (2) the two alleles of a heterozygote may be unequally amplified, a condition known as ***preferential amplification*** in which a smaller sized allele is amplified more efficiently than larger ones.

#### *18.5.2.4 Allelic Dropout*

Allelic dropout occurs when an allele, usually one of the heterozygote alleles, fails to be detected. To date, our understanding of what causes the dropout is very limited. Some believe allelic dropout is the result of an extreme situation of preferential amplification or heterozygote imbalance.

### 18.5.3 Electrophoretic Artifacts

#### *18.5.3.1 Pull-Up Peaks*

A pull-up peak occurs when a minor peak of one color on an electropherogram is pulled up from a major allelic peak in another color (Figure 18.17) when the colors have overlapping spectra. For example, a green peak may pull up a yellow peak, or a blue peak may pull up a green peak. A pull-up peak may contribute to the inaccuracy of a profile if it corresponds to a position of an

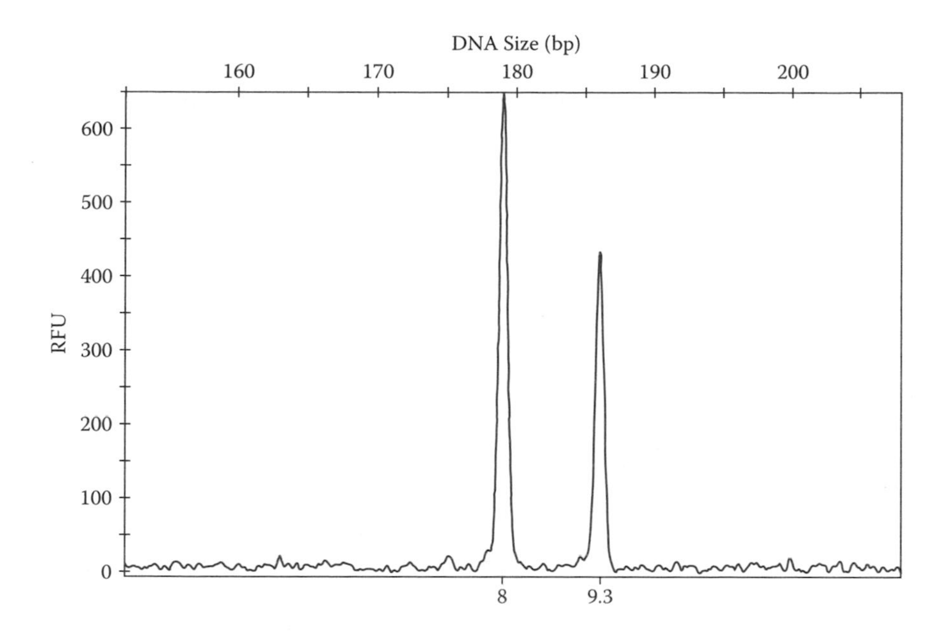


**Figure 18.16** Heterozygote imbalance. The signal intensity of one allele is greater than the other allele within the same locus.

allele. A pull-up peak often occurs when a sample is overloaded or a matrix file (a spectral calibration) is not updated. Thus, loading a smaller sample or installing a new matrix file can reduce the occurrence of pull-up peaks.

#### *18.5.3.2 Spikes*

Spikes are sharp peaks in all colors of an electropherogram with similar signal intensities (Figure 18.18). They are caused by air bubbles and urea crystals in the capillary of an electrophoretic platform. Voltage spikes can also contribute to spike peaks. The spikes are electrophoretic artifacts and are not reproducible. Thus, electrophoresis can be repeated to verify that the spikes occurred on a previous run.

## 18.6 Genotyping of Challenging Forensic Sample

### 18.6.1 Degraded DNA

Environmental exposure of biological evidence can lead to DNA degradation (breaking of DNA molecules into smaller fragments). The more severe the degradation, the fewer intact DNA templates are available in a sample. The lack of an intact DNA template will result in failure of PCR amplification.

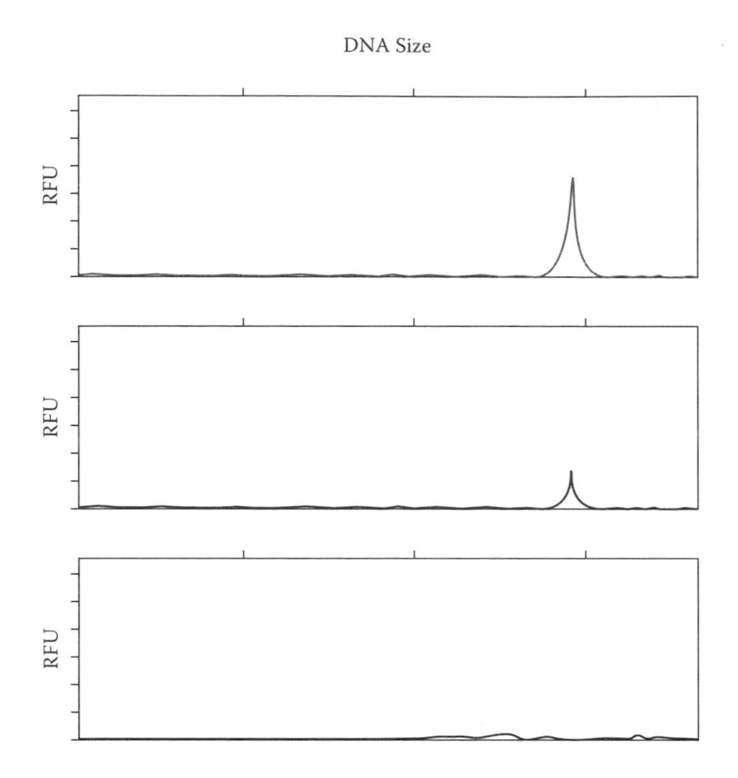


**Figure 18.17** Pull-up peaks. The peaks with overlapping spectra observed in the top and middle panels are not observed in the bottom panel.

An average STR amplicon is 100 to 400 bp in length. It seems that smaller alleles are more likely to be amplified when a sample has some degradation. In degraded samples, the larger alleles are usually the ones that fail to be amplified (Figure 18.19).

Redesigned PCR primers have been developed recently for a miniSTR multiplex kit. These primers are located more proximally to the STR core repeat region to yield reduced size amplicons. The miniSTR kit perform better with degraded samples than existing commercial STR kits.

### 18.6.2 Low Copy Number DNA Testing

***Low copy number (LCN)*** DNA analysis involves the testing of very small amounts of DNA (less than 100 pg) in a sample. Such low levels are often encountered in samples such as fingerprints and from tools and weapons handled by perpetrators. STR analysis of extremely low levels of human DNA can be achieved by increasing the number of PCR cycles (for example,

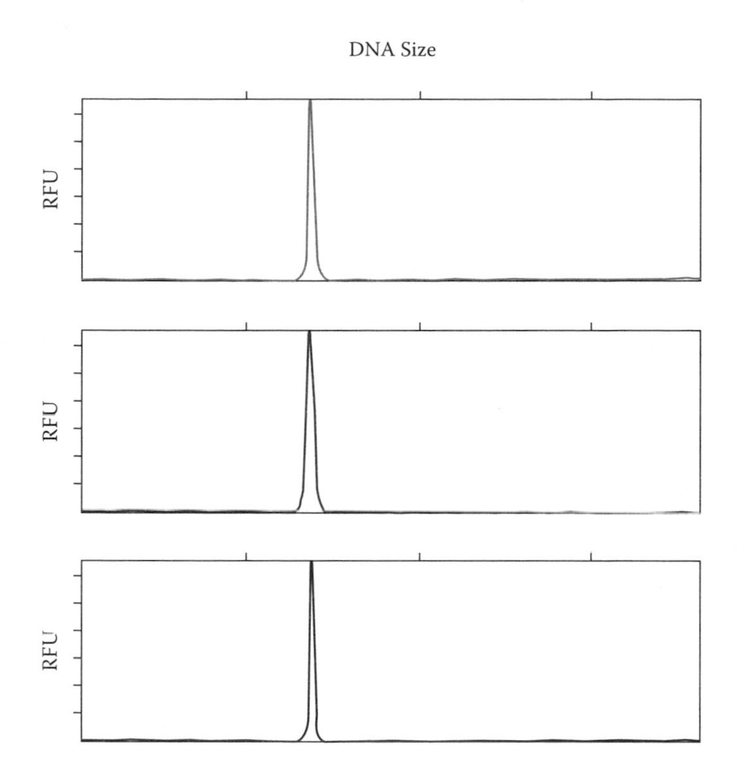


**Figure 18.18** Spike peaks can be observed in various intensities.

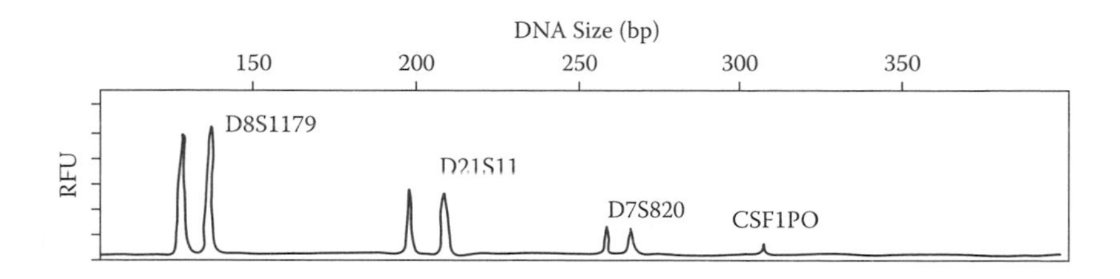


**Figure 18.19** DNA profile of degraded sample.

from 28 to 34 cycles) to improve the yield of amplicons, thus increasing the sensitivity of the analysis.

However, this approach also increases the appearance of artifacts that can make interpretation difficult. For instance, allele dropout, heterozygote imbalance, and stuttering are frequently observed in such cases. Additionally, allele drop-in can arise from contamination. The allele drop-in phenomenon is usually not reproducible. Thus, with LCN testing, genotypes can be determined if the alleles can be identified in two independent amplification reactions.

### 18.6.3 Mixtures

Samples of DNA from two or more contributors are commonly encountered in forensic cases such as sexual assaults in which the evidence recovered from a victim is mixed with a suspect's biological fluids (Figure 18.20). The interpretation of DNA profiles of mixed stains is known as ***mixture interpretation***. The procedures for analyzing mixed stains using STR typing results are described below.

**Determine the presence of a mixture** — First, determine whether the source of the DNA in the sample came from one or more individuals by examination of alleles at multiple loci. The characteristics listed below usually indicate a mixture, but caution should be taken not to confuse various artifacts such as stutters and non-template adenylation with true alleles.

- Severe heterozygote imbalance
- Stutter ratio above 0.15
- Presence of three or more alleles per locus at multiple loci

**Determine genotypes of all the alleles and identify the number of contributors** — Note that the maximum number of alleles at any given locus is four for a two-person mixture. In case of homozygous or allele overlap, the number of alleles observed can be less than four.

**Estimate the ratios of the contributions**— Determine the relative ratios of the contributions to the mixture made by each individual by comparing the peak areas or amplitudes. Amelogenin, a gender type marker, is useful in determining the genders of contributors.

**Consider all possible genotype combinations** — This may be done by pair-wise comparisons to determine the allele combinations that belong to the minor contributor and those that belong to the major contributor.

**Compare reference samples** — The final step is to compare the genotype profiles with the genotypes of reference samples from a suspect and/or victim. If the DNA profile of the suspect's reference sample matches a major or minor component of the mixture, the suspect cannot be excluded as a contributor.

## 18.7 Interpretation of STR Profiling Results

The guidelines for interpretation and reporting of STR profile results should be followed. General guidelines were set by the Scientific Working Group

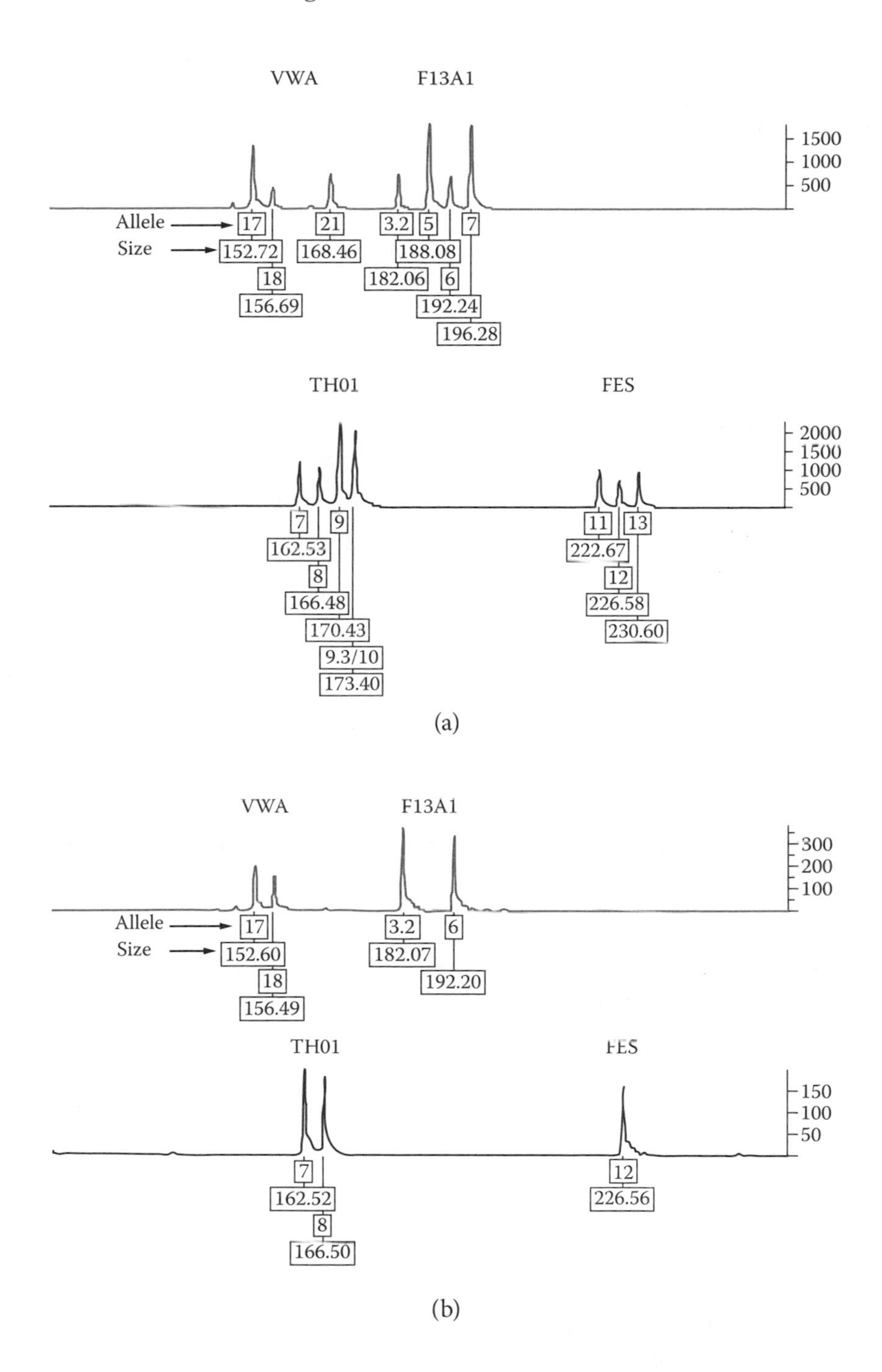


**Figure 18.20** DNA profiles of mixed biological fluids (a) DNA profile of mixed stains from evidence (b) DNA profile of victim. (c) DNA profile of suspect.

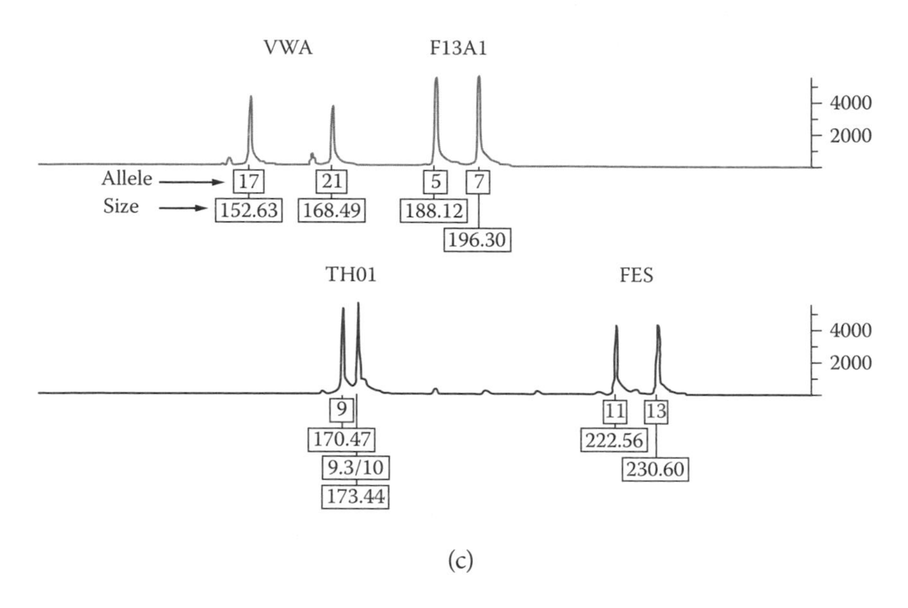


**Figure 18.20** (continued)

on DNA Analysis Methods (SWGDAM) and the DNA Commission of the International Society of Forensic Genetics (ISFG). Conclusions are typically categorized as inclusion (match), exclusion, or inconclusive result.

**Inclusion (match)** — Peaks of compared STR loci show identical genotypes. The strength of the conclusion can be evaluated via statistical analysis and is usually cited in the case report.

**Exclusion** — The genotypes of two or more samples differ and the profile of the sample is determined to be an exclusion, meaning that the profiles originated from different sources.

**Inconclusive result** — The data does not support a conclusion of inclusion or exclusion. In other words, insufficient information is available to reach a conclusion.

## Bibliography

Andersen, J. et al. 1996. Report on third EDNAP collaborative STR exercise. *Forensic Sci Int* 78 (2):83.

Anker, R., T. Steinbrueck, and H. Donis-Keller. 1992. Tetranucleotide repeat polymorphism at the human thyroid peroxidase (hTPO) locus. *Hum Mol Genet* 1 (2):137.

Bar, W. et al. 1997. DNA recommendations: Further report of the DNA Commission of the International Society for Forensic Haemogenetics regarding the use of short tandem repeat systems. *Int J Legal Med* 110 (4):175.

Barber, M. D., B. J. McKeown, and B. H. Parkin. 1996. Structural variation in the alleles of a short tandem repeat system at the human alpha fibrinogen locus. *Int J Legal Med* 108 (4):180.

Barber, M. D. and B. H. Parkin. 1996. Sequence analysis and allelic designation of the two short tandem repeat loci D18S51 and D8S1179. *Int J Legal Med* 109 (2):62.

Bill, M. et al. 2005. Pendulum: A guideline-based approach to the interpretation of STR mixtures. *Forensic Sci Int* 148 (2-3):181.

Brinkmann, B., E. Meyer, and A. Junge. 1996. Complex mutational events at the HumD21S11 locus. *Hum Genet* 98 (1):60.

Budowle, B. et al. 1997. Validation studies of the CTT STR multiplex system. *J Forensic Sci* 42 (4):701.

Budowle, B. et al. 2000. *DNA Typing Protocols: Molecular Biology and Forensic Analysis*. Natick, MA: Eaton Publishing.

Budowle, B. et al. 2001. STR primer concordance study. *Forensic Sci Int* 124 (1):47.

Buse, E.L. et al. 2003. Performance evaluation of two multiplexes used in fluorescent short tandem repeat DNA analysis. *J Forensic Sci* 48 (2):348.

Butler, J. M. 2005. Constructing STR multiplex assays. *Methods Mol Biol* 297:53.

Butler, J. M. 2005. *Forensic DNA Typing: Biology, Technology, and Genetics of STR Markers*, 2nd ed. Burlington, MA: Elsevier.

_____. 2006. Genetics and genomics of core short tandem repeat loci used in human identity testing. *J Forensic Sci* 51 (2):253.

Butler, J. M., Y. Shen, and B. R. McCord. 2003. The development of reduced size STR amplicons as tools for analysis of degraded DNA. *J Forensic Sci* 48 (5):1054.

Chakraborty, R. et al. 1999. The utility of short tandem repeat loci beyond human identification: Implications for development of new DNA typing systems. *Electrophoresis* 20 (8):1682.

Clayton, T. M. et al. 2004. A genetic basis for anomalous band patterns encountered during DNA STR profiling. *J Forensic Sci* 49 (6):1207.

Coble, M. D., and J. M. Butler. 2005. Characterization of new miniSTR loci to aid analysis of degraded DNA. *J Forensic Sci* 50 (1):43.

Collins, J. R. et al. 2003. An exhaustive DNA micro-satellite map of the human genome using high performance computing. *Genomics* 82 (1):10.

Collins, P. J. et al. 2004. Developmental validation of a single-tube amplification of the 13 CODIS STR loci, D2S1338, D19S433, and amelogenin: The AmpFlSTR Identifiler PCR Amplification Kit. *J Forensic Sci* 49 (6):1265.

Cotton, E. A. et al. 2000. Validation of the AMPFlSTR SGM plus system for use in forensic casework. *Forensic Sci Int* 112 (2-3):151.

Crouse, C. A., and J. Schumm. 1995. Investigation of species specificity using nine PCR-based human STR systems. *J Forensic Sci* 40 (6):952.

Dib, C. et al. 1996. A comprehensive genetic map of the human genome based on 5,264 microsatellites. *Nature* 380 (6570):152.

Drabek, J. et al. 2004. Concordance study between Miniplex assays and a commercial STR typing kit. *J Forensic Sci* 49 (4):859.

Duewer, D. L. et al. 2004. NIST mixed stain study 3: Signal intensity balance in commercial short tandem repeat multiplexes. *Anal Chem* 76 (23):6928.

Edwards, A. et al. 1991. DNA typing and genetic mapping with trimeric and tetrameric tandem repeats. *Am J Hum Genet* 49 (4):746.

Edwards, A. et al. 1992. Genetic variation at five trimeric and tetrameric tandem repeat loci in four human population groups. *Genomics* 12 (2):241.

Ellegren, H. 2004. Microsatellites: Simple sequences with complex evolution. *Nat Rev Genet* 5 (6):435.

Fregeau, C. J. et al. 2003. AmpFlSTR profiler Plus short tandem repeat DNA analysis of casework samples, mixture samples, and nonhuman DNA samples amplified under reduced PCR volume conditions (25 microL). *J Forensic Sci* 48 (5):1014.

Fregeau, C. J., and R. M. Fourney. 1993. DNA typing with fluorescently tagged short tandem repeats: A sensitive and accurate approach to human identification. *Biotechniques* 15 (1):100.

Gill, P. 2002. Role of short tandem repeat DNA in forensic casework in the UK: past, present, and future perspectives. *Biotechniques* 32 (2):366.

Gill, P. et al. 2006. The evolution of DNA databases—recommendations for new European STR loci. *Forensic Sci Int* 156 (2-3):242.

Griffiths, R. A. et al. 1998. New reference allelic ladders to improve allelic designation in a multiplex STR system. *Int J Legal Med* 111 (5):267.

Hammond, H. A. et al. 1994. Evaluation of 13 short tandem repeat loci for use in personal identification applications. *Am J Hum Genet* 55 (1):175.

Heinrich, M. et al. 2004. Allelic drop-out in the STR system ACTBP2 (SE33) as a result of mutations in the primer binding region. *Int J Legal Med* 118 (6):361.

Hering, S. et al. 2002. Sequence variations in the primer binding regions of the highly polymorphic STR system SE33. *Int J Legal Med* 116 (6):365.

Hering, S. et al. 2007. Complex variability of intron 40 of the von Willebrand factor (vWF) gene. *Int J Legal Med.*

Holt, C. L. et al. 2002. TWGDAM validation of AmpFlSTR PCR amplification kits for forensic DNA casework. *J Forensic Sci* 47 (1):66.

Hou, Y. P. et al. 1999. D20S161 data for three ethnic populations and forensic validation. *Int J Legal Med* 112 (6):400.

Jobling, M. A. 2001. In the name of the father: Surnames and genetics. *Trends Genet* 17 (6):353.

Kadash, K. et al. 2004. Validation study of the TrueAllele automated data review system. *J Forensic Sci* 49 (4):660.

Kimpton, C. et al. 1994. Evaluation of an automated DNA profiling system employing multiplex amplification of four tetrameric STR loci. *Int J Legal Med* 106 (6):302.

Kimpton, C., A. Walton, and P. Gill. 1992. A further tetranucleotide repeat polymorphism in the vWF gene. *Hum Mol Genet* 1 (4):287.

Kimpton, C. P. et al. 1993. Automated DNA profiling employing multiplex amplification of short tandem repeat loci. *PCR Methods Appl* 3 (1):13.

Krenke, B. E. et al. 2002. Validation of a 16-locus fluorescent multiplex system. *J Forensic Sci* 47 (4):773.

Lareu, M. V. et al. 1996. A highly variable STR at the D12S391 locus. *Int J Legal Med* 109 (3):134.

Leclair, B. et al. 2004. Systematic analysis of stutter percentages and allele peak height and peak area ratios at heterozygous STR loci for forensic casework and database samples. *J Forensic Sci* 49 (5):968.

Leibelt, C. et al. 2003. Identification of a D8S1179 primer binding site mutation and the validation of a primer designed to recover null alleles. *Forensic Sci Int* 133 (3):220.

Levedakou, E. N. et al. 2002. Characterization and validation studies of PowerPlex 2.1, a nine-locus short tandem repeat (STR) multiplex system and penta D monoplex. *J Forensic Sci* 47 (4):757.

Li, H. et al. 1993. Three tetranucleotide polymorphisms for loci: D3S1352; D3S1358; D3S1359. *Hum Mol Genet* 2 (8):1327.

Margolis-Nunno, H. et al. 2001. A new allele of the short tandem repeat (STR) locus, CSF1PO. *J Forensic Sci* 46 (6):1480.

Masibay, A., T. J. Mozer, and C. Sprecher. 2000. Promega Corporation reveals primer sequences in its testing kits. *J Forensic Sci* 45 (6):1360.

Mills, K. A., D. Even, and J. C. Murray. 1992. Tetranucleotide repeat polymorphism at the human alpha fibrinogen locus (FGA). *Hum Mol Genet* 1 (9):779.

Moller, A., E. Meyer, and B. Brinkmann. 1994. Different types of structural variation in STRs: HumFES/FPS, HumVWA and HumD21S11. *Int J Legal Med* 106 (6):319.

Mornhinweg, E. et al. 1998. D3S1358: Sequence analysis and gene frequency in a German population. *Forensic Sci Int* 95 (2):173.

Nelson, M. S. et al. 2002. Detection of a primer-binding site polymorphism for the STR locus D16S539 using the Powerplex 1.1 system and validation of a degenerate primer to correct for the polymorphism. *J Forensic Sci* 47 (2):345.

Opel, K. L. et al. 2006. The application of miniplex primer sets in the analysis of degraded DNA from human skeletal remains. *J Forensic Sci* 51 (2):351.

Poltl, R. et al. 1997. Typing of the short tandem repeat D8S347 locus with different fluorescence markers. *Electrophoresis* 18 (15):2871.

Poltl, R., C. Luckenbach, and H. Ritter. 1998. The short tandem repeat locus D3S1359. *Forensic Sci Int* 95 (2):163.

Polymeropoulos, M. H. et al. 1991. Tetranucleotide repeat polymorphism at the human tyrosine hydroxylase gene (TH). *Nucleic Acids Res* 19 (13):3753.

Puers, C. et al. 1993. Identification of repeat sequence heterogeneity at the polymorphic short tandem repeat locus HUMTH01[AATG]n and reassignment of alleles in population analysis by using a locus-specific allelic ladder. *Am J Hum Genet* 53 (4):953.

Reichenpfader, B., R. Zehner, and M. Klintschar. 1999. Characterization of a highly variable short tandem repeat polymorphism at the D2S1242 locus. *Electrophoresis* 20 (3):514.

Ruitberg, C. M., D. J. Reeder, and J. M. Butler. 2001. STRBase: A short tandem repeat DNA database for the human identity testing community. *Nucleic Acids Res* 29 (1):320.

Schneider, H. R. et al. 1998. ACTBP2-nomenclature recommendations of GEDNAP. *Int J Legal Med* 111 (2):97.

Seidl, C. et al. 1999. Sequence analysis and population data of short tandem repeat polymorphisms at loci D8S639 and D11S488. *Int J Legal Med* 112 (6):355.

Sharma, V., and M. Litt. 1992. Tetranucleotide repeat polymorphism at the D21S11 locus. *Hum Mol Genet* 1 (1):67.

Smith, R. N. 1995. Accurate size comparison of short tandem repeat alleles amplified by PCR. *Biotechniques* 18 (1):122.

Tamaki, K. et al. 1996. Evaluation of tetranucleotide repeat locus D7S809 (wg1g9) in the Japanese population. *Forensic Sci Int* 81 (2-3):133.

Thangaraj, K., A. G. Reddy, and L. Singh. 2002. Is the amelogenin gene reliable for gender identification in forensic casework and prenatal diagnosis? *Int J Legal Med* 116 (2):121.

Urquhart, A. et al. 1994. Variation in short tandem repeat sequences—a survey of twelve microsatellite loci for use as forensic identification markers. *Int J Legal Med* 107 (1):13.

Urquhart, A., C. P. Kimpton, and P. Gill. 1993. Sequence variability of the tetranucleotide repeat of the human beta-actin related pseudogene H-beta-Ac-psi-2 (ACTBP2) locus. *Hum Genet* 92 (6):637.

Vanderheyden, N. et al. 2007. Identification and sequence analysis of discordant phenotypes between AmpFlSTR SGM Plustrade mark and PowerPlex(R). *Int J Legal Med* 16.

Walsh, P. S., H. A. Erlich, and R. Higuchi. 1992. Preferential PCR amplification of alleles: Mechanisms and solutions. *PCR Methods Appl* 1 (4):241.

Walsh, P. S., N. J. Fildes, and R. Reynolds. 1996. Sequence analysis and characterization of stutter products at the tetranucleotide repeat locus vWA. *Nucleic Acids Res* 24 (14):2807.

Weber, J. L., and P. E. May. 1989. Abundant class of human DNA polymorphisms which can be typed using the polymerase chain reaction. *Am J Hum Genet* 44 (3):388.

Wiegand, P. et al. 1993. Forensic validation of the STR systems SE 33 and TC 11. *Int J Legal Med* 105 (6):315.

Wiegand, P. et al. 1999. D18S535, D1S1656 and D10S2325: Three efficient short tandem repeats for forensic genetics. *Int J Legal Med* 112 (6):360.

Yeung, S. H. et al. 2006. Rapid and high-throughput forensic short tandem repeat typing using a 96-lane microfabricated capillary array electrophoresis microdevice. *J Forensic Sci* 51 (4):740.

## Study Questions

1. For moderately degraded DNA samples:
   (a) Smaller PCR products work better for STR testing.
   (b) Bigger PCR products work better for STR testing.

2. For paternity testing, which of the following are preferred?
   (a) Markers with low mutation rate
   (b) Markers with high mutation rate

3. Forensic STR markers usually are located in the:
   (a) Coding region of genomic DNA
   (b) Noncoding region of genomic DNA

4. The most commonly used STRs are tetrameric repeats. Why?
   (a) Because dimeric and trimeric repeats are less common
   (b) Because pentameric and hexameric repeats have more stutter effect
   (c) Because pentameric and hexameric repeats are less variable
   (c) None of the above

5. Alleles of the same length that have different sequences:
   (a) Can be detected by current STR analysis methods
   (b) Cannot be detected by current STR analysis methods

6. A stutter peak is usually:
   (a) One repeat unit shorter
   (b) One repeat unit longer

7. The STR loci currently used in forensic labs:
   (a) Are found only in humans
   (b) May be found in humans and other primates

8. Which of the following is the more sensitive method?
   (a) RFLP
   (b) STR

9. The length of STR fragments used for current forensic DNA testing is approximately:
   (a) 10,000 to 50,000 base pairs
   (b) 1000 to 5000 base pairs
   (c) 100 to 500 base pairs

10. Which of the following may cause someone to mistake a profile as a DNA mixture?
    (a) Stutter products
    (b) Triallelic patterns due to extra chromosomal occurrences
    (c) Both of the above

11. To identify the presence of a mixture, more than __________ real peaks per locus must be observed.
    (a) Two
    (b) Three
    (c) Four

12. With STR testing, which of the following are fluorescently labeled for PCR amplification?
    (a) dNTPs
    (b) PCR primers

13. Which of the following have lower mutation rates?
    (a) STRs
    (b) SNPs

14. Which of the following are used to estimate the size of an STR fragment?
    (a) Size standards
    (b) Allelic ladders

15. You have obtained alleles 15 and 16 at one locus from an STR kit. Another STR kit reveals only allele 16 during testing of the same sample. How can this be explained?
    (a) Sample mix-up
    (b) Mixture
    (c) Null allele due to a mutation at the primer binding sequence

16. What conclusion can be made if only a partial DNA profile is obtained?
    (a) The sample may have some degradation
    (b) Preferential amplification
    (c) All of the above

17. Mixture profiles can be identified using:
    (a) RFLP
    (b) STR
    (c) All of the above

18. Your STR testing showed size standards but no PCR products. What can cause this?
    (a) An electrophoresis problem
    (b) An amplification problem
    (c) A detection problem

19. Identical quantities of stutter products should be observed among various loci.
    (a) True
    (b) False

20. Which of the following is true?
    (a) A PCR product is usually a +adenine product.
    (b) A template product is usually a +adenine product .
    (c) A PCR product is usually a –adenine product.
    (d) A template product is usually a –adenine product.

21. During capillary electrophoresis, the size standards should:
    (a) Be mixed with the DNA sample
    (b) Not be mixed with the DNA sample

22. The allelic ladder should:
    (a) Be mixed with the DNA sample
    (b) Not be mixed with the DNA sample

23. Your STR testing shows neither size standards nor PCR products. What is a possible explanation?
    (a) An electrophoresis problem
    (b) An amplification problem
    (c) A detection problem

# Y Chromosome Profiling and Gender Typing 19

The Y chromosome is inherited from the father and is passed on to all male offspring (Figure 19.1). Thus, the Y chromosome is unique to males. The chromosome encodes dozens of genes required for male-specific functions, including sex determination and spermatogenesis.

Y chromosome loci are very important for forensic DNA profiling and this chapter will discuss such applications. For instance, the Y chromosome STR (Y-STR) used in forensic DNA testing is male specific (for humans and certain higher primates) and is thus useful in investigations of sexual assault cases involving male suspects. The evidence gathered in such cases usually consists of mixtures of high levels of female DNA and low levels of male DNA. The Y chromosome-specific loci can be examined without interference from large amounts of female DNA; differential extraction of sperm and nonsperm cells may not be needed. Furthermore, the Y-STR system is useful for determining numbers of male perpetrators in sexual assault cases involving more than one male. The Y-STR loci used for forensic applications are located in the non-recombining section of the Y chromosome so that paternal lineages can be established. The technique can be used for paternity testing and identification of missing persons. Finally, data interpretation can be simplified by the use of a single allele per Y-STR locus profile.

The major disadvantage of Y chromosome loci is that their discriminating power is low compared to the discriminating power of autosomal loci. Because Y chromosome loci are linked, the product rule for statistical calculations for profile probability does not apply. Chapter 23 discusses statistical evaluation of the strengths of matches. Also, the Y chromosome loci test cannot distinguish individuals with the same paternal lineage.

## 19.1 Human Y Chromosome Genome

The human Y chromosome genome contains approximately 60 million bp and the chromosome can be divided into two regions: the pseudo-autosomal region (PAR) and the male-specific Y (MSY) region.

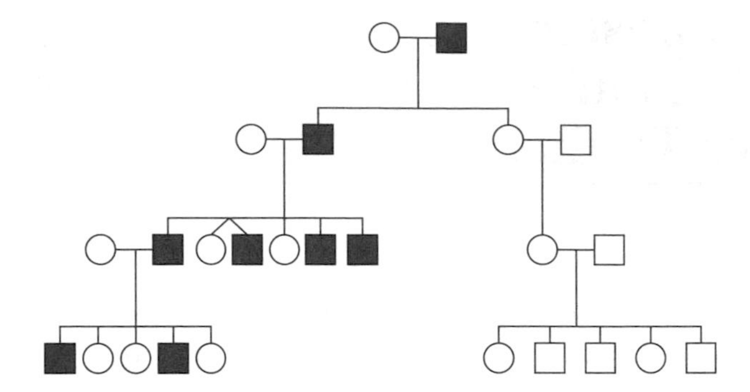

**Figure 19.1** Human family pedigree showing inheritance of Y chromosome. Females and males are denoted by circles and squares, respectively. Red symbols indicate individuals who inherited the same Y chromosome.

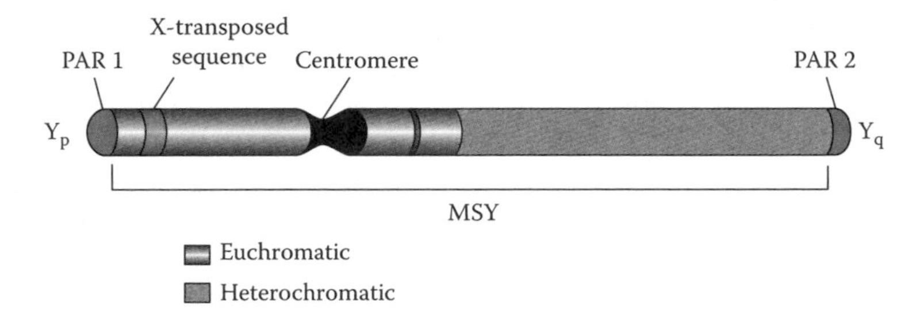


**Figure 19.2** Human Y-STR chromosome structure. PAR = pseudo-autosomal region. MSY = male-specific Y region. Yp = short arm of Y chromosome. Yq = long arm of Y chromosome.

### 19.1.1 Pseudo-Autosomal Region

Approximately 5% of the Y chromosome sequence is located at the telomeres of the chromosome. In particular, PAR1 is located on the tip of the short arm and PAR2 is located at the tip of the long arm (Figure 19.2). This region undergoes recombination with homologous regions on the X chromosome during meiosis in males.

### 19.1.2 Male-Specific Y Region

The remainder of the Y chromosome is known as the MSY region. It was previously called the non-recombining Y (NRY) region (Figure 19.2). It does not participate in homologous recombination. However, certain sections involve intrachromosomal gene conversion.

About 40 megabases (Mb) within the MSY region are heterochromatic (highly repetitive sequences) including the centromeric region and the bulk of the distal long arm. The euchromatic region is about 23 Mb and most of it has been sequenced. Certain sections of the euchromatic region share some homology with the X chromosome. For instance, X-transposed sequences of the Y chromosome are 99% identical to sequences within Xq21 (a band in the long arm of the X chromosome). Additionally, dozens of genes located in the euchromatic region share 60% to 96% homology with their X chromosome counterparts. These X-homologous regions should be avoided when selecting Y chromosome-specific markers for forensic DNA profiling.

### 19.1.3 Polymorphic Sequences

The Y chromosome contains an abundance of repetitive elements, namely STRs, Alu, and LINE elements. Many of these are highly polymorphic. To date Y-STRs are usually used for Y chromosome DNA testing. Single nucleotide polymorphisms (SNPs) at the Y chromosome are also useful for forensic applications and will be discussed in Chapter 20.

## 19.2 Profiling Systems

### 19.2.1 Y-STR

More than 400 STR loci have been identified in the Y chromosome genome. The precise locations of these loci have been sequentially mapped using human genome sequencing data. The distribution of Y-STR loci at the Y chromosome has also been analyzed. Most Y-STR loci, approximately 60% of the 400 identified, are located at the long arm of the chromosome; about 22% are located at the short arm and a few are found in the centromeric region. Y-STRs in the telomeric region have yet to be identified. Only about 5% of Y-STRs are located within 5′ untranslated or intron regions of protein coding genes. The repeat unit length of identified Y-STRs have been analyzed. Among the 400 Y-STRs, 6% are dimeric repeats, 39% are trimeric, 45% are tetrameric, 9% are pentameric, and 1% are hexameric (Figure 19.3).

Fewer than half the STRs have been characterized. Some loci are polymorphic and are useful for forensic applications and developing new Y-STR multiplex systems. The STR loci at the Y chromosome are usually referred to as ***haplotypes***. A haplotype is a collection of alleles that are usually linked (inherited together) since homologous recombination does not occur on the majority of the Y chromosome. The most commonly used Y-STR loci for forensic testing are described below (Figure 19.4 and Table 19.1).

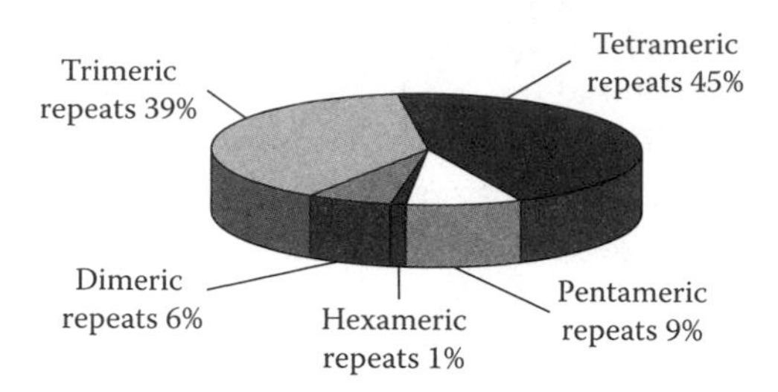


**Figure 19.3** Human Y-STRs with different repeat unit length. About 400 Y-STRs have been identified and categorized according to repeat unit length (Source: Adapted from Hanson, E. K., and J. Ballantyne. 2006. *Leg Med (Tokyo)* 8 (2):110.)

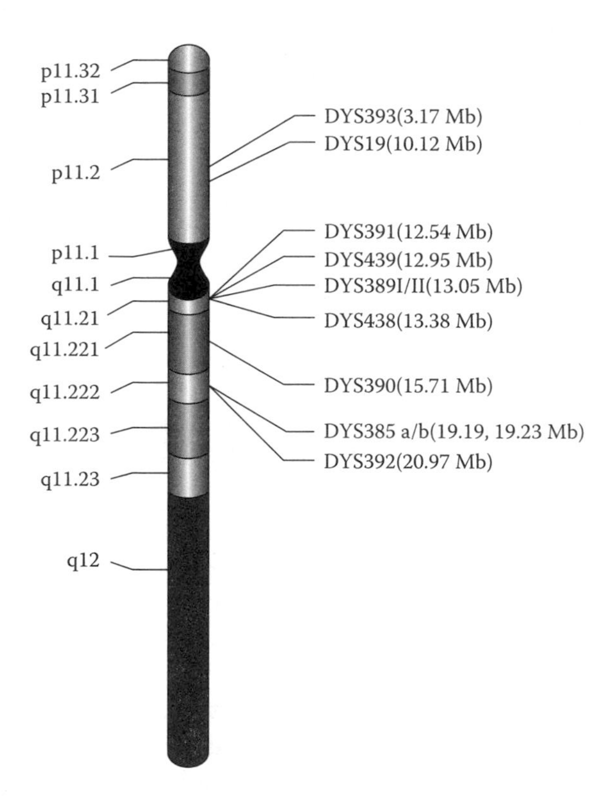


**Figure 19.4** Human cytogenetic map of Y chromosome. The Y-STRs and positions are shown. Mb = megabase. Cytogenetic patterns with alternating dark and light bands are shown.

### 19.2.2 Core Y-STR Loci

In 1997, the ***European minimal haplotype (EMH)*** locus set was recommended by the International Y-STR User Group for forensic applications. This haplotype set includes a core set of nine Y-STR loci: DYS19, DYS385a

**Table 19.1 Common Y-STR Core Loci**

| Locus | EMH | U.S. Haplotype Loci | Repeat Motif |
|---|---|---|---|
| DYS19 | ☑ | ☑ | TAGA |
| DYS385a/b | ☑ | ☑ | GAAA |
| DYS389 I | ☑ | ☑ | TCTA |
| DYS389 II | ☑ | ☑ | [TCTG][TCTA] |
| DYS390 | ☑ | ☑ | [TCTG] [TCTA] |
| DYS391 | ☑ | ☑ | TCTA |
| DYS392 | ☑ | ☑ | TAT |
| DYS393 | ☑ | ☑ | AGAT |
| DYS438 | | ☑ | TTTTC |
| DYS439 | | ☑ | GATA |

and b, DYS389I, DYS389II, DYS390, DYS391, DYS392, and DYS393. In 2003, the ***U.S. haplotype loci*** were recommended by the Scientific Working Group on DNA Analysis Methods (SWGDAM) for forensic DNA analysis. The U.S. haplotype loci includes the EMH loci set plus two additional loci, DYS438 and DYS439.

DYS385 and DYS389 are ***multi-local Y-STR loci (MLL)***. The MLL designation refers to a presence of a particular STR at more than one site on the Y chromosome DNA due to duplication. To date, about 50 such MLL Y-STRs have been identified. Further MLL subdivisions are designated bi-local, tri-local, etc. DYS385 and DYS389 are bi-local.

The DYS385 locus has two inverted duplicated clusters and is separated by a $4 \times 10^4$ bp interstitial region (Figure 19.5). It can be amplified by a single set of primers. One allele is observed if the duplicates are the same length. If the duplicated clusters have different lengths, they can generate two different alleles when amplified. The smaller sized allele is designated “a” and the larger sized allele is designated “b.”

The DYS389 locus has two duplicated clusters with the same orientation (Figure 19.6). In a single set of PCR primers, there are two binding sites for the same forward primer at each 5′ flanking sequence of the core repeat region of DYS389. These binding sites between DYS389I and DYS389II are about 120 bp apart. Therefore, two amplicons are produced. DYS389I is designed for the smaller allele and DYS389II is designated for the larger allele.

The average mutation rate for the core Y-STR loci is approximately $10^{-3}$ per generation—similar to the mutation rate of autosomal STR loci. Mutations can exert major impacts on the interpretation of paternity test results.

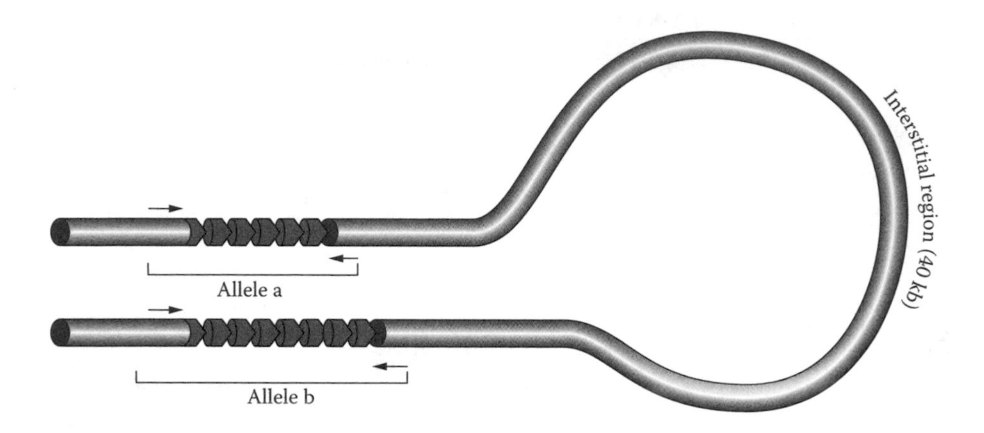


**Figure 19.5** MLL Y-STR locus DYS385. At the DYS385 locus, note the two inverted duplicated regions of the Y chromosome with an interstitial region of 40 kilobases (Kb). These inverted regions can be amplified with a single pair of primers (indicated by arrows) in one PCR reaction. The allele designations are described in the text.

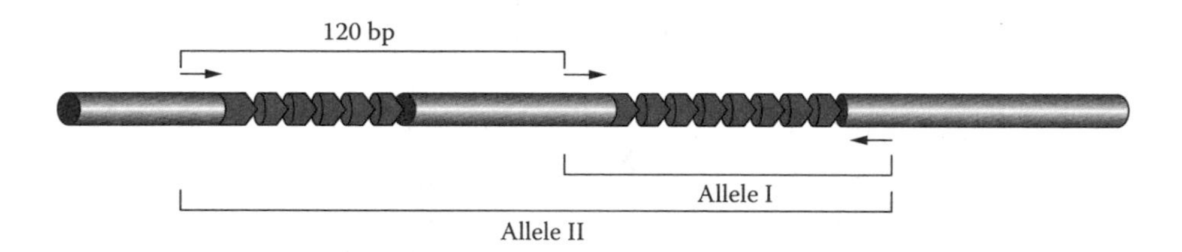


**Figure 19.6** MLL Y-STR locus DYS389. At the DYS389 locus, note the two duplicated regions of the Y-chromosome with the same orientation. These duplicated regions can be amplified with a single pair of primers (indicated by arrows) in one PCR reaction. The allele designations are described in the text.

## 19.2.3 Multiplex Y-STR

The application Y-STR for forensic casework was initiated in Europe. In the U.S., the laboratory of the Office of Chief Medical Examiner in New York City was the first to perform Y-STR testing of four loci (DYS19, DYS390, DYS389I and II) for casework (Figure 19.7). The use of Y-STR loci has been facilitated by various commercially available PCR amplification kits in multiplex systems (Table 19.2).

ReliaGene Technologies developed the first commercial multiplex Y-STR system, the Y-PLEX™6. The kit includes DYS19, DYS385a and b, DYS389II, DYS390, DYS391, and DYS393. Additional commercially available kits with more Y-STR loci are now available and have been validated

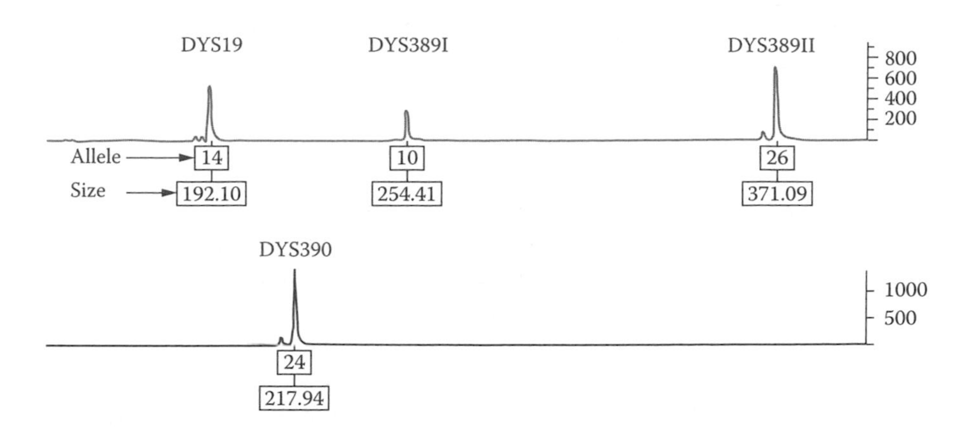


**Figure 19.7** Electropherogram of Y-STR profile using a multiplex of four loci.

**Table 19.2 Core Y-STR Loci Covered by Common Commercial Kits**

| Locus | genRES® DYSplex-1 (Serac, Germany) | genRES® DYSplex-2 (Serac, Germany) | MenPlex® Argus Y-MH (Biotype, | PowerPlex® Y (Promega, U.S.) | Yfiler™ (Applied Biosystems, U.S.) | Y-PLEX™ 12 (ReliaGene Technologies, |
|---|---|---|---|---|---|---|
| DYS19 | | ☑ | ☑ | ☑ | ☑ | ☑ |
| DYS385a and b | ☑ | | ☑ | ☑ | ☑ | ☑ |
| DYS389 I | ☑ | ☑ | ☑ | ☑ | ☑ | ☑ |
| DYS389 II | ☑ | ☑ | ☑ | ☑ | ☑ | ☑ |
| DYS390 | ☑ | | ☑ | ☑ | ☑ | ☑ |
| DYS391 | ☑ | | ☑ | ☑ | ☑ | ☑ |
| DYS392 | | ☑ | ☑ | ☑ | ☑ | ☑ |
| DYS393 | | ☑ | ☑ | ☑ | ☑ | ☑ |
| DYS438 | | | | ☑ | ☑ | ☑ |
| DYS439 | | | | ☑ | ☑ | ☑ |

for forensic use. To improve discriminating power, multiplex systems including new Y-STR loci are desired. Many new Y-STR loci are being characterized for developing new multiplex systems.

## 19.3 Gender Typing

Gender typing of a biological sample is useful in forensic investigation, for example, for victim identification in disaster cases and suspect identification in sexual assaults. One commonly used gender typing marker is the amelogenin (*AMEL*) locus.

### 19.3.1 Amelogenin Locus

This region encodes extracellular matrix proteins involved in tooth enamel formation (Table 19.3). Mutations in the *AMEL* gene can lead to an enamel defect known as amelogenesis imperfecta. The *AMEL* locus has two homologous genes: *AMELX* (Xp22.1–Xp22.3) is located on the human X chromosome (Figure 19.8) and *AMELY* (Yp11.2) is located on the human Y chromosome (Figure 19.9 and Figure 19.10). Although the genes constitute a homologous pair, they differ in size and sequence. Gender typing can be performed using various primers designed specifically for the sequences of the homologous region on these genes, followed by amplification. Different sizes of amplicons are obtained.

The most commonly used gender typing method at the *AMEL* locus is the detection of a 6-bp deletion at intron 1 of *AMELX* (Figure 19.11). This deletion is not present in *AMELY.* Primer sets were developed to amplify both alleles in a single PCR by Forensic Science Service in the United Kingdom in 1993. The amplicons generated from *AMELY* and *AMELX* are separated by

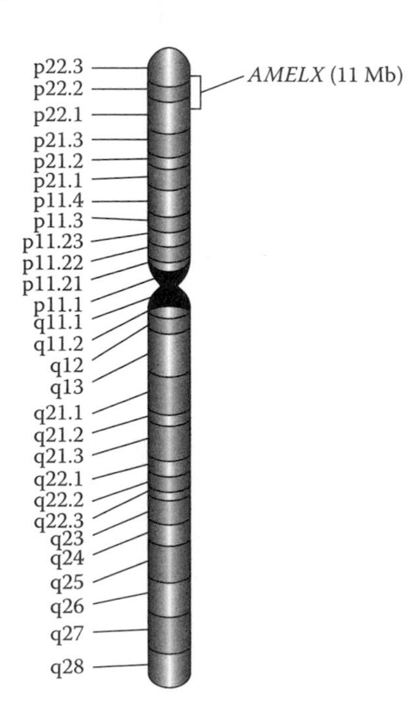


**Figure 19.8** Cytogenetic map of the human X chromosome and *AMELX* location shown with physical positions (Mb = megabases). Cytogenetic patterns with alternating dark and light bands are shown.

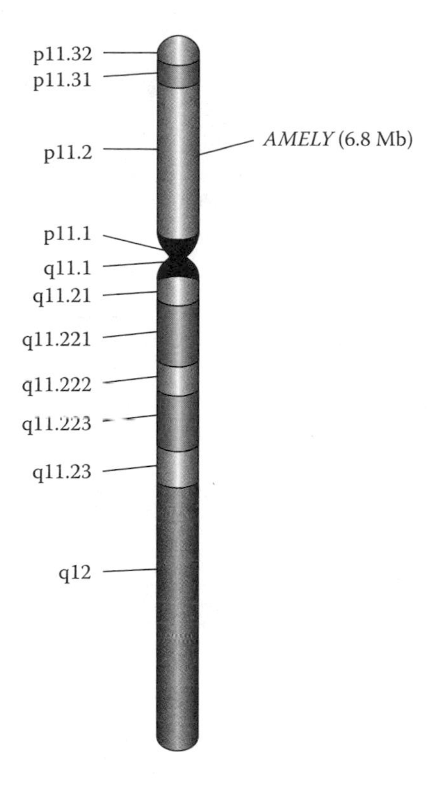


**Figure 19.9** Cytogenetic map of the human Y chromosome and *AMELY* location shown with physical positions (Mb = megabases). Cytogenetic patterns with alternating dark and light bands are shown.

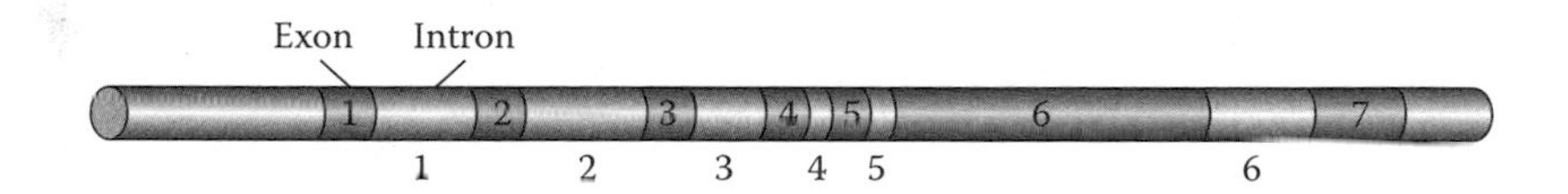


**Figure 19.10** Structure of human *AMELY* gene. Exons 1 through 7 and introns 1 through 6 are numbered.

electrophoresis. The observation of the *AMELX* fragment alone indicates a female, whereas the observation of both *AMELX* and *AMELY* indicates a male. Nevertheless, primate and some rudiment DNA can be amplified as well but the amplicon sizes vary.

The *AMEL* locus has been co-amplified with other markers to provide a combined gender and identity test. Such combined tests have been used in D1S80 AFLP and various STR multiplex analyses.

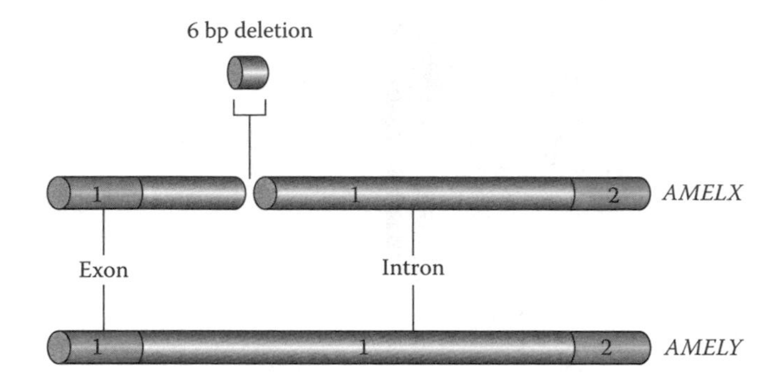


**Figure 19.11** Gender typing using *AMEL* markers. A 6-base pair deletion in intron 1 is present in *AMELX* but not in *AMELY* and can be resolved using electrophoresis as described in the text.

**Table 19.3 Genomic Locations of *AMEL* Loci**

| Gene Symbol | Gene Name | Chromosomal Location | Physical Location |
|---|---|---|---|
| *AMELX* | Amelogenin, amelogenesis imperfecta 1, X-linked | Xp22.31–p22.1 | 11.2 Mb |
| *AMELY* | Amelogenin, Y-linked | Yp11.2 | 6.8 Mb |

### 19.3.2 AMELY Null Mutations

Several cases of *AMELY* null mutations have been reported. Only the *AMELX* fragment was detected in these *AMELY* null males. Many of them are phenotypically normal but present the *AMEL* gender types of females. Various interstitial deletions at the Y chromosome short arm have been identified as the cause of some *AMELY* null gender typing. The frequency of *AMELY* null males is rare, but is higher in Sri Lanka and India.

## Bibliography

### Y-STR

Alves, C. et al. 2003. Evaluating the informative power of Y-STRs: A comparative study using European and new African haplotype data. *Forensic Sci Int* 134 (2-3):126.

Ayub, Q. et al. 2000. Identification and characterisation of novel human Y-chromosomal microsatellites from sequence database information. *Nucleic Acids Res* 28 (2):e8.

Ballard, D. J. et al. 2005. A study of mutation rates and the characterisation of intermediate, null and duplicated alleles for 13 Y chromosome STRs. *Forensic Sci Int* 155 (1):65.

Beleza, S. et al. Extending STR markers in Y chromosome haplotypes. *Int J Legal Med* 117 (1):27.

Bosch, E. et al. 2002. High resolution Y chromosome typing: 19 STRs amplified in three multiplex reactions. *Forensic Sci Int* 125 (1):42.

Butler, J. M. 2005. *Forensic DNA Typing: Biology, Technology, and Genetics of STR Markers*, 2nd ed. Burlington, MA: Elsevier.

Caglia, A. et al. 1998. Increased forensic efficiency of a STR-based Y-specific haplotype by addition of the highly polymorphic DYS385 locus. *Int J Legal Med* 111 (3):142.

Carracedo, A. et al. 2001. Results of a collaborative study of the EDNAP group regarding the reproducibility and robustness of the Y-chromosome STRs DYS19, DYS389 I and II, DYS390 and DYS393 in a PCR pentaplex format. *Forensic Sci Int* 119 (1):28.

Cerri, N. et al. 2003. Mixed stains from sexual assault cases: Autosomal or Y-chromosome short tandem repeats? *Croat Med J* 44 (3):289.

Corach, D. et al. 2001. Routine Y-STR typing in forensic casework. *Forensic Sci Int* 118 (2-3):131.

de Knijff, P. et al. 1997. Chromosome Y microsatellites: Population genetic and evolutionary aspects. *Int J Legal Med* 110 (3):134.

Dekairelle, A. F., and B. Hoste. 2001. Application of a Y-STR-pentaplex PCR (DYS19, DYS389I and II, DYS390 and DYS393) to sexual assault cases. *Forensic Sci Int* 118 (2-3):122.

Delfin, F. C. et al. 2005. Y-STR analysis for detection and objective confirmation of child sexual abuse. *Int J Legal Med* 119 (3):158.

Dupuy, B. M. et al. 2004. Y-chromosomal microsatellite mutation rates: Differences in mutation rate between and within loci. *Hum Mutat* 23 (2):117.

Foster, E. A. et al. 1998. Jefferson fathered slave's last child. *Nature* 396 (6706):27.

Gill, P. et al. 2001. DNA commission of the International Society of Forensic Genetics: Recommendations on forensic analysis using Y-chromosome STRs. *Int J Legal Med* 114 (6):305.

Gonzalez-Neira, A. et al. 2001. Sequence structure of 12 novel Y chromosome microsatellites and PCR amplification strategies. *Forensic Sci Int* 122 (1):19.

Gusmao, L., C. Alves, and A. Amorim. 2001. Molecular characteristics of four human Y-specific microsatellites (DYS434, DYD437, DYS438, DYS439) for poulation and forensic studies. *Ann Hum Genet* 65 (Pt 3):285.

Gusmao, L. et al. 1999. Robustness of the Y STRs DYS19, DYS389 I and II, DYS390 and DYS393: Optimization of a PCR pentaplex. *Forensic Sci Int* 106 (3):163.

Gusmao, L. et al. 2000. Alternative primers for DYS391 typing: Advantages of their application to forensic genetics. *Forensic Sci Int* 112 (1):49.

Hall, A., and J. Ballantyne. 2003. Novel Y-STR typing strategies reveal the genetic profile of the semen donor in extended interval post-coital cervicovaginal samples. *Forensic Sci Int* 136 (1-3):58.

Hammer, M. F. et al. 2001. Hierarchical patterns of global human Y-chromosome diversity. *Mol Biol Evol* 18 (7):1189-203.

Hanson, E. K., and J. Ballantyne. 2006. Comprehensive annotated STR physical map of the human Y chromosome: Forensic implications. *Leg Med (Tokyo)* 8 (2):110.

Hartzell, B., K. Graham, and B. McCord. 2003. Response of short tandem repeat systems to temperature and sizing methods. *Forensic Sci Int* 133 (3):228.

Henke, J. et al. 2001. Application of Y-chromosomal STR haplotypes to forensic genetics. *Croat Med J* 42 (3):292.

Heyer, E. et al. 1997. Estimating Y chromosome specific microsatellite mutation frequencies using deep rooting pedigrees. *Hum Mol Genet* 6 (5):799.

Holtkemper, U. et al. 2001. Mutation rates at two human Y-chromosomal microsatellite loci using small pool PCR techniques. *Hum Mol Genet* 10 (6):629.

Hurles, M. E., and M. A. Jobling. 2003. A singular chromosome. *Nat Genet* 34 (3):246.

Jobling, M. A. et al. 1999. Y-chromosome-specific microsatellite mutation rates reexamined using a minisatellite, MSY1. *Hum Mol Genet* 8 (11):2117.

Jobling, M. A., A. Pandya, and C. Tyler-Smith. 1997. The Y chromosome in forensic analysis and paternity testing. *Int J Legal Med* 110 (3):118.

Jobling, M. A. and C. Tyler-Smith. 1995. Fathers and sons: The Y chromosome and human evolution. *Trends Genet* 11 (11):449.

_____. 2000. New uses for new haplotypes the human Y chromosome, disease and selection. *Trends Genet* 16 (8):356.

_____. 2003. The human Y chromosome: An evolutionary marker comes of age. *Nat Rev Genet* 4 (8):598.

Junge, A. et al. 2003. Validation of the multiplex kit genRESMPX-2 for forensic casework analysis. *Int J Legal Med* 117 (6):317.

Kayser, M. et al. 2000. Characteristics and frequency of germline mutations at microsatellite loci from the human Y chromosome, as revealed by direct observation in father/son pairs. *Am J Hum Genet* 66 (5):1580.

Kayser, M. et al. 2004. A comprehensive survey of human Y-chromosomal microsatellites. *Am J Hum Genet* 74 (6):1183.

Kayser, M. and A. Sajantila. 2001. Mutations at Y-STR loci: Implications for paternity testing and forensic analysis. *Forensic Sci Int* 118 (2-3):116.

Krawczak, M. 2001. Forensic evaluation of Y-STR haplotype matches: A comment. *Forensic Sci Int* 118 (2–3):114.

Krenke, B. E. et al. 2005. Validation of a male-specific, 12-locus fluorescent short tandem repeat (STR) multiplex. *Forensic Sci Int* 148 (1):1.

Lahn, B. T., N. M. Pearson, and K. Jegalian. 2001. The human Y chromosome, in the light of evolution. *Nat Rev Genet* 2 (3):207.

Lim, S. K. et al. 2007. Variation of 52 new Y-STR loci in the Y Chromosome Consortium worldwide panel of 76 diverse individuals. *Int J Legal Med* 121 (2):124.

Mulero, J. J., C. W. Chang, and L. K. Hennessy. 2006. Characterization of the N+3 stutter product in the trinucleotide repeat locus DYS392. *J Forensic Sci* 51 (5):1069.

Nebel, A. et al. 2001. Haplogroup-specific deviation from the stepwise mutation model at the microsatellite loci DYS388 and DYS392. *Eur J Hum Genet* 9 (1):22.

Niederstatter, H. et al. 2005. Separate analysis of DYS385a and b versus conventional DYS385 typing: Is there forensic relevance? *Int J Legal Med* 119 (1):1.

Park, M. J. et al. 2007. Y-STR analysis of degraded DNA using reduced-size amplicons. *Int J Legal Med* 121 (2):152.

Parson, W. et al. 2003. Improved specificity of Y-STR typing in DNA mixture samples. *Int J Legal Med* 117 (2):109.

Parson, W. et al. 2001. When autosomal short tandem repeats fail: Optimized primer and reaction design for Y-chromosome short tandem repeat analysis in forensic casework. *Croat Med J* 42 (3):285.

Pascali, V. L., M. Dobosz, and B. Brinkmann. 1999. Coordinating Y-chromosomal STR research for the Courts. *Int J Legal Med* 112 (1):1.

Prinz, M. et al. 1997. Multiplexing of Y chromosome specific STRs and performance for mixed samples. *Forensic Sci Int* 85 (3):209.

Prinz, M. et al. 2001. Validation and casework application of a Y chromosome specific STR multiplex. *Forensic Sci Int* 120 (3):177.

Prinz, M., and M. Sansone. 2001. Y chromosome-specific short tandem repeats in forensic casework. *Croat Med J* 42 (3):288.

Redd, A. J. et al. 2002. Forensic value of 14 novel STRs on the human Y chromosome. *Forensic Sci Int* 130 (2-3):97.

Redd, A. J., S. L. Clifford, and M. Stoneking. 1997. Multiplex DNA typing of short-tandem-repeat loci on the Y chromosome. *Biol Chem* 378 (8):923.

Roewer, L. et al. 2000. A new method for the evaluation of matches in non-recombining genomes: Application to Y-chromosomal short tandem repeat (STR) haplotypes in European males. *Forensic Sci Int* 114 (1):31.

Roewer, L. et al. 2001. Online reference database of European Y-chromosomal short tandem repeat (STR) haplotypes. *Forensic Sci Int* 118 (2-3):106.

Rolf, B. et al. 2001. Paternity testing using Y-STR haplotypes: Assigning a probability for paternity in cases of mutations. *Int J Legal Med* 115 (1):12.

Rozen, S. et al. 2003. Abundant gene conversion between arms of palindromes in human and ape Y chromosomes. *Nature* 423 (6942):873.

Sanchez-Diaz, P. et al. 2003. Results of the GEP-ISFG collaborative study on two Y-STRs tetraplexes: GEPY I (DYS461, GATA C4, DYS437 and DYS438) and GEPY II (DYS460, GATA A10, GATA H4 and DYS439). *Forensic Sci Int* 135 (2):158.

Santos, F. R., N. O. Bianchi, and S. D. Pena. 1996. Worldwide distribution of human Y-chromosome haplotypes. *Genome Res* 6 (7):601.

Santos, F. R., S. D. Pena, and J. T. Epplen. 1993. Genetic and population study of a Y-linked tetranucleotide repeat DNA polymorphism with a simple non-isotopic technique. *Hum Genet* 90 (6):655.

Saxena, R. et al. The DAZ gene cluster on the human Y chromosome arose from an autosomal gene that was transposed, repeatedly amplified and pruned. *Nat Genet* 14 (3):292.

Schneider, P. M. et al. 1999. Results of collaborative study regarding the standardization of the Y-linked STR system DYS385 by the European DNA Profiling (EDNAP) group. *Forensic Sci Int* 102 (2-3):159.

Schoske, R. et al. 2003. Multiplex PCR design strategy used for the simultaneous amplification of 10 Y chromosome short tandem repeat (STR) loci. *Anal Bioanal Chem* 375 (3):333.

Schoske, R. et al. 2004. High-throughput Y-STR typing of U.S. populations with 27 regions of the Y chromosome using two multiplex PCR assays. *Forensic Sci Int* 139 (2-3):107.

Seo, Y. et al. 2003. A method for genotyping Y chromosome-linked DYS385a and DYS385b loci. *Leg Med (Tokyo)* 5 (4):228.

Sibille, I. et al. 2002. Y-STR DNA amplification as biological evidence in sexually assaulted female victims with no cytological detection of spermatozoa. *Forensic Sci Int* 125 (2-3):212.

Skaletsky, H. et al. 2003. The male-specific region of the human Y chromosome is a mosaic of discrete sequence classes. *Nature* 423 (6942):825.

Szibor, R., M. Kayser, and L. Roewer. 2000. Identification of the human Y-chromosomal microsatellite locus DYS19 from degraded DNA. *Am J Forensic Med Pathol* 21 (3):252.

Tilford, C. A. et al. 2001. A physical map of the human Y chromosome. *Nature* 409 (6822):943.

Tsuji, A. et al. 2001. Personal identification using Y-chromosomal short tandem repeats from bodily fluids mixed with semen. *Am J Forensic Med Pathol* 22 (3):288.

Underhill, P. A. et al. 2000. Y chromosome sequence variation and the history of human populations. *Nat Genet* 26 (3):358.

White, P. S. et al. 1999. New, male-specific microsatellite markers from the human Y chromosome. *Genomics* 57 (3):433.

Zhivotovsky, L. A. et al. 2004. The effective mutation rate at Y chromosome short tandem repeats, with application to human population–divergence time. *Am J Hum Genet* 74 (1):50.

## Gender Typing

Buel, E., G. Wang, and M. Schwartz. 1995. PCR amplification of animal DNA with human X–Y amelogenin primers used in gender determination. *J Forensic Sci* 40 (4):641.

Cadenas, A. M. et al. 2007. Male amelogenin dropouts: Phylogenetic context, origins and implications. *Forensic Sci Int* 166 (2–3):155.

Chang, Y. M., L. A. Burgoyne, and K. Both. 2003. Higher failures of amelogenin sex test in an Indian population group. *J Forensic Sci* 48 (6):1309.

Chang, Y. M. et al. 2007. A distinct Y-STR haplotype for Amelogenin negative males characterized by a large Y(p)11.2 (DYS458-MSY1-AMEL-Y) deletion. *Forensic Sci Int* 166 (2-3):115.

Fincham, A. G. et al. 1991. Human developing enamel proteins exhibit a sex-linked dimorphism. *Calcif Tissue Int* 48 (4):288.

Haas-Rochholz, H., and G. Weiler. 1997. Additional primer sets for an amelogenin gene PCR-based DNA-sex test. *Int J Legal Med* 110 (6):312.

Kashyap, V. K. et al. 2006. Deletions in the Y-derived amelogenin gene fragment in the Indian population. *BMC Med Genet* 7:37.

Lattanzi, W. et al. 2005. A large interstitial deletion encompassing the amelogenin gene on the short arm of the Y chromosome. *Hum Genet* 116 (5):395.

Lau, E. C. et al. 1989. Human and mouse amelogenin gene loci are on the sex chromosomes. *Genomics* 4 (2):162.

Lau, E. C., H. C. Slavkin, and M. L. Snead. 1990. Analysis of human enamel genes: Insights into genetic disorders of enamel. *Cleft Palate J* 27 (2):121.

Mannucci, A. et al. 1994. Forensic application of a rapid and quantitative DNA sex test by amplification of the X-Y homologous gene amelogenin. *Int J Legal Med* 106 (4):190.

Mitchell, R. J. et al. 2006. An investigation of sequence deletions of amelogenin (AMELY), a Y-chromosome locus commonly used for gender determination. *Ann Hum Biol* 33 (2):227.

Nakahori, Y., O. Takenaka, and Y. Nakagome. 1991. A human X-Y homologous region encodes amelogenin. *Genomics* 9 (2):264.

Roffey, P. E., C. I. Eckhoff, and J. L. Kuhl. 2000. A rare mutation in the amelogenin gene and its potential investigative ramifications. *J Forensic Sci* 45 (5):1016.

Sire, J. Y., S. Delgado, and M. Girondot. 2006. The amelogenin story: Origin and evolution. *Eur J Oral Sci* 114 Suppl 1:64.

Snead, M. L. et al. 1989. Of mice and men: Anatomy of the amelogenin gene. *Connect Tissue Res* 22 (1–4):101.

Steinlechner, M. et al. 2002. Rare failures in the amelogenin sex test. *Int J Legal Med* 116 (2):117-20.

Thangaraj, K., A. G. Reddy, and L. Singh. 2002. Is the amelogenin gene reliable for gender identification in forensic casework and prenatal diagnosis? *Int J Legal Med* 116 (2):121.

Umeno, M. et al. 2006. A rapid and simple system of detecting deletions on the Y chromosome related with male infertility using multiplex PCR. *J Med Invest* 53 (1-2):147.

## Study Questions

1. The amelogenin locus of the:
   (a) X chromosome has a 6-base pair deletion at intron 1.
   (b) Y chromosome has a 6-base pair deletion at intron 1.

2. Most Y chromosome STRs used for forensic cases are:
   (a) Trimeric repeats
   (b) Tetrameric repeats
   (c) Pentameric repeats
   (d) Hexameric repeats

3. The Y chromosome should be identical among:
   (a) Brothers from the same father
   (b) Half brothers from the same father
   (c) Both of the above

4. Which test is most preferred for gender identification?
   (a) Y-STR
   (b) Y-SNP
   (c) Amelogenin marker

5. Which of the following has a lower mutation rate?
   (a) Y-STR
   (b) Y-SNP

6. The Y chromosome is always located in the:
   (a) Mitochondria
   (b) Nucleus
   (c) Both of the above
   (d) None of the above

# Single Nucleotide Polymorphism Profiling 20

## 20.1 Basic Characteristics of SNPs

Sequence polymorphisms are sequence variations in the human genome. One type is called a ***single-nucleotide polymorphism (SNP)*** and constitutes single base pair change originating from spontaneous mutation. SNPs can result from base substitutions, insertions, or deletions at a single site. They account for most human DNA polymorphisms. An estimated 10 million SNPs exist in the human genome and approximately 1.4 million SNPs have been identified. Most appear in noncoding regions of the genome, although some are found in coding regions as well (Figure 20.1).

Most SNPs are biallelic, although very rare triallelic and tetraallelic SNPs also occur. As noted earlier, an SNP originates from a spontaneous mutation in the genome. If it is a germ-line mutation, it can be inherited by offspring and spread in the population. As a result, both the parent and mutant alleles are produced (biallelic SNP). Subsequently, if a third mutation occurs at the same nucleotide site, a rare triallelic SNP is produced.

SNP loci have the advantages in that (1) they are abundant within the human genome and can be used as markers for forensic applications; (2) SNP amplicon sizes (usually 50 to 100 base pairs in length) are smaller than STR amplicons and thus can be useful when dealing with degraded DNA samples in which many STR loci are not successfully amplified by PCR; (3) SNPs have low mutation rates and are thus useful for paternity testing; and (4) many SNP analysis methods including multiplex systems are available.

The technique also has some disadvantages in that (1) SNP loci are not as polymorphic for forensic identity testing as STR loci. It is estimated that 50 to 60 SNP loci are required to achieve a similar level of the population match probability (*P*m; the lower the *P*m, the less likely a match will occur between two randomly chosen individuals) using 13 STR loci in CODIS; (2) it is difficult to resolve mixed DNA profiles because most SNPs are biallelic; and (3) most DNA databases contain STR profiles instead of SNP profiles.

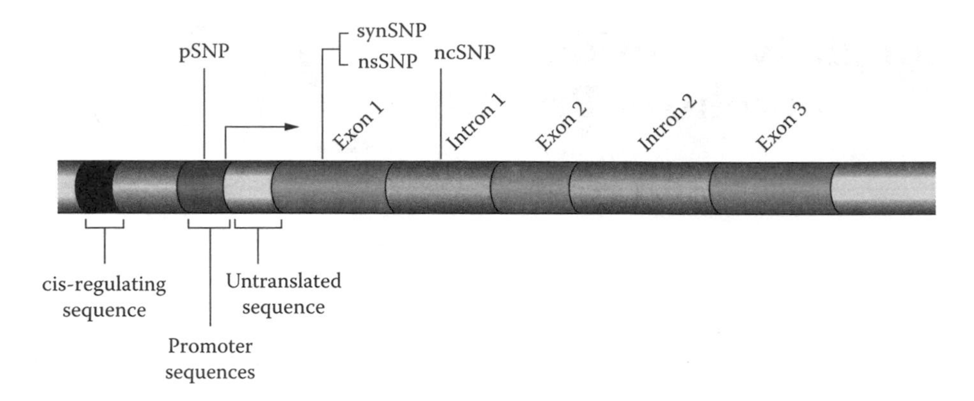


**Figure 20.1** SNPs fall into several classes. Most reside in the noncoding regions of DNA and are designated as noncoding SNPs (ncSNPs). A subset of ncSNPs can also be found in introns. SNPs residing in exons are further divided into two types: the synonymous type (synSNP) is an exonic SNP that does not change the amino acid composition of the encoded polypeptide. Conversely, a nonsynonymous type (nsSNP) changes the encoded amino acid. SNPs in the promoter regions of the genome are known as promoter SNPs (pSNPs). Arrow = transcription start site.

## 20.2 Forensic Applications of SNP Profiling

### 20.2.1 HLA-DQA1 LOCUS

The first use of SNP-based profiling for forensic application involved sequence polymorphisms at the HLA-DQA1 locus (Table 20.1). The HLA-DQA1 gene is a highly polymorphic member of the human leukocyte antigen (HLA) family involved in the immune response. The HLA-DQA1 locus is located within human HLA gene clusters on chromosome 6. The region tested for forensic use is located at the second exon of the gene.

#### *20.2.1.1 Commercial Kits*

The DQα AmpliType® kit (HLA-DQA1 was formerly called the DQα locus) was the first commercial kit developed in the late 1980s by Cetus Corporation (Emeryville, CA). It can distinguish among seven allelic genotypes at the DQA1 locus designated 1.1, 1.2, 1.3, 2, 3, and 4. The *Pm* is approximately $5 \times 10^{-2}$. Although the *Pm* of this panel is high, it is useful as a preliminary test to quickly exclude innocent suspects.

In addition to the HLA-DQA1 locus, five additional loci, LDLR, GYPA, HBGG, D7S8, and GC, were utilized for forensic application in 1993 (Table 20.1). They are included in the second generation of the kit known as the AmpliType® PM PCR amplification and typing kit (also known as Polymarker) manufactured by Perkin-Elmer (Norwalk, CT). It consists of two panels for the testing of HLA-DQAl and other loci (Figure 20.2).

**Table 20.1 Chromosomal Locations of SNP Loci Used in AmpliType® PM PCR Amplification and Typing Kit**

| Locus | Gene Product | Chromosome Location | Number of Alleles | Further Reading |
|---|---|---|---|---|
| HLA-DQA1 | HLA-DQA1 | 6p21.3 | 7 | Gyllensten and Erlich (1988) |
| LDLR | Low density lipoprotein receptor | 19p 13.1-13.3 | 2 | Yamamoto et al. (1984) |
| GYPA | Glycophorin A | 4q28-31 | 2 | Siebert and Fukuda (1987) |
| HBGG | Hemoglobin G gammaglobin | 11 p 15.5 | 3 | Slightom et al. (1980) |
| D7S8 | Anonymous | 7q22-31.1 | 2 | Horn et al. (1990) |
| GC | Group specific component | 4q11-13 | 3 | Yang et al. (1985) |

*Source*: Adapted from Budowle, B. et al. 2000. DNA *Typing Protocols: Molecular Biology and Forensic Analysis.* Natick, MA: Eaton Publishing.

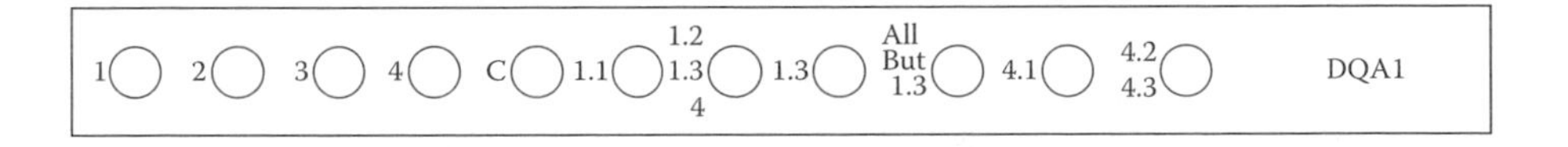




**Figure 20.2** Panels of immobilized probes in the Polymarker kit. Top: DQA1. Bottom: additional five loci (LDLR, GYPA, HBGG, D7S8, and GC). C and S = threshold control dots.

The DQA1 panel can distinguish the following: the 1 allele (subtyped as 1.1, 1.2, or 1.3), 2, 3, and 4 alleles (subtyped as 4.1 and 4.2/4.3 in which the 4.2 and 4.3 alleles are combined and cannot be distinguished). A total of 28 genotypes were possible from combinations of these alleles. LDLR, GYPA, and D7S8 each have two detectable alleles (designated A and B), while HBGG and GC each have three alleles (A, B, and C) that can be typed. As a result, the *Pm* decreases to $10^{-4}$. Polymarker profiles have been accepted in U.S. courts. The Polymarker was utilized in many forensic laboratories until replaced by STR profiling in the late 1990s. Both DQα AmpliType® and Polymarker kits utilize the assay described below.

### 20.2.1.2 Allele-Specific Oligonucleotide Hybridization

Also known as ***ASO hybridization***, this technique analyzes single nucleotide variations such as SNPs at a given locus. It is based on the principle that ASO probes, usually 14 to 17 bases in length, hybridize to their complementary DNA sequences to distinguish known polymorphic alleles. ASO probes for multiple alleles at several loci can be arranged on the same panel to establish the presence or absence of specific alleles in PCR-amplified fragments of a DNA sample (Figure 20.3). Thus, the genotypes can be determined.

In the DQα AmpliType® and Polymarker kits, the oligonucleotides representing different alleles are immobilized to a solid matrix consisting of nylon membrane strips. Each immobilized probe at a particular site on the membrane is utilized to detect corresponding SNPs. Since the probe rather than the target DNA (as with regular blot format) is immobilized to the solid phase, the configuration used here is known as a ***reverse blot*** format. The regions of the DNA in question are amplified by PCR. One of each pair of the primers is conjugated with biotin at the 5′ end (Figure 20.4). Thus, the amplified products are biotinylated for purposes of detection. This kit established multiplexing of a six-loci system allowing simultaneous amplification of the HLA-DQA1 locus along with LDLR, GYPA, HBGG, D7S8, and GC in a single reaction for each sample. Following the denaturation of the PCR product to separate the two DNA strands of the PCR products, the biotinylated strands are hybridized to the immobilized probes. Hybridization and washing conditions are established to ensure perfect hybridization of the ASO probe and its target sequence. Unbound amplified products are washed away.

The presence of a PCR product bound to a specific probe can be detected by a colorimetric detection system (Figure 20.5). Since the amplified PCR product is biotinylated, a horseradish peroxidase-conjugated streptavidin

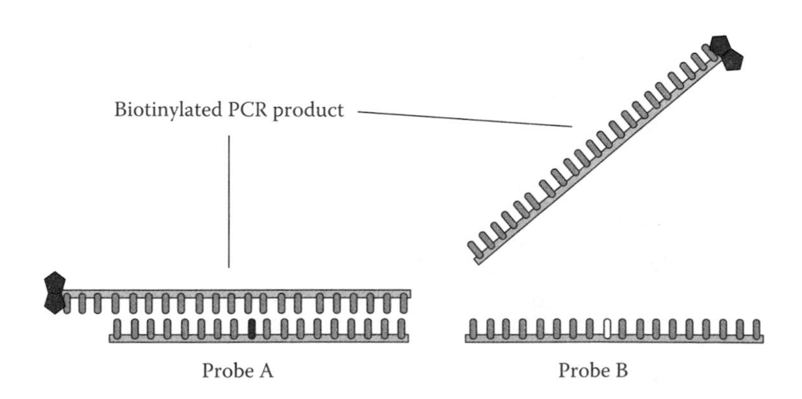


**Figure 20.3** Hybridization with ASO. Two probes are incubated with the target DNA containing the SNP. Only the perfectly matched probe–target DNA can hybridize under optimal conditions.

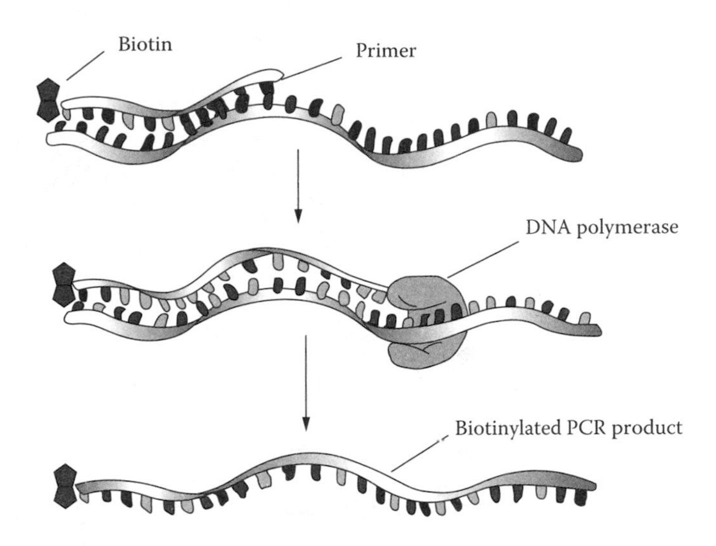


**Figure 20.4** Amplification of DNA using biotinylated primers. The biotin is conjugated at the 5′ end of the primer. The amplified products are biotinylated.

complex is allowed to bind to the biotinylated PCR products. The horseradish peroxidase then catalyzes the oxidation reaction of the colorless substrate, tetramethyl benzidine (TMB), into a blue precipitate at a designated location on the membrane strip, allowing the genotype of the sample to be determined (Figure 20.6 and Figure 20.7).

The kit used a threshold control (C dot in the HLA-DQA1 panel and S dot in the panel of the other five loci) to distinguish between signal and background noise and determined whether a sufficient amount of DNA was amplified to detect all the alleles present in a sample. If the signal intensities of allele dots were greater then or equal to the threshold control, the alleles were considered true. If the allele dots were less intense than the threshold control, they were considered inconclusive for determination of full genotypes because an allele may not have been detected due to a low level of DNA template.

The Polymarker system demonstrated a number of advantages compared to the previous RFLP methods:

- It is a PCR-based method and is thus more sensitive (sensitivity is approximately 2 ng) than RFLP and capable of successfully analyzing less amounts of a DNA sample.
- The amplicons of all loci tested are much smaller than that of RFLP so that degraded DNA can be analyzed.
- It is more rapid and less laborious than RFLP.
- All the allele amplicons at a given locus have identical lengths and are thus not amenable to preferential amplification.

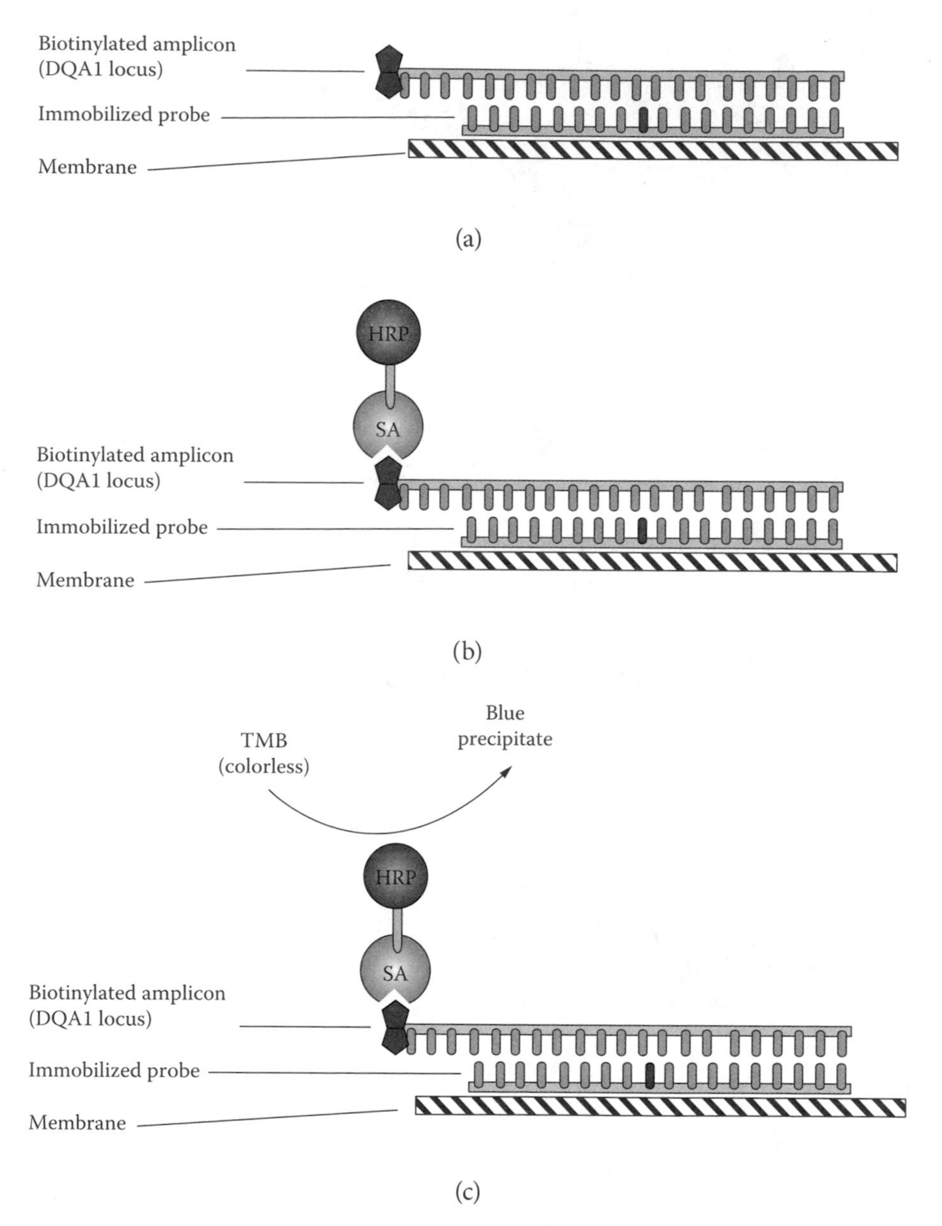


**Figure 20.5** Reverse blot assay. (a) Probe is immobilized onto a solid phase membrane and hybridized with a biotinylated PCR product having the target sequence. (b) Detection of hybridization is carried out by streptavidin (SA) and horseradish peroxidase (HRP) conjugate. (c) Colorimetric reaction is catalyzed by HRP using TMB as substrate.

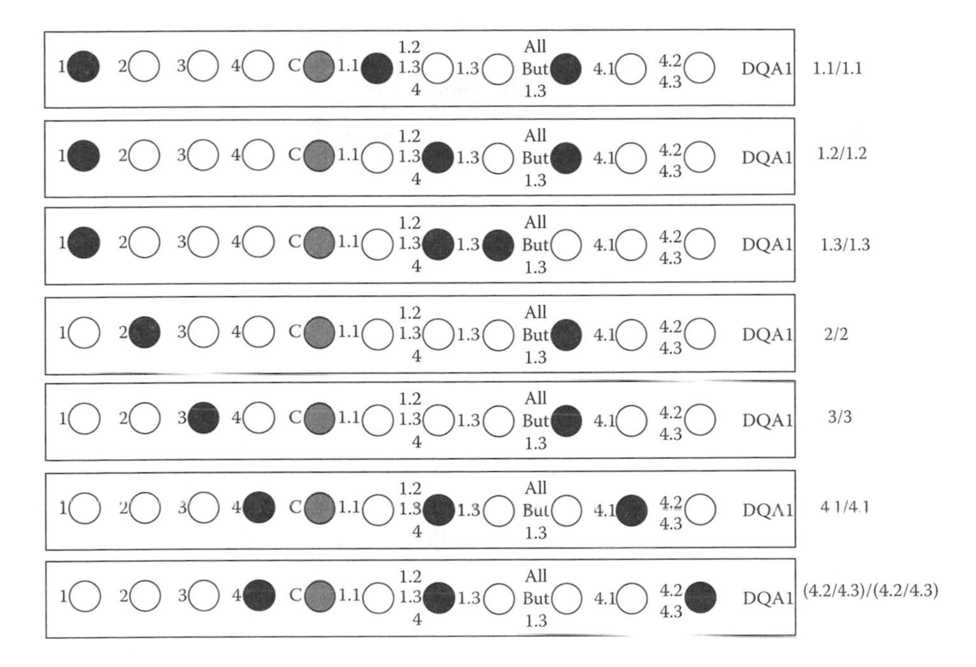


**Figure 20.6** DNA typing of homozygous alleles at DQA1 locus using reverse blot assay (Polymarker kit). The genotype of each sample is noted at the right.

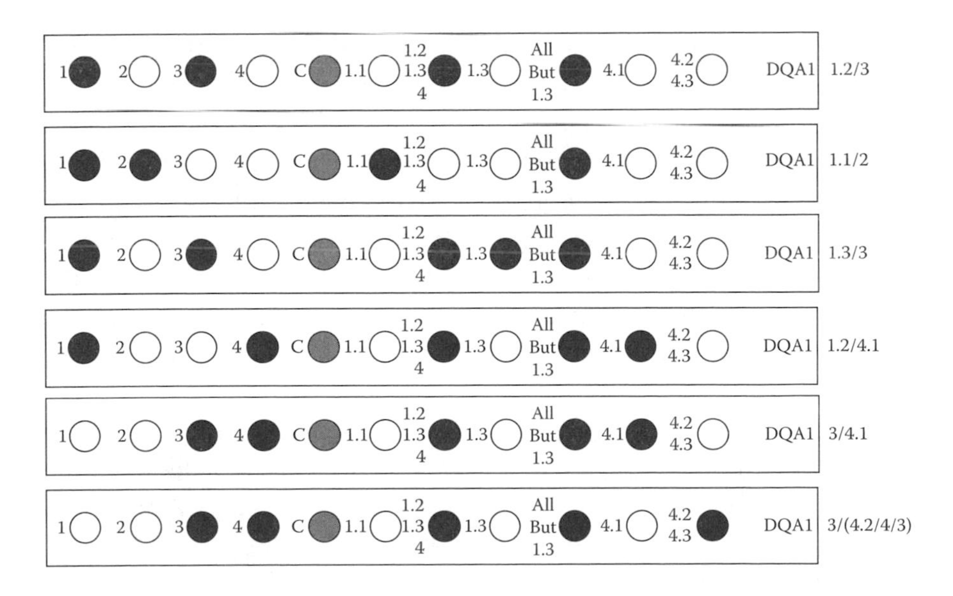


**Figure 20.7** DNA typing of heterozygous alleles at DQA1 locus using reverse blot assay (Polymarker kit). The representative genotypes of samples are noted at the right.

However, the Polymarker system has several limitations:

- The *Pm* values of two randomly selected individuals are higher in the Polymarker system compared to the values in the RFLP method.
- The limited number of alleles per locus makes identification of components of mixtures more difficult than with RFLP. Polymarker loci are less variable than VNTR loci and it may be possible that contributors share common alleles in a mixed sample.

## 20.2.2 Existing and Potential Applications

### *20.2.2.1 Application of SNPs for Forensic Identification*

The applications of SNPs for forensic identity applications are summarized in Table 20.2. Autosomal SNP panels can be used for many types of forensic testing including analysis of degraded DNA samples. The SNP panels for mtDNA profiling are under development and may serve as alternative methods to direct sequencing which is time consuming and laborious. SNP loci on the Y chromosome are also potentially useful markers for paternity testing because of low mutation rates. SNP loci such as ancestry informative markers (AIM) can be used to determine ethnic origins of questioned samples to generate leads for investigations.

### *20.2.2.2 Potential Application of SNP for Phenotyping*

One potential DNA analysis application is determining phenotypic information, also known as ***phenotyping*** (Table 20.2). The relevant SNP loci usually are ***nonsynonymous SNPs*** (***nsSNPs***); they reside in the exon and change the encoded amino acid and lead to an altered phenotype. Phenotyping of a questioned sample can reveal physical characteristics such as hair and eye colors to provide leads for investigations. A number of SNPs residing within the melanocortin 1 receptor gene (*MC1R)* are associated with red hair, fair skin, and freckles, while SNPs residing within the *P* gene that play a role in pigmentation are associated with eye color variations.

Phenotyping can also be employed in the area of forensic pathology. Cardiac arrhythmia long QT syndrome (LQTS) can cause sudden death. A number of LQTS-associated SNPs, for example, SNPs in *KCNH2* and *SCN5A* genes, have been shown to correlate to such deaths. Thus, these SNPs are potentially useful for investigating the causes of death. Finally, phenotyping also has applications in forensic toxicology. A number of SNPs in the genes such as *CYP2D6* that are responsible for metabolizing drugs can serve as potential markers for postmortem investigations of drug overdose cases.

**Table 20.2 Examples of Forensic Applications of SNP Profiling**

| Testing | Candidate SNP Loci | Application | Further Reading |
|---|---|---|---|
| Identity | Autosomal SNPs | Human identification | Sanchez et al., 2006 Kidd et al., 2006 |
| | | Human identification via degraded DNA | Budowle et al., 2004 |
| | mtDNA SNPs | Human identification | Grignani et al., 2006 |
| | Y chromosome SNPs | Paternity testing | Hammer et al., 2006 |
| Biogeographical origin | Ancestry informative markers (AIMs) | Ethnic group identification | Frudakis et al., 2003; Shriver and Kittles, 2004 |
| Physical characteristics | *MC1R* (melanocortin 1 receptor gene) | Hair color identification (investigative lead) | Grimes et al., 2001 |
| | *P* (gene has role in pigmentation) | Eye color identification (investigative lead) | Rebbeck et al., 2002 |
| Pathology | *KCNH2* (cardiac potassium channel gene) | Determining cause of sudden death from cardiac arrhythmia long QT syndrome | Lunetta et al., 2002 |
| | *SCN5A* (encodes cardiac sodium channel gene) | Determining cause of sudden death from cardiac arrhythmia long QT syndrome | Burke et al., 2005 |
| Toxicology | *CYP2C19, CYP2D6, CYP3A4, CYP2E1* (drug metabolizing enzyme genes) | Investigation of drug overdose (including death) | Kupiec et al., 2006 |

#### *20.2.2.3 Techniques*

Over the years, various SNP analysis techniques have been developed and can be divided into four groups based on mechanisms used: allele-specific hybridization, primer extension, oligonucleotide ligation, and invasive cleavage. In ***allele-specific hybridization***, allele discrimination is based on an optimal condition allowing only the perfectly matched probe–target hybridization to form. ***Primer extension*** methods are based on the ability of DNA polymerase to incorporate specific dNTPs (deoxynucleotides) complementary to the sequence of the template DNA. ***Allele-specific oligonucleotide ligation*** is based on the condition that only the allelic probe perfectly matched to the target will be ligated. In the ***invasive cleavage*** method, allelic discrimination is based on DNA sequence-specific cleavage by endonucleases.

A number of detection methods can be utilized in SNP analysis such as the measurements of fluorescence, luminescence, and molecular mass. Most assays are carried out in solution or on a solid matrix support such as glass slide, chip, or bead. Table 20.3 summarizes the representative assays for SNP typing.

**Table 20.3 Techniques for SNP Assay**

| Basis of Technique | Representative Assay | Detection | Format | Notes | Further Reading |
|---|---|---|---|---|---|
| Allele-specific hybridization | Reverse Blot | Colorimetry | Membrane--based | Useful for screen assay | Saiki et al., 1989 |
| | LightCycler® (Roche) | Fluorescence | Solution-based | Applicable for automation; limited multiplexing capability | Lareu et al., 2001 |
| | TaqMan® (Applied Biosystems) | Fluorescence | Solution-based | Applicable for automation; limited multiplexing capability | Livak, 1999 |
| | Molecular beacons | Fluorescence | Solution-based | Applicable for automation; limited multiplexing capability | Tyagi et al., 1998 |
| | GeneChip® (Affymetrix) | Fluorescence | Chip-based | Designed for genotyping large number of SNPs; exceeds forensic needs | Mei et al., 2000 |
| Primer extension | SNaPshot™ (Applied Biosystems) | Fluorescence | Solution-based | Applicable for capillary electrophoresis and multiplexing | Sanchez et al., 2006 |
| | PinPoint (Applied Biosystems) | Mass spectrometry | Solution-based | Multiplex capability lower than SNaPshot™ | Haff and Smirnov, 1997 |
| | Arrayed primer extension (APEX) | Fluorescence | Chip-based | Low reproducibility | Pastinen et al., 1997 |
| | Pyrosequencing | Fluorescence | Solution-based | May be applicable for analysis of mixed profiles; not applicable for automation; limited multiplexing capability | Ronaghi et al., 1998 |

**Table 20.3 Techniques for SNP Assay** (continued)

| Allele-specific oligonucleotide ligation | SNPstream® UHT (Orchid-Gene Screen) | Fluorescence | Chip-based | Developed for mass disaster cases | Bell et al., 2002 |
|---|---|---|---|---|---|
| | SNPlex™ (Applied Biosystems) | Fluorescence | Solution-based | Applicable for multiplexing; requires higher amount of target DNA template than other PCR-based methods | Delahunty et al., 1996 |
| Invasive cleavage | Invader® (Third Wave Technology) | Fluorescence | Solution-based | PCR not required; large amount of target DNA required | Lyamichev et al., 1999 |

*Source:* Adapted from Sobrino, B., M. Brion, and A. Carracedo. 2005. *Forensic Sci Int* 154 (2–3):181; Budowle, B. 2004. *Forensic Sci Int* 146 Suppl:S139.

## Bibliography

### Polymarker Loci

Blake, E. et al. 1992. Polymerase chain reaction (PCR) amplification and human leukocyte antigen (HLA)-DQ alpha oligonucleotide typing on biological evidence samples: Casework experience. *J Forensic Sci* 37 (3):700.

Budowle, B., F. S. Baechtel, and R. Fejeran. 1998. Polymarker, HLA-DQA1, and D1S80 allele frequency data in Chamorro and Filipino populations from Guam. *J Forensic Sci* 43 (6):1195.

Budowle, B. et al. Validation and population studies of the loci LDLR, GYPA, HBGG, D7S8, and Gc (PM loci), and HLA-DQ alpha using a multiplex amplification and typing procedure. *J Forensic Sci* 40 (1):45.

Budowle, B. et al. 2000. *DNA Typing Protocols: Molecular Biology and Forensic Analysis*. Natick, MA: Eaton Publishing.

Comey, C. T., and B. Budowle. 1991. Validation studies on the analysis of the HLA DQα locus using the polymerase chain reaction. *J Forensic Sci* 36 (6):1633.

Comey, C. T. et al. 1993. PCR amplification and typing of the HLA DQ alpha gene in forensic samples. *J Forensic Sci* 38 (2):239.

Cowland, J. B., H. O. Madsen, and N. Morling. 1995. HLA-DQA1 typing in Danes by two polymerase chain reaction (PCR) based methods. *Forensic Sci Int* 73 (1):1.

Dimo-Simonin, N., and C. Brandt-Casadevall. 1996. Evaluation and usefulness of reverse dot blot DNA PolyMarker typing in forensic case work. *Forensic Sci Int* 81 (1):61.

Erlich, H. A. 1992. HlA-DQalpha typing of forensic specimens. *Forensic Sci Int* 53 (2):227.

Erlich, H. A. et al. 1990. Reliability of the HLA-DQ alpha PCR-based oligonucleotide typing system. *J Forensic Sci* 35 (5):1017.

Erlich, H. A., E. L. Sheldon, and G. Horn. 1986. HLA typing using DNA probes. *Biotechnology* 4:7.

Fildes, N., and R. Reynolds. 1995. Consistency and reproducibility of AmpliType PM results between seven laboratories: Field trial results. *J Forensic Sci* 40 (2):279.

Gene, M. et al. 2000. Population study of Aymara Amerindians for the PCR-DNA polymorphisms HUMTH01, HUMVWA31A, D3S1358, D8S1179, D18S51, D19S253, YNZ22 and HLA-DQalpha. *Int J Legal Med* 113 (2):126.

Gino, S. et al. 1999. LDLR, GYPA, HBGG, D7S8 and GC allele and genotype frequencies in the northwest Italian population. *J Forensic Sci* 44 (1):171.

Giroti, R., and V. K. Kashyap. 1998. Detection of the source of mislabeled biopsy tissue paraffin block and histopathological section on glass slide. *Diagn Mol Pathol* 7 (6):331.

Giroti, R. I., R. Biswas, and K. Mukherjee. 2002. Restriction fragment length polymorphism and polymerase chain reaction: HLA-DQA1 and polymarker analysis of blood samples from transfusion recipients. *Am J Clin Pathol* 118 (3):382.

Gross, A. M., and R. A. Guerrieri. 1996. HLA DQA1 and Polymarker validations for forensic casework: standard specimens, reproducibility, and mixed specimens. *J Forensic Sci* 41 (6):1022.

Gyllensten, U. B., and H. A. Erlich. 1988. Generation of single-stranded DNA by the polymerase chain reaction and its application to direct sequencing of the HLA-DQA locus. *Proc Natl Acad Sci USA* 85 (20):7652.

Hallenberg, C., and N. Morling. 2001. Report of the 1997, 1998 and 1999 Paternity Testing Workshops of the English Speaking Working Group of the International Society for Forensic Genetics. *Forensic Sci Int* 116 (1):23.

Helmuth, R. et al. 1990. HLA-DQ alpha allele and genotype frequencies in various human populations, determined by using enzymatic amplification and oligonucleotide probes. *Am J Hum Genet* 47 (3):515.

Herrin, G., N. Fildes, and R. Reynolds. 1994. Evaluation of the AmpliType PM DNA test system on forensic case samples. *J Forensic Sci* 39:9.

Hochmeister, M. N. et al. 1995. A method for the purification and recovery of genomic DNA from an HLA DQA1 amplification product and its subsequent amplification and typing with the AmpliType PM PCR amplification and typing kit. *J Forensic Sci* 40 (4):649.

Horn, G. T. et al. 1990. Characterization and rapid diagnostic analysis of DNA polymorphisms closely linked to the cystic fibrosis locus. *Clin Chem* 36 (9):1614.

Hromadnikova, I. et al. 2001. Analysis of paternal alleles in nucleated red blood cells enriched from maternal blood. *Folia Biol (Praha)* 47 (1):36.

Jung, J. M. et al. 1991. Extraction strategy for obtaining DNA from bloodstains for PCR amplification and typing of the HLA-DQα gene. *Int J Legal Med* 104 (3):145.

Kubo, S. et al. 2002. Personal identification from skeletal remain by D1S80, HLA DQA1, TH01 and polymarker analysis. *J Med Invest* 49 (1-2):83-6.

Pai, C. Y. et al. 1995. Flow chart HLA-DQA1 genotyping and its application to a forensic case. *J Forensic Sci* 40 (2):228.

Romero, R. L. et al. 1997. Applicability of formalin-fixed and formalin fixed paraffin embedded tissues in forensic DNA analysis. *J Forensic Sci* 42 (4):708.

Rudin, N. and K. Inman. 2002. Procedures for forensic DNA analysis, *in Introduction to Forensic DNA Analysis*, 2nd ed. Boca Raton, FL: CRC Press, LLC.

Sachidanandam, R. et al. 2001. A map of human genome sequence variation containing 1.42 million single nucleotide polymorphisms. *Nature* 409 (6822):928.

Saiki, R. K. et al. 1986. Analysis of enzymatically amplified beta-globin and HLA-DQα DNA with allele-specific oligonucleotide probes. *Nature* 324 (6093):163.

Saiki, R. K. et al. 1989. Genetic analysis of amplified DNA with immobilized sequence-specific oligonucleotide probes. *Proc Natl Acad Sci USA* 86 (16):6230.

Sato, Y. et al. 2003. HLA typing of aortic tissues from unidentified bodies using hot start polymerase chain reaction-sequence specific primers. *Leg Med (Tokyo)* 5 Suppl 1:S191.

Schneider, P. M., and C. Rittner. 1993. Experience with the PCR-based HLA-DQα DNA typing system in routine forensic casework. *Int J Legal Med* 105 (5):295.

Siebert, P. D., and M. Fukuda. 1987. Molecular cloning of a human glycophorin B cDNA: Nucleotide sequence and genomic relationship to glycophorin A. *Proc Natl Acad Sci USA* 84 (19):6735.

Slightom, J. L., A. E. Blechl, and O. Smithies. 1980. Human fetal Gγ and Aγ globin genes: Complete nucleotide sequences suggest that DNA can be exchanged between these duplicated genes. *Cell* 21 (3):627.

Tahir, M. A. et al. 2000. DNA typing of samples for polymarker, DQA1, and nine STR loci from a human body exhumed after 27 years. *J Forensic Sci* 45 (4):902.
Taylor, M. S., A. Challed-Spong, and E. A. Johnson. 1997. Co-amplification of the amelogenin and HLA DQ alpha genes: Optimization and validation. *J Forensic Sci* 42 (1):130.
Tsongalis, G. J., and M. M. Berman. 1997. Application of forensic identity testing in a clinical setting. Specimen identification. *Diagn Mol Pathol* 6 (2):111.
Vu, N. T., A. K. Chaturvedi, and D. V. Canfield. 1999. Genotyping for DQA1 and PM loci in urine using PCR-based amplification: Effects of sample volume, storage temperature, preservatives, and aging on DNA extraction and typing. *Forensic Sci Int* 102 (1):23.
Wilson, R. B. et al. 1994. Guidelines for internal validation of the HLA-DQα DNA typing system. *Forensic Sci Int* 66 (1):9.
Yamamoto, T. et al. 1984. The human LDL receptor: A cysteine-rich protein with multiple Alu sequences in its mRNA. *Cell* 39 (1):27.
Yang, F. et al. 1985. Human group-specific component (Gc) is a member of the albumin family. *Proc Natl Acad Sci USA* 82 (23):7994.

## Other SNPs

Ackerman, M. J., D. J. Tester, and D. J. Driscoll. 2001. Molecular autopsy of sudden unexplained death in the young. *Am J Forensic Med Pathol* 22 (2):105.
Altshuler, D. et al. 2000. An SNP map of the human genome generated by reduced representation shotgun sequencing. *Nature* 407 (6803):513.
Arnestad, M. et al. 2007. Prevalence of long-QT syndrome gene variants in sudden infant death syndrome. *Circulation* 115 (3):361.
Arnestad, M. et al. 2007. A mitochondrial DNA polymorphism associated with cardiac arrhythmia investigated in sudden infant death syndrome. *Acta Paediatr* 96 (2):206.
Brandstatter, A., A. Salas, H. Niederstatter, C. Gassner, A. Carracedo, and W. Parson. 2006. Dissection of mitochondrial superhaplogroup H using coding region SNPs. *Electrophoresis* 27 (13):2541.
Braun, A., D. P. Little, and H. Koster. 1997. Detecting CFTR gene mutations by using primer oligo base extension and mass spectrometry. *Clin Chem* 43 (7):1151.
Brenner, C. H., and B. S. Weir. 2003. Issues and strategies in the DNA identification of World Trade Center victims. *Theor Popul Biol* 63 (3):173.
Brion, M. et al. 2005. Hierarchical analysis of 30 Y-chromosome SNPs in European populations. *Int J Legal Med* 119 (1):10.
Budowle, B. 2004. SNP typing strategies. *Forensic Sci Int* 146 Suppl:S139.
Budowle, B. et al. 2000. PCR-based analysis: Allele-specific oligonucleotide assays, in *DNA Typing Protocols: Molecular Biology and Forensic Analysis*. Natick, MA: Eaton Publishing.
Burke, A. et al. 2005. Role of SCN5A Y1102 polymorphism in sudden cardiac death in blacks. *Circulation* 112 (6):798.
Chen, X., K. J. Livak, and P. Y. Kwok. 1998. A homogeneous, ligase-mediated DNA diagnostic test. *Genome Res* 8 (5):549.

Consortium, International HapMap. 2005. A haplotype map of the human genome. *Nature* 437 (7063):1299.

Dixon, L. A. et al. 2005. Validation of a 21-locus autosomal SNP multiplex for forensic identification purposes. *Forensic Sci Int* 154 (1):62.

Doktycz, M .J. et al. 1995. Analysis of polymerase chain reaction-amplified DNA products by mass spectrometry using matrix-assisted laser desorption and electrospray. *Anal Biochem* 230 (2):205.

Duffy, D. L., N. F. Box, W. Chen, J. S. Palmer, G. W. Montgomery, M. R. James, N. K. Hayward, N. G. Martin, J. and R.A. Sturm. 2004. Interactive effects of MC1R and OCA2 on melanoma risk phenotypes. *Hum Mol Genet* 13 (4):447.

Eiberg, H., and J. Mohr. 1996. Assignment of genes coding for brown eye colour (BEY2) and brown hair colour (HCL3) on chromosome 15q. *Eur J Hum Genet* 4 (4):237.

Fan, J. B. et al. 2000. Parallel genotyping of human SNPs using generic high-density oligonucleotide tag arrays. *Genome Res* 10 (6):853.

Fei, Z., T. Ono, and L. M. Smith. 1998. MALDI-TOF mass spectrometric typing of single nucleotide polymorphisms with mass-tagged ddNTPs. *Nucleic Acids Res* 26 (11):2827.

Flanagan, N. et al. 2000. Pleiotropic effects of the melanocortin 1 receptor (MC1R) gene on human pigmentation. *Hum Mol Genet* 9 (17):2531.

Frudakis, T. et al. 2003. Sequences associated with human iris pigmentation. *Genetics* 165 (4):2071.

Gabriel, S. B. et al. 2002. Structure of haplotype blocks in the human genome. *Science* 296 (5576):2225.

Gill, P. et al. 2004. An assessment of whether SNPs will replace STRs in national DNA databases—joint considerations of the DNA working group of European Network of Forensic Science Institutes (ENFSI) and Scientific Working Group on DNA Analysis Methods (SWGDAM). *Sci Justice* 44 (1):51.

Graf, J., R. Hodgson, and A. van Daal. 2005. Single nucleotide polymorphisms in the MATP gene are associated with normal human pigmentation variation. *Hum Mutat* 25 (3):278.

Grignani, P. et al. 2006. Subtyping mtDNA haplogroup H by SNaPshot minisequencing and its application in forensic individual identification. *Int J Legal Med* 120 (3):151.

Grimes, E. A. et al. 2001. Sequence polymorphism in the human melanocortin 1 receptor gene as an indicator of the red hair phenotype. *Forensic Sci Int* 122 (2-3):124.

Grossman, P. D. et al. 1994. High-density multiplex detection of nucleic acid sequences: Oligonucleotide ligation assay and sequence-coded separation. *Nucleic Acids Res* 22 (21):4527-34.

Haff, L. A., and I. P. Smirnov. 1997. Multiplex genotyping of PCR products with MassTag-labeled primers. *Nucleic Acids Res* 25 (18):3749.

Hall, J. G. et al. 2000. Sensitive detection of DNA polymorphisms by the serial invasive signal amplification reaction. *Proc Natl Acad Sci USA* 97 (15):8272.

Hammer, M. F. et al. 2006. Population structure of Y chromosome SNP haplogroups in the United States and forensic implications for constructing Y chromosome STR databases. *Forensic Sci Int* 164 (1):45.

Hardenbol, P. et al. 2003. Multiplexed genotyping with sequence-tagged molecular inversion probes. *Nat Biotechnol* 21 (6):673.

Hirschhorn, J. N. et al. 2000. SBE-TAGS: An array-based method for efficient single-nucleotide polymorphism genotyping. *Proc Natl Acad Sci USA* 97 (22):12164.

Hsu, T. M. et al. 2001. Genotyping single-nucleotide polymorphisms by the invader assay with dual-color fluorescence polarization detection. *Clin Chem* 47 (8):1373.

Inagaki, S. et al. 2004. A new 39-plex analysis method for SNPs including 15 blood group loci. *Forensic Sci Int* 144 (1):45.

Jobling, M. A. 2001. Y-chromosomal SNP haplotype diversity in forensic analysis. *Forensic Sci Int* 118 (2-3):158.

Jobling, M. A., and C. Tyler-Smith. 2003. The human Y chromosome: An evolutionary marker comes of age. *Nat Rev Genet* 4 (8):598.

Kidd, K. K. et al. 2006. Developing a SNP panel for forensic identification of individuals. *Forensic Sci Int* 164 (1):20.

Kostrikis, L. G. et al. 1998. Spectral genotyping of human alleles. *Science* 279 (5354):1228.

Krawczak, M. 1999. Informativity assessment for biallelic single nucleotide polymorphisms. *Electrophoresis* 20 (8):1676.

Kupiec, T. C., V. Raj, and N. Vu. 2006. Pharmacogenomics for the forensic toxicologist. *J Anal Toxicol* 30 (2):65.

Kuppuswamy, M. N. et al. 1991. Single nucleotide primer extension to detect genetic diseases: Experimental application to hemophilia B (factor IX) and cystic fibrosis genes. *Proc Natl Acad Sci USA* 88 (4):1143.

Lai, E. 2001. Application of SNP technologies in medicine: Lessons learned and future challenges. *Genome Res* 11 (6):927.

Lamason, R. L. et al. 2005. SLC24A5, a putative cation exchanger, affects pigmentation in zebrafish and humans. *Science* 310 (5755):1782.

Landegren, U. et al. A ligase-mediated gene detection technique. *Science* 241 (4869):1077.

Lareu, M. et al. 2001. The use of the LightCycler for the detection of Y chromosome SNPs. *Forensic Sci Int* 118 (2-3):163.

Latif, S. et al. 2001. Fluorescence polarization in homogeneous nucleic acid analysis II: 5' nuclease assay. *Genome Res* 11 (3):436.

Levo, A. et al. 2003. Postmortem SNP analysis of CYP2D6 gene reveals correlation between genotype and opioid drug (tramadol) metabolite ratios in blood. *Forensic Sci Int* 135 (1):9.

Lunetta, P. et al. 2002. Death in bathtub revisited with molecular genetics: A victim with suicidal traits and a LQTS gene mutation. *Forensic Sci Int* 130 (2-3):122.

Lyamichev, V. et al. 1999. Polymorphism identification and quantitative detection of genomic DNA by invasive cleavage of oligonucleotide probes. *Nat Biotechnol* 17 (3):292.

Mei, R. et al. 2000. Genome-wide detection of allelic imbalance using human SNPs and high-density DNA arrays. *Genome Res* 10 (8):1126.

Mein, C. A. et al. 2000. Evaluation of single nucleotide polymorphism typing with invader on PCR amplicons and its automation. *Genome Res* 10 (3):330.

Muddiman, D. C. et al. 1997. Length and base composition of PCR-amplified nucleic acids using mass measurements from electrospray ionization mass spectrometry. *Anal Chem* 69 (8):1543.

Myakishev, M. V. et al. 2001. High-throughput SNP genotyping by allele-specific PCR with universal energy-transfer-labeled primers. *Genome Res* 11 (1):163.

Paracchini, S. et al. 2002. Hierarchical high-throughput SNP genotyping of the human Y chromosome using MALDI-TOF mass spectrometry. *Nucleic Acids Res* 30 (6):e27.

Pastinen, T. et al. 1997. Minisequencing: A specific tool for DNA analysis and diagnostics on oligonucleotide arrays. *Genome Res* 7 (6):606.

Pastinen, T. et al. 2000. A system for specific, high-throughput genotyping by allele-specific primer extension on microarrays. *Genome Res* 10 (7):1031.

Quintana, B. et al. 2004. Typing of mitochondrial DNA coding region SNPs of forensic and anthropological interest using SNaPshot minisequencing. *Forensic Sci Int* 140 (2–3):251.

Rebbeck, T. R. et al. 2002. P gene as an inherited biomarker of human eye color. *Cancer Epidemiol Biomarkers Prev* 11 (8):782.

Rees, J. L. 2003. Genetics of hair and skin color. *Annu Rev Genet* 37:67.

Reich, D. E. et al. 2002. Human genome sequence variation and the influence of gene history, mutation and recombination. *Nat Genet* 32 (1):135.

Ronaghi, M. et al. 1996. Real-time DNA sequencing using detection of pyrophosphate release. *Anal Biochem* 242 (1):84.

Ronaghi, M., M. Uhlen, and P. Nyren. 1998. A sequencing method based on real-time pyrophosphate. *Science* 281 (5375):363.

Ross, P. et al. 1998. High level multiplex genotyping by MALDI-TOF mass spectrometry. *Nat Biotechnol* 16 (13):1347.

Sanchez, J. J. et al. 2003. Multiplex PCR and minisequencing of SNPs: Model with 35 Y chromosome SNPs. *Forensic Sci Int* 137 (1):74.

Sanchez, J. J. et al. 2006. A multiplex assay with 52 single nucleotide polymorphisms for human identification. *Electrophoresis* 27 (9):1713.

Sauer, S. et al. 2000. A novel procedure for efficient genotyping of single nucleotide polymorphisms. *Nucleic Acids Res* 28 (5):E13.

Shriver, M. D., and R. A. Kittles. 2004. Genetic ancestry and the search for personalized genetic histories. *Nat Rev Genet* 5 (8):611.

Sobrino, B., M. Brion, and A. Carracedo. 2005. SNPs in forensic genetics: A review on SNP typing methodologies. *Forensic Sci Int* 154 (2-3):181.

Sokolov, B. P. 1990. Primer extension technique for the detection of single nucleotide in genomic DNA. *Nucleic Acids Res* 18 (12):3671.

Syvanen, A. C. et al. 1990. A primer-guided nucleotide incorporation assay in the genotyping of apolipoprotein E. *Genomics* 8 (4):684.

Tyagi, S., D. P. Bratu, and F. R. Kramer. 1998. Multicolor molecular beacons for allele discrimination. *Nat Biotechnol* 16 (1):49.

Tyagi, S., and F. R. Kramer. 1996. Molecular beacons: Probes that fluoresce upon hybridization. *Nat Biotechnol* 14 (3):303.

Tyagi, S., S. A. Marras, and F.R. Kramer. 2000. Wavelength-shifting molecular beacons. *Nat Biotechnol* 18 (11):1191.

Underhill, P. A. et al. 2001. The phylogeography of Y chromosome binary haplotypes and the origins of modern human populations. *Ann Hum Genet* 65 (Pt 1):43.

Vallone, P. M., and J. M. Butler. 2004. Y-SNP typing of U.S. African American and Caucasian samples using allele-specific hybridization and primer extension. *J Forensic Sci* 49 (4):723.

Wang, X. et al. 2005. Single nucleotide polymorphism in transcriptional regulatory regions and expression of environmentally responsive genes. *Toxicol Appl Pharmacol* 207 (2 Suppl):84.

## Study Questions

1. In the reverse blot method of HLA DQA1 testing:
   (a) The probe is labeled with the biotin.
   (b) The primer is labeled with the biotin.

2. Single nucleotide polymorphisms:
   (a) Can be useful for degraded DNA samples
   (b) Can be useful for mixed samples from more than one donor
   (c) Can be useful for forensic identification because of high discriminating power

3. Which of the following is an example of a single nucleotide poly morphism?
   (a) A base substitution on Y chromosome
   (b) A base substitution on autosomal DNA
   (c) A base substitution on mitochondrial DNA
   (d) All of the above

4. Most SNPs are:
   (a) Biallelic
   (b) Triallelic
   (c) Tetraallelic
   (d) All of the above

5. Most SNPs for human identification reside in:
   (a) Noncoding regions of DNA
   (b) Exons that do not change the amino acid compositions of proteins
   (c) Exons that change the amino acid compositions of proteins
   (d) Promoter regions of genes

6. SNPs for phenotyping reside in:
   (a) Noncoding regions of DNA
   (b) Exons that do not change the amino acid compositions of proteins
   (c) Exons that change the amino acid compositions of proteins
   (d) Promoter regions of genes

# Mitochondrial DNA Profiling

# 21

Forensic mitochondrial DNA (mtDNA) analysis is a useful tool for human identification. Because mtDNA is maternally inherited, it is especially useful for identifying victims. Additionally, the mtDNA genome produces much higher numbers of copies per cell than the nuclear genome. Thus, mtDNA testing is frequently used when nuclear DNA in samples is insufficient. For example, hair shafts, bones, and decomposed samples may be tested with mtDNA analysis.

## 21.1 Human Mitochondrial Genome

Mitochondria are subcellular organelles that serve as the energy-generating components of cells (Figure 21.1). Each cell contains hundreds of mitochondria that have their own extrachromosomal genomes separate from the nuclear genome. Although each human mitochondrion contains several copies of the mtDNA genome, the exact copy number varies for each cell. However, it is estimated that hundreds of copies of mtDNA genome exist in most cell types. Recombination has not been observed in mtDNA. Thus, the mtDNA type, also referred to as the ***mitotype***, is considered a haplotype treated as a single locus. The mitochondrial genome has a higher mutation rate (up to 10 times higher) than its nuclear counterpart.

### 21.1.1 Genetic Contents of Mitochondrial Organelle Genomes

Organelle genomes are usually much smaller than their nuclear counterparts. The much smaller mitochondrial genome has been sequenced and is known as the Cambridge reference sequence. It was established in the early 1980s, later modified, and is now known as the ***revised Cambridge reference sequence*** (***rCRS***) that is presently used as the standard for sequence comparisons.

The human mitochondrial genome is a circular DNA molecule consisting of 16,569 bp containing 37 genes (Figure 21.2). Thirteen of these genes code for proteins involved in the respiratory complex, a main energy-generating component in mitochondria. The other 24 specify noncoding RNA molecules required for expression of the mitochondrial genome. The genes in the human mitochondrial genome are much more closely packed than in the nuclear genome and contain no introns. A ***control region***, also known as

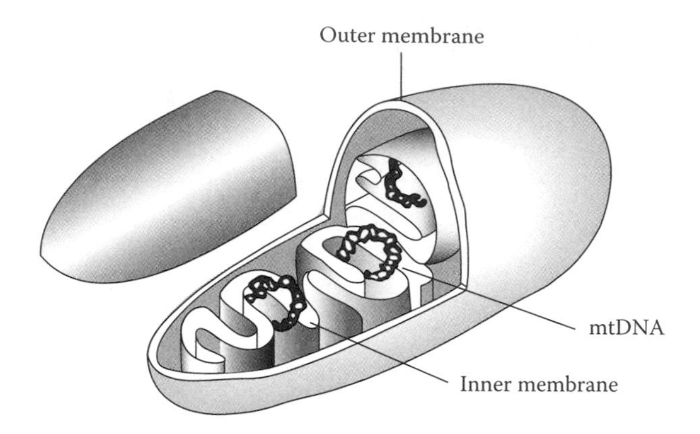


**Figure 21.1** Mitochondrion. mtDNA = mitochrondrial DNA.

a ***displacement loop*** (***D loop***), contains the origin of replication for one of the mtDNA strands but does not code for any gene products (Figure 21.2).

An asymmetric distribution of nucleotides gives rise to light (L) and heavy (H) strands when mtDNA molecules are separated in alkaline CsCl gradients. The ***H strand*** that contains a greater number of guanine nucleotides has a higher molecular weight in comparison to the ***L strand***.

### 21.1.2 Maternal Inheritance of mtDNA

Maternal inheritance is typically observed for the mtDNA genome. mtDNA is inherited differently from nuclear genes; it does not obey the rules of Mendelian inheritance and is thus called nonMendelian inheritance. The mitochondria of the spermatozoa are located at the midpieces of spermatozoa. At conception, only the head portion of a spermatozoon (containing a nucleus but no mitochondria) enters the egg. The fertilized egg contains the maternal mitochondria which is transmitted to progeny. The mtDNA sequence is identical for relatives within the same maternal lineage (Figure 21.3). This characteristic of maternal inheritance is useful for identifying samples by comparing them with samples from maternal relatives.

## 21.2 mtDNA Polymorphic Regions

### 21.2.1 Hypervariable Regions

The most polymorphic region of mtDNA is located within the D loop (Figure 21.4). The three hypervariable regions in the D-loop region are designated

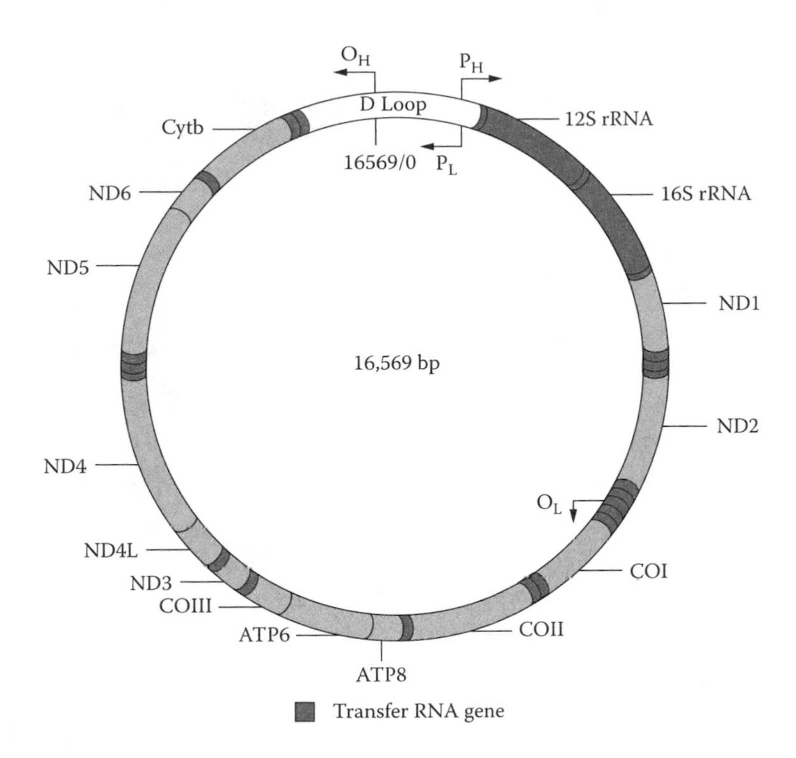


**Figure 21.2** Human circular mitochondrial genome. The transcription direction for the H (heavy) and L (light) strands are indicated by arrows ($P_H$, $P_L$). The origins of replication are labeled $O_H$ for heavy strand and $O_L$ for light strand, respectively. The mitochondrial DNA genome encodes genes. ND = NADH coenzyme Q oxidoreductase complex. CO = cytochrome c oxidase complex. Cytb = cytochrome b. ATP = ATP synthase. rRNA, ribosomal RNA. Transfer RNA genes are shown as indicated.

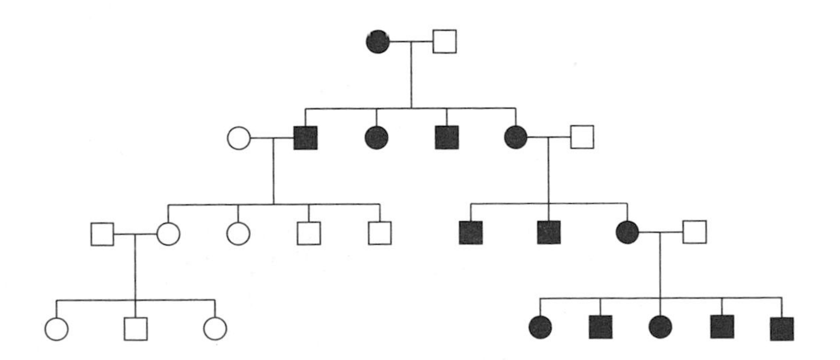

**Figure 21.3** Pedigree of a human family showing inheritance of mtDNA. Females and males are denoted by circles and squares, respectively. Red symbols indicate individuals who inherited the same mtDNA.

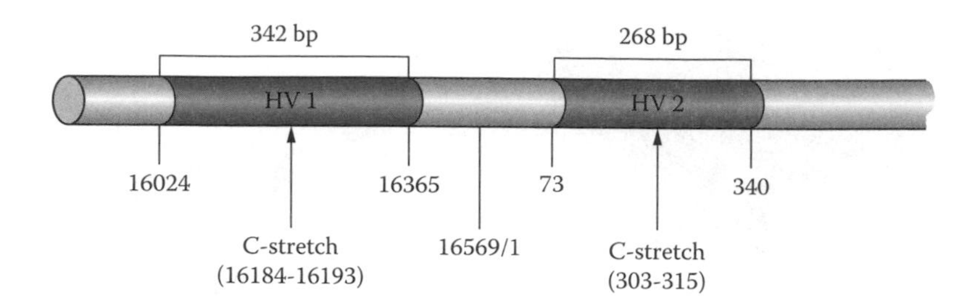


**Figure 21.4** Hypervariable regions of the D loop in mtDNA (with nucleotide positions).

HV1 (16024–16365; 342 bp), HV2 (73–340; 268 bp), and HV3 (438–574; 137 bp). The most common polymorphic regions of the human mtDNA genome analyzed for forensic purposes are the two hypervariable regions within the D loop known as ***hypervariable region I (HV1)*** and ***hypervariable region II (HV2)***.

### 21.2.2 Heteroplasmy

***Heteroplasmy*** occurs when an individual carries more than one mtDNA haplotype. Heteroplasmy may be observed with one type of tissue and be absent in other tissue types; for example, it is commonly observed in hair samples. Several instances of heteroplasmy may be observed in different tissue types. An individual may exhibit one mitotype in one tissue and a different mitotype in another. Thus, it is necessary to obtain and process additional samples to confirm the heteroplasmy when it is observed in a questioned sample but not in a known sample or vice versa. The two types are sequence and length heteroplasmies.

#### *21.2.2.1 Sequence Heteroplasmy*

Sequence heteroplasmy is defined as the presence of two nucleotides at a single position shown as overlapping peaks in a sequence electropherogram (Figure 21.5). Heteroplasmy usually occurs at one position, but on rare occasions can be observed at more than one position. Hot spots for heteroplasmy have been documented at both HV1 and HV2 regions. Heteroplasmy may complicate the interpretation of mtDNA results, but its presence can also improve the strength of a match.

#### *21.2.2.2* Length Heteroplasmy

Both HV1 and HV2 of the human mtDNA D-loop region contain homopolymeric cytosine sequences known as ***C stretches***. The HVI region contains a C stretch between positions 16184 and 16193, interrupted by a thymine

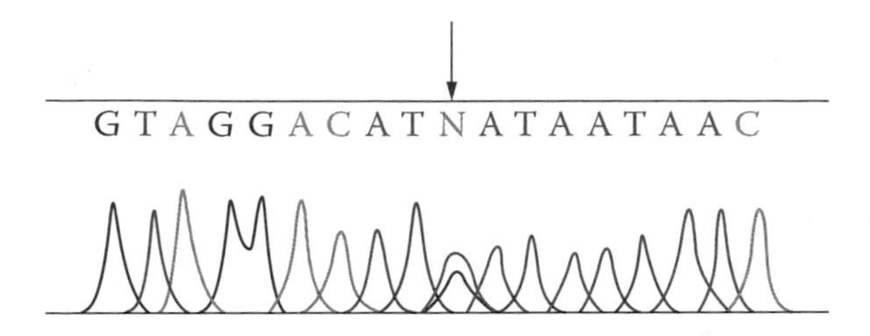


**Figure 21.5** Electropherogram showing mtDNA sequence heteroplasmy at position 234R (A/G) as indicated by arrow. N = unresolved sequence.

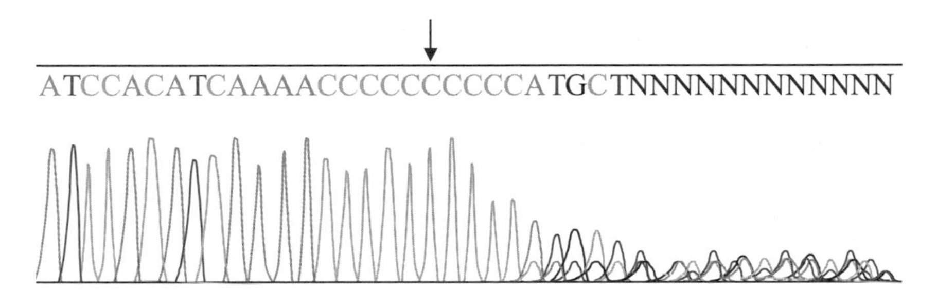


**Figure 21.6** Electropherogram showing mtDNA length heteroplamy at C stretch of HV1 region where position 16189 is a C as indicated by arrow. N = unresolved sequence.

at position 16189. If a base transition from T to C occurs at position 16189 (a variant present in approximately 20% of the population), it results in an uninterrupted C stretch. A similar C stretch resides between positions 303 and 315 of the HV2 region.

Length heteroplasmies are often observed at the uninterrupted C stretches, which create serious problems with sequence analysis downstream from the homopolymeric regions (Figure 21.6). It is not clear whether the length heteroplasmy is due to replication slippage at the C stretches or results from a mixture of length variants in the cells. If length heteroplasmy occurs, alternative sequencing primers can be used to obtain the downstream sequences of the C stretches.

## 21.3 Forensic mtDNA Testing

### 21.3.1 General Considerations

mtDNA analysis is often used on samples derived from skeletal or decomposed remains. The surface of the sample should be cleaned to remove any adhering debris or contaminants. Bones and teeth are pulverized to facilitate extraction of the mtDNA. Duplicate extractions (e.g., two sections of a single hair) are recommended if sufficient sample material is available. mtDNA is extracted using a similar method to nuclear DNA (nuclear DNA is co-extracted with mtDNA). The amount of mtDNA can therefore be estimated from the quantity of nuclear DNA obtained. Alternatively, mtDNA-specific quantization methods using real-time PCR can also be used to directly obtain measurements of mtDNA extracted.

For mtDNA sequencing, analysis of both strands of the mtDNA in a given region must be performed to ensure accuracy. Due to the high sensitivity of mtDNA analysis, it is essential to minimize risks of contamination during the procedure. Contamination must be strictly monitored using proper controls such as reagent blanks (see Section 14.44) and negative controls (samples containing all reagents except DNA template).

Finally, a positive control must also be used to monitor the success of the analysis. It should be introduced at the amplification step and remain through the sequencing process. A positive control consists of a DNA template of known sequence such as DNA purified from an HL60 cell line.

### 21.3.2 mtDNA Screen Assay

One of the most common assays for screening mtDNA variations is the allele-specific oligonucleotide (ASO) assay. It allows rapid screening of mtDNA sequence polymorphisms and has the potential to reduce the number of samples required for mtDNA sequencing. This method is also useful for excluding or eliminating suspects from a case. However, HV1 and HV2 sequencing should be performed to obtain complete sequence information for the targeted HV regions to confirm a match.

The commercial Linear Array™ mtDNA HV1/HV2 region sequence typing kit (Roche Applied Sciences, Indianapolis, Indiana) utilizes reverse ASO configuration with a panel of immobilized ASO probes that detect common polymorphic sites (Figure 21.7). The mtDNA is amplified at both HV1 (444 bp amplicon) and HV2 (415 bp amplicon) regions and the forward primers are biotin labeled at the 5′ ends of the oligonucleotides. Thus, the amplified PCR product is biotinylated. A horseradish peroxidase-conjugated streptavidin complex is then allowed to bind to the biotinylated PCR product. Finally, colorimetric detection is carried out with tetramethylbenzidine (TMB) as the substrate to produce a colored precipitate at the designated location. The typing kit detects sequence variation in 19 positions within the HV1 and HV2 regions.

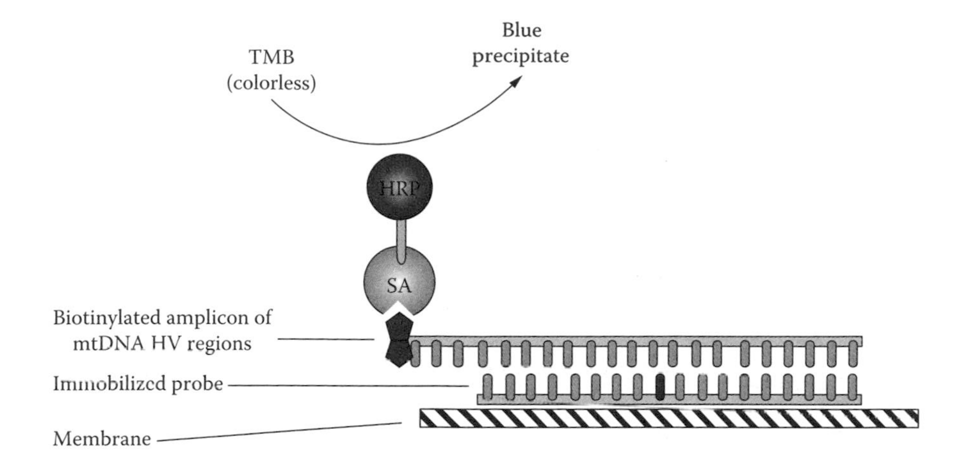


**Figure 21.7** Reverse blot assay employed in mtDNA screen. A probe is immobilized onto a solid phase membrane, then hybridized with a biotinylated amplicon of the mtDNA HV sequences. Hybridization is detected by a streptavidin (SA) and horseradish peroxidase (HRP) conjugate. A colorimetric reaction is catalyzed by HRP using tetramethylbenzidine (TMB) as a substrate.

## 21.3.3 mtDNA Sequencing

To sequence a specific region of mtDNA, a combination of PCR amplification and DNA sequencing techniques is employed that reduce the time and labor needed to obtain DNA sequences from genomic DNA templates. mtDNA sequencing usually consists of (1) PCR amplification, (2) DNA sequencing reactions, (3) separation using electrophoresis, and (4) data collection and sequence analysis (Figure 21.8). Chapter 23 describes the evaluation of the strength of the results via statistical analysis.

### *21.3.3.1 PCR Amplification*

The extracted DNA samples must be amplified to yield sufficient quantities of template for sequencing reactions. PCR amplification of all or a part of the D loop region can be carried out with various primer sets. If a sample contains high quality and high copy number mtDNA, the HV1, and HV2 regions can be amplified as two amplicons, each of about 350 to 400 bp in length.

If a sample is degraded or contains low copy number mtDNA, the hypervariable regions can be amplified as smaller PCR products. PCR amplification of mtDNA is usually done in 34 to 38 cycles. Protocols for highly degraded DNA specimens sometimes require 42 cycles. The use of higher PCR cycle numbers can improve the yield of the amplicon.

Following mtDNA amplification, a purification step is necessary to remove excess primers and deoxynucleotide triphosphates (dNTPs). This step can be

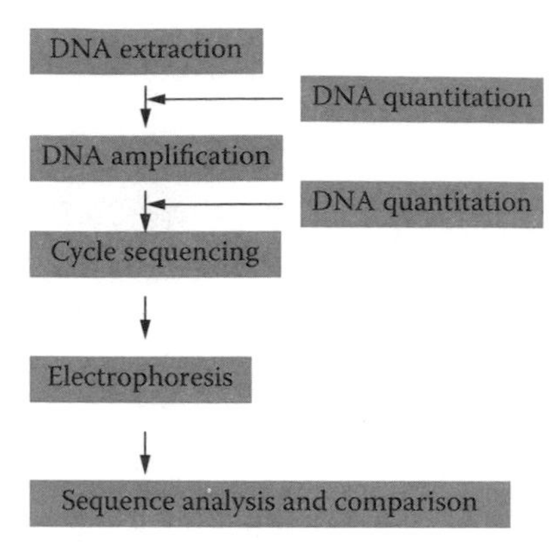


**Figure 21.8** Overview of usual steps for mtDNA sequencing.

performed by using filtration devices such as a Microcon® to remove small molecules from the sample or by using nuclease digestion with shrimp alkaline phosphatase or exonuclease I to degrade remaining primers and dNTPs. The concentration of the PCR product is important for an optimal sequencing reaction in the next phase of mtDNA sequencing. The quality and quantity of the mtDNA amplicon must be evaluated to confirm the presence or absence of PCR products and their concentrations. This can be done using an agarose yield gel to visualize the PCR products of the sample or via capillary electrophoresis, a more informative method, for quantifying PCR products.

### *21.3.3.2 DNA Sequencing Reactions*

The best known DNA sequencing techniques are the ***chain termination*** method and the ***chemical degradation*** method developed, respectively, by Sanger and Gilbert (who shared the Nobel Prize for their work) in 1977. Over the years, the chain termination method became more common because it was applicable for automation and did not require the toxic chemicals necessary for the chemical degradation method.

**21.3.3.2.1 Chain Termination or Sanger Method** An oligonucleotide primer that can anneal to a single stranded DNA template is employed. A sequencing reaction contains a DNA polymerase and the four dNTPs. The reaction also contains small quantities of dideoxynucleotide triphosphates (ddNTPs, Figure 21.9). If a ddNTP molecule is incorporated into a growing DNA chain, the absence of a 3′ OH group prevents formation of a phosphodiester bond and thus prevents the extension of the oligonucleotide chain. A sequencing reaction involves competition between extension and termination of the chain (Figure 21.10).

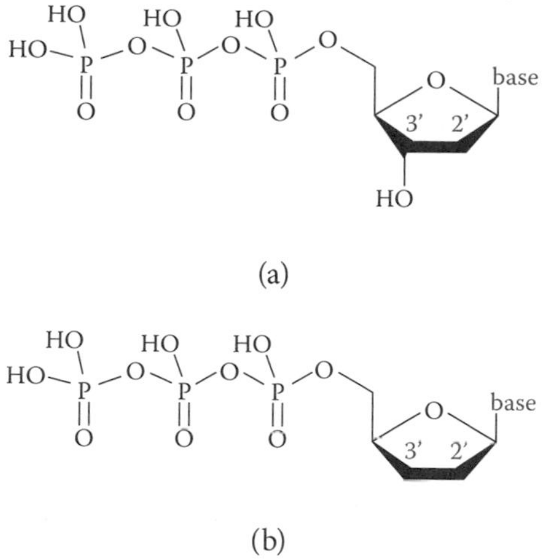

(a)

(b)

**Figure 21.9** Chemical structures of dNTP and ddNTP analog. (a) dNTP. (b) ddNTP. Both hydroxyl groups attached to the 2′ and 3′ carbons of ribose are replaced by hydrogens.

The ratio of ddNTPs to dNTPs has been optimized to result in a collection of DNA fragments varying in length by one nucleotide from primer length to the full length of the sequencing reaction product. As a result, the products of the sequencing reaction consist of a population of various lengths of oligonucleotide chains terminated by ddNTPs. By using the four different ddNTPs, populations of DNA fragments are generated that terminate at positions occupied by every A, C, G, or T in the template strand. The labeled products of sequencing reactions are then resolved during electrophoresis and the sequencing data can be collected.

**21.3.3.2.2 Cycle Sequencing** The chain termination reaction is carried out using a ***cycle sequencing*** technique commonly used in forensic laboratories for mtDNA sequencing. Cycle sequencing is a technique developed in the late 1980s in which PCR is employed to generate a single stranded template for chain termination sequencing reactions. The application of PCR in a sequencing reaction greatly increases the signal intensity and thus the sensitivity of the sequencing.

The sequencing reactions are carried out with multiple rounds of thermal cycling. Each cycle consists of three steps: denaturation of the double-stranded DNA template, annealing of a sequencing primer to its target sequence, and the extension of the annealed primer by DNA polymerase. However, cycle sequencing employs only a single primer per reaction, and during the extension of cycle sequencing, the extension of the strand is terminated with the incorporation of a ddNTP (Figure 21.11). The resulting partially double stranded hybrid, consisting of the full-length template strand

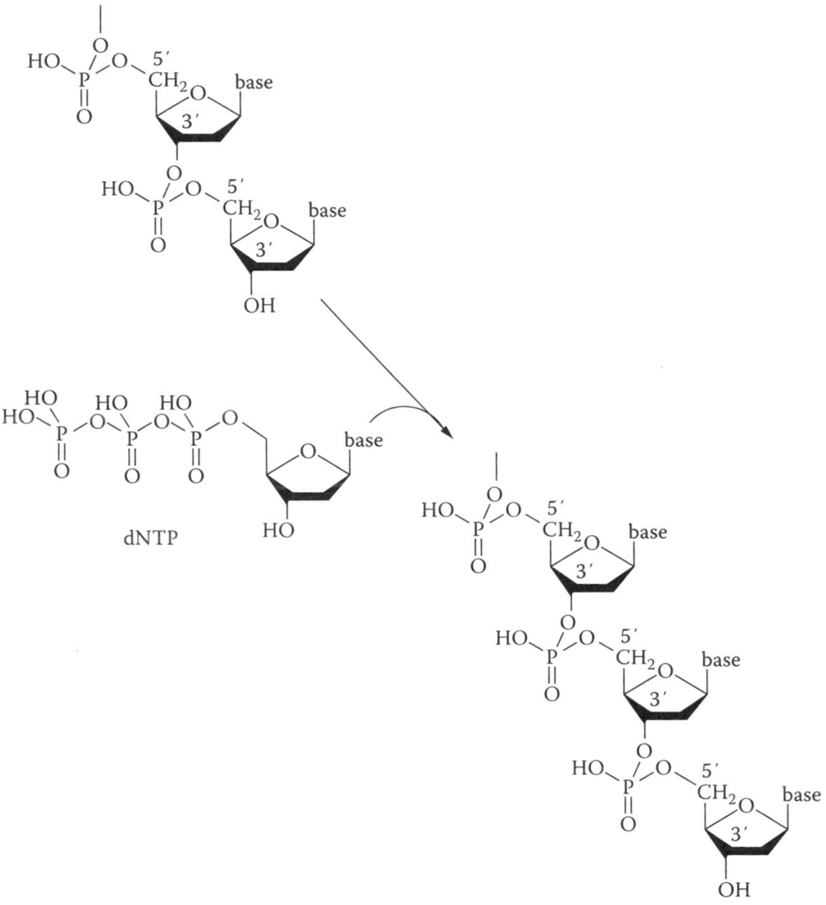

(a)

**Figure 21.10** Competition between extension and termination. (a) Growing chain in extension. (b) DNA strand with labeled ddNTP incorporated into a growing DNA chain that interrupts additions of new nucleotides. The ddNTP usually is a fluorescent dye labeled for detection.

and its complementary chain-terminated product, is denatured during the first step of the next cycle, thereby liberating the template strand for another round of annealing, extension, and termination.

**21.3.3.2.3 Labeling Methods** The sequencing product of chain termination can be labeled (1) with the ***dye terminator*** system in which the terminator is labeled or (2) ***dye primer*** system in which the primer is labeled. The dye terminator system is commonly used for mtDNA sequencing in forensic laboratories. With the dye terminator system, the ddNTPs are labeled with four

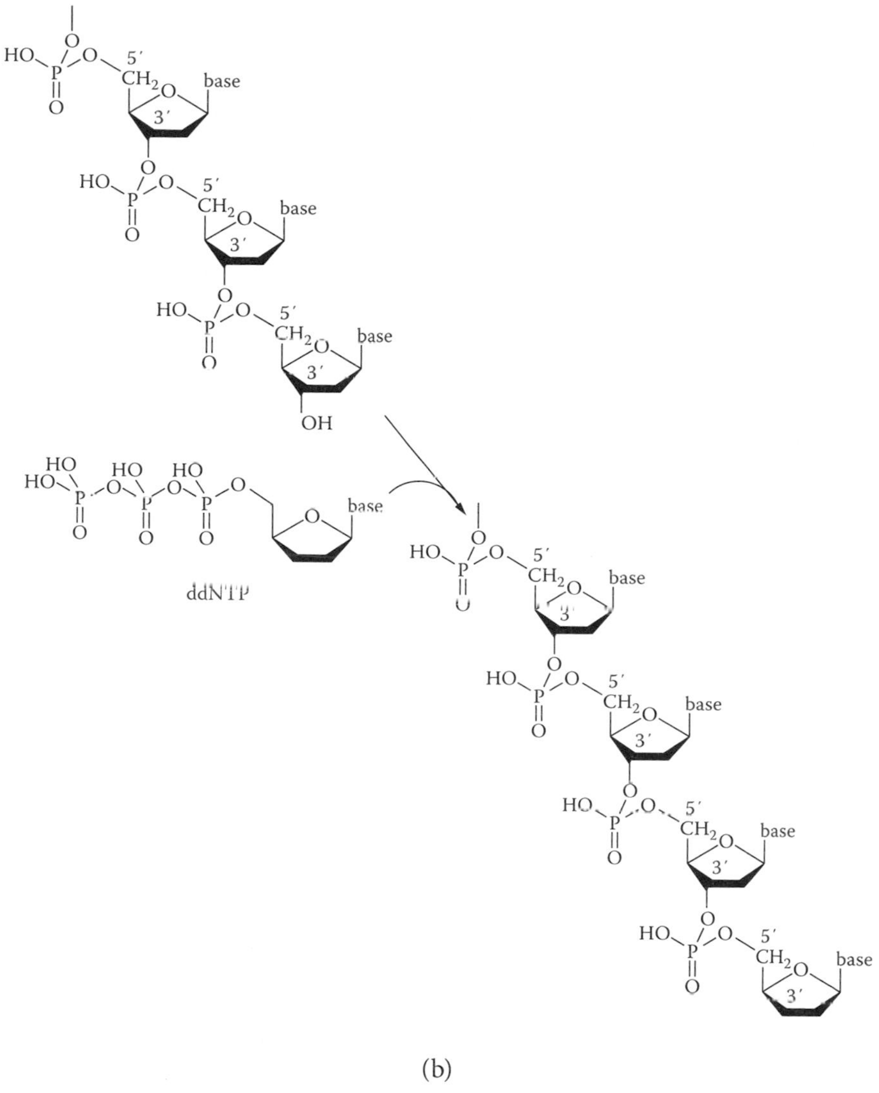

(b)

**Figure 21.10** (continued)

different fluorescent dyes, each with a distinct spectrum. Thus, the sequencing with all four dideoxy nucleotides can be carried out in a single reaction.

### *21.3.3.3 Electrophoresis and Sequence Analysis*

The cycle sequencing products can be separated using electrophoresis in a 4% polyacrylamide denatured gel or a POP-6 polymer (Applied Biosystems) as the matrix for capillary electrophoresis (Figure 21.12). Following data collection, sequence data analysis can be performed with the Sequencher™

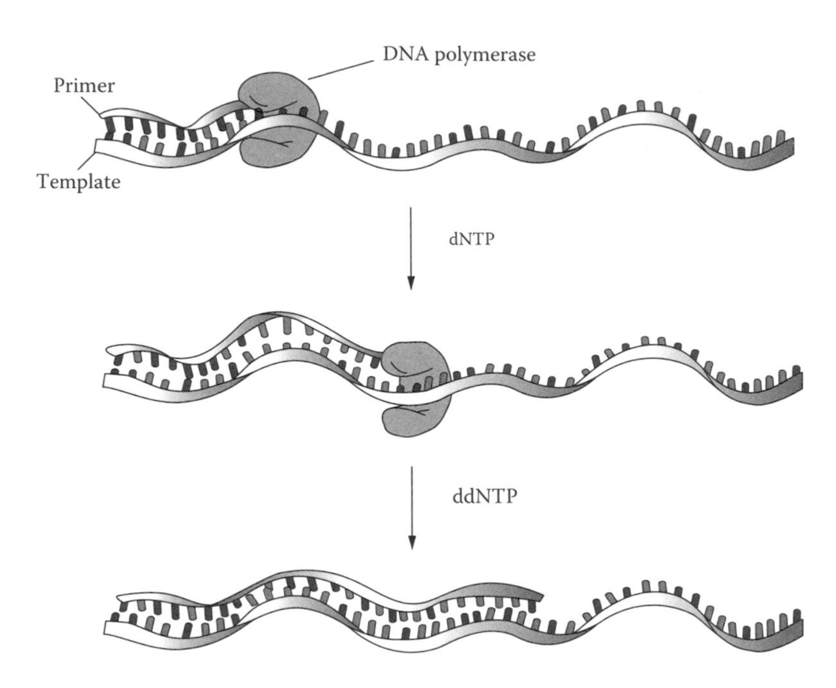


**Figure 21.11** Cycle sequencing reaction that includes DNA polymerase, a template, and a primer. During DNA synthesis, the dNTP is incorporated by a new phosphodiester bond with the primer. The incorporation of ddNTP blocks further DNA synthesis of the growing chain.

software (Gene Codes Corporation, Ann Arbor, MI, USA). Figure 21.13 shows sequence data.

### *21.3.3.4 Mitotype Designations*

Sequencing of a mtDNA region should be performed twice. Sequencing both strands of a mtDNA region is preferable to reduce ambiguities in sequence determination. The sequences of both the reference and evidence samples are compared. The nomenclature should be compatible with International Union of Pure and Applied Chemistry (IUPAC) codes.

**Reporting Format** — The rCRS is used as a reference standard to facilitate the designation of mitotypes. For reporting purposes, sequence differences relative to the rCRS are listed in data format. When a difference between an individual's sequence and that of the rCRS sequence is observed, only the position designated by a number and the nucleotide differing from the reference standard are recorded. In this format, nucleotides identical to the rCRS are not listed. For example, at position 228 (HV2), the rCRS has a G. If a mitotype carries an A at position 228, the individual's mtDNA sequence is described as 228A. If an unresolved sequence ambiguity is observed at a

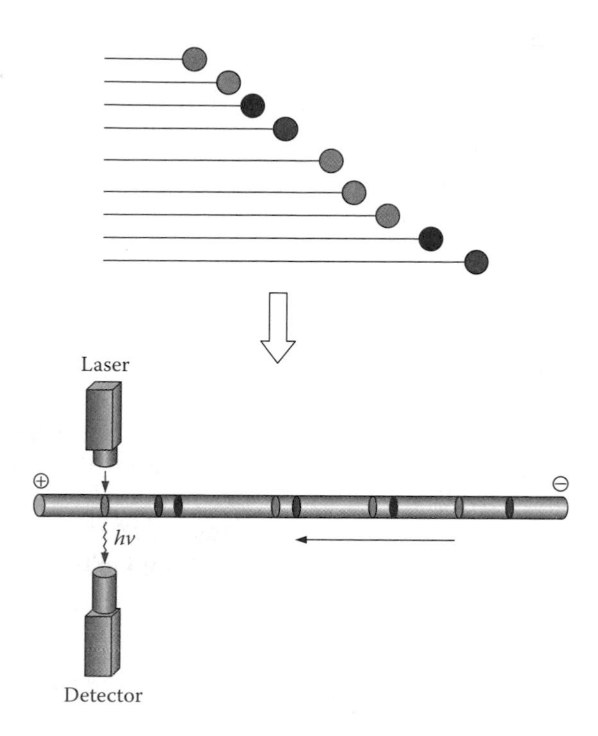


**Figure 21.12** Separation of cycle sequencing products using capillary electrophoresis. The DNA chain lengths are determined by the addition of ddNTP at different positions. The fluorescent dye-labeled cycle sequencing products are separated using electrophoresis. The fluorescent dyes are resolved by a detector and the peaks corresponding to each DNA fragment are identified.

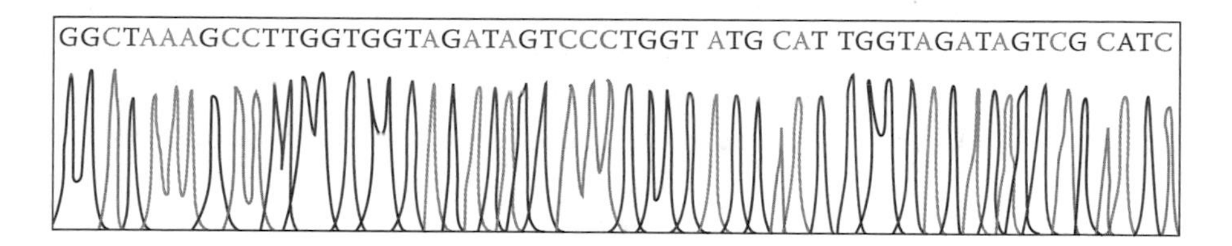


**Figure 21.13** Electropherogram containing DNA sequence presented as peaks with different colors. The bases associated with each peak are noted. The sequence is read from left to right (5′ to 3′).

position, the base number for the position is listed followed by an N. For example, 228N means an unresolved sequence ambiguity was observed at position 228.

**Insertions** — The insertion site is described by noting the position (5′ to the insertion) followed by a decimal point and a number. The number indicates the order of the insertions (e.g., 1 indicates the first insertion, 2 indicates the second, etc.) The base calling following the number indicates the inserted nucleotide (for example, 524.1A, 524.2C).

**Deletions** — The deletion site designation is followed by a d. For example, a deletion at position 16296 is recorded as 16296d.

**Heteroplasmic Sites** — The IUPAC codes for base calling can be applied to sequence heteroplasmic sites. For example, an A/G heteroplasmy can be designated as R, and a C/T heteroplasmy can be designated as Y.

### 21.3.4 Interpretation of mtDNA Profiling Results

Interpretation guidelines are used when the evaluation of sequencing results from evidence and reference samples are necessary. General guidelines were set forth by the Scientific Working Group on DNA Analysis Methods (SWGDAM) and the DNA Commission of the International Society of Forensic Genetics (ISFG). The limitations of mtDNA technology should be taken into account as should the higher mutation rates found with the mtDNA genome than found with the nuclear genome. Mutations seem to be more common in certain tissues. For that reason the sources of the tissues investigated should be taken into consideration as well. In reporting mtDNA profiling results, the most common categories of conclusions are: cannot exclude, exclusion, and inconclusive result.

**Exclusion** — If the sequences are different, then the samples can be excluded as originating from the same source. Additionally, the SWGDAM's guidelines define that this conclusion can be made if there are two or more nucleotide differences between the questioned and known samples.

**Cannot exclude** — If the sequences are the same, the reference sample and evidence cannot be excluded as potentially arising from the same source. When a mtDNA profile cannot be excluded, it is desirable to evaluate the weight of the evidence. In cases where the same heteroplasmy is observed in both questioned and known samples, its presence increases the strength of the evidence. However, if heteroplasmy is observed in a questioned sample but not in a known sample or vice versa, a common maternal lineage still cannot be excluded.

**Inconclusive result** — If the questioned and known samples differ by a single nucleotide, and no evidence of heteroplasmy is present, the interpretation may be that the results are inconclusive.

## Bibliography

Anderson, S. et al. 1981. Sequence and organization of the human mitochondrial genome. *Nature* 290 (5806):457.

Andrews, R. et al. 1999. Reanalysis and revision of the Cambridge reference sequence for human mitochondrial DNA. *Nat Genet* 23 (2):147.
Bendall, K. E. et al. 1996. Heteroplasmic point mutations in the human mtDNA control region. *Am J Hum Genet* 59 (6):1276.
Bendall, K. E., V. A. Macaulay, and B. C. Sykes. 1997. Variable levels of a heteroplasmic point mutation in individual hair roots. *Am J Hum Genet* 61 (6):1303.
Bendall, K. E., and B. C. Sykes. 1995. Length heteroplasmy in the first hypervariable segment of the human mtDNA control region. *Am J Hum Genet* 57 (2):248.
Budowle, B. et al. 1999. Mitochondrial DNA regions HVI and HVII population data. *Forensic Sci Int* 103 (1):23.
Budowle, B. et al. 2000. *DNA Typing Protocols: Molecular Biology and Forensic Analysis*. Natick, MA: Eaton Publishing.
Budowle, B. et al. 2003. Forensics and mitochondrial DNA: Applications, debates, and foundations. *Annu Rev Genomics Hum Genet* 4:119.
Butler, J. M. 2005. *Forensic DNA Typing: Biology, Technology, and Genetics of STR Markers*, 2nd ed. Burlington, MA: Elsevier.
Butler, J. M., and B. C. Levin. 1998. Forensic applications of mitochondrial DNA. *Trends Biotechnol* 16 (4):158.
Calloway, C. D. et al. 2000. The frequency of heteroplasmy in the HVII region of mtDNA differs across tissue types and increases with age. *Am J Hum Genet* 66 (4):1384.
Carracedo, A. et al. 2000. DNA commission of the international society for forensic genetics: Guidelines for mitochondrial DNA typing. *Forensic Sci Int* 110 (2):79.
Cassandrini, D. et al. 2006. A new method for analysis of mitochondrial DNA point mutations and assess levels of heteroplasmy. *Biochem Biophys Res Commun* 342 (2):387.
Chakraborty, R. et al. 1999. The utility of short tandem repeat loci beyond human identification: Implications for development of new DNA typing systems. *Electrophoresis* 20 (8):1682.
Clayton, D. A. 1992. Structure and function of the mitochondrial genome. *J Inherit Metab Dis* 15 (4):439.
Comas, D., S. Paabo, and J. Bertranpetit. 1995. Heteroplasmy in the control region of human mitochondrial DNA. *Genome Res* 5 (1):89.
Danielson, P. B. et al. 2005. Separating human DNA mixtures using denaturing high-performance liquid chromatography. *Expert Rev Mol Diagn* 5 (1):53-63.
D'Eustachio, P. 2002. High levels of mitochondrial DNA heteroplasmy in human hairs by Budowle et al. *Forensic Sci Int* 130 (1):63.
Grignani, P. et al. 2006. Subtyping mtDNA haplogroup H by SNaPshot minisequencing and its application in forensic individual identification. *Int J Legal Med* 120 (3):151.
Hauswirth, W. W., and D. A. Clayton. 1985. Length heterogeneity of a conserved displacement loop sequence in human mitochondrial DNA. *Nucleic Acids Res* 13 (22):8093.
Linch, C. A., D. A. Whiting, and M. M. Holland. 2001. Human hair histogenesis for the mitochondrial DNA forensic scientist. *J Forensic Sci* 46 (4):844.

Miller, K. W., and B. Budowle. 2001. A compendium of human mitochondrial DNA control region: Development of an international standard forensic database. *Croat Med J* 42 (3):315.

Obata, M. et al. 2006. Sperm mitochondrial DNA transmission to both male and female offspring in the blue mussel *Mytilus galloprovincialis*. *Dev Growth Differ* 48 (4):253.

Pakendorf, B., and M. Stoneking. 2005. Mitochondrial DNA and human evolution. *Annu Rev Genomics Hum Genet* 6:165.

Parsons, T. J., and M. D. Coble. 2001. Increasing the forensic discrimination of mitochondrial DNA testing through analysis of the entire mitochondrial DNA genome. *Croat Med J* 42 (3):304.

Rew, D. A. 2001. Mitochondrial DNA, human evolution and the cancer genotype. *Eur J Surg Oncol* 27 (2):209.

Tully, G. et al. 2001. Considerations by European DNA profiling (EDNAP) group on working practices, nomenclature and interpretation of mitochondrial DNA profiles. *Forensic Sci Int* 124 (1):83.

Tully, G. et al. 2004. Results of a collaborative study of the EDNAP group regarding mitochondrial DNA heteroplasmy and segregation in hair shafts. *Forensic Sci Int* 140 (1):1.

Tully, L. A., and B. C. Levin. 2000. Human mitochondrial genetics. *Biotechnol Genet Eng Rev* 17:147.

Umetsu, K., and I. Yuasa. 2005. Recent progress in mitochondrial DNA analysis. *Leg Med (Tokyo)* 7 (4):259.

Wilson, M. R. et al. 2002. Recommendations for consistent treatment of length variants in the human mitochondrial DNA control region. *Forensic Sci Int* 129 (1):35.

Wilson, M. R. et al. 1995. Validation of mitochondrial DNA sequencing for forensic casework analysis. *Int J Legal Med* 108 (2):68.

Wilson, M. R. et al. 1995. Extraction, PCR amplification and sequencing of mitochondrial DNA from human hair shafts. *Biotechniques* 18 (4):662.

## Study Questions

1. Which of the following has the lowest mutation rate?
   (a) STRs
   (b) Autosomal SNPs
   (c) Mitochondrial HV sequences

2. Which of the following works better for DNA samples with degradation?
   (a) STR
   (b) RFLP
   (c) mtDNA sequencing

3. mtDNA testing may be used to determine:
   (a) Maternal inheritance
   (b) Paternity

4. Heteroplasmy is usually observed in:
   (a) Nuclear DNA
   (b) Mitochondrial DNA
   (c) The Y chromosome
   (d) All of the above

5. Human mtDNA consists of approximately how many base pairs?
   (a) About 1,650
   (b) About 16,500
   (c) About 165,000
   (d) About 1,650,000

6. What is the Cambridge reference sequence?
   (a) The human mtDNA reference sequence
   (b) The human and animal mtDNA reference sequence
   (c) A reference sequence with the human hypervariable region only
   (d) A reference sequence with human and animal hypervariable regions only

7. Which of the following is true?
   (a) An individual may have more than one mtDNA type in a single tissue.
   (b) An individual may exhibit one mtDNA type in one tissue and a different type in another tissue.
   (c) An individual may be heteroplasmic in one tissue sample and homoplasmic in another tissue sample.
   (d) All of the above

8. Which of the following is true for mtDNA?
   (a) The heavy strand has higher GC content.
   (b) The light strand has higher GC content.

9. What region of the mtDNA D loop is used in forensic analysis?
   (a) The heavy strand
   (b) The light strand
   (c) Hypervariable regions
   (d) tRNA

10. Which of the following is true?
    (a) Signal dropout is common after the C stretch of HV1 when polymorphic T is absent.
    (b) Signal dropout is common after the C stretch of HV1 when polymorphic T is present.

11. Mitochondrial DNA should be identical among:
    (a) Half-brothers from the same mother
    (b) Half-sisters from the same mother
    (c) Sisters and brothers from the same mother
    (d) All of the above

12. mtDNA usually can be found in:
    (a) Shed hair
    (b) Pulled hair
    (c) All of the above

13. An electropherogram of a mtDNA sequence has a C/T at position 16093 but not at any other position. Which of the following is most likely?
    (a) Contamination
    (b) Heteroplasmy
    (c) DNA mixture

14. Which part of the mtDNA genome is currently used with forensic casework?
    (a) The D loop
    (b) The coding region
    (c) About 75% of the human mitochondrial genome
    (d) The STR region

15. How long is the HV1 of mtDNA?
    (a) About 1,600 base pairs
    (b) About 800 base pairs
    (c) About 400 base pairs
    (d) About 200 base pairs

16. Which of the following is utilized in DNA sequencing?
    (a) 3′ dNTP
    (b) 2′3′ ddNTP
    (c) 3′ dNTP and 2′3′ ddNTP

17. Which of the following is preferable?
    (a) Perform mtDNA sequencing before mtDNA screening
    (b) Perform mtDNA screening prior to mtDNA sequencing

# VI

# Forensic Issues

# DNA Databases

# 22

DNA databases were developed to help solve crimes by creating a network for exchanges of information among law enforcement agencies. More specifically, DNA databases allow forensic laboratories to compare DNA profiles electronically to identify perpetrators. The United Kingdom established the world's first national DNA database (the ***National DNA Database***® or ***NDNAD*** in 1995. It demonstrated initial success in solving crimes. Three years later, the United States introduced its national ***COMBINED DNA INDEX SYSTEM*** or ***CODIS***. By the end of 1998, other countries (Austria, Germany, Netherlands, New Zealand, and Slovenia) had also introduced national DNA databases. Table 22.1 describes some national DNA databases.

## 22.1 U.K. National Database

### 22.1.1 Brief History

NDNAD was established in 1995 in England and Wales. Scotland and Northern Ireland have their own databases but also submit their profiles to NDNAD.

### 22.1.2 Management

Three entities handle the management and operation of NDNAD:

- A Strategic Board consisting of the Home Office, the Association of Chief Police Officers (ACPO), the Association of Police Authorities (APA), and a member of the Human Genetics Commission (HGC) is responsible for oversight.
- A Custodian Unit of the Home Office unit is responsible for setting standards of performance for forensic science laboratories to provide DNA profiles to NDNAD by monitoring participating laboratories and ensuring that profiles are submitted and maintained.
- The Forensic Science Service (FSS) carries out day-to-day operations of the NDNAD under contract to the Home Office. The FSS is responsible

**Table 22.1 Characteristics of National DNA Databases**

| Country | Year Established | Suspect Entry Criteria | Convicted Offender Entry Criteria | Removal Criteria |
|---|---|---|---|---|
| United Kingdom | 1995 | Any recordable offense that leads to imprisonment | Entered as suspect | Never removed |
| New Zealand | 1996 | No suspects entered | Relevant offense (≥7 years in prison) | Never removed unless conviction quashed |
| Austria | 1997 | Any recordable offense that leads to imprisonment | Entered as suspect | Only after acquittal |
| Netherlands | 1997 | No suspects entered except when suspect's DNA is tested for case | Offense leading to >4 years in prison | 20 to 30 years after conviction |
| Germany | 1998 | Offense leading to>1 year in prison | After court decision | After acquittal or 5 to 10 years after conviction if prognosis good |
| Slovenia | 1998 | Any recordable offense that leads to imprisonment | Entered as suspect | Depends on severity of crime |
| United States | 1998 | No suspects entered; under revision | Depends on state law | Depends on state law |
| Finland | 1999 | offense leading to >1 year in prison | Entered as suspect | Only after acquittal |
| Sweden | 2000 | No suspects entered | Offense leading to >2 years in prison | 10 years after release from prison |
| Switzerland | 2000 | Any recordable offense that leads to imprisonment | Entered as suspect | After acquittal or 5 to 30 years after conviction |
| France | 2001 | No suspects entered | Sexual assaults and serious crimes | 40 years after conviction |

*Source:* Adapted from Jobling, M. A., and P. Gill. 2004. *Nat Rev Genet* 5 (10):739.

for loading DNA profiles onto the system, conducting searches for matches, and reporting the results back to police. The FSS is the largest provider of DNA profiles to NDNAD. The FSS's delivery of NDNAD contract services is overseen by the custodian unit of the Home Office.

### 22.1.3 Index

Two types of samples are stored in the database: crime scene samples and samples taken from offenders.

### 22.1.4 Entries

NDNAD provides that DNA samples can be taken from any individual arrested and detained in police custody in connection with a recordable offense. Most offenses (other than traffic offenses) committed must be recorded in the Police National Computer system as part of an individual's criminal record. Samples can be nonintimate (typically mouth swabs) taken without consent or intimate samples (blood) taken with consent and known as criminal justice or CJ samples.

The NDNAD contains the world's largest number of DNA profiles in proportion to population—approximately 5%. As of February 2006, the NDNAD held approximately 3.4 million offender profiles and 290,000 crime scene profiles. Additionally, more than 721,495 investigations were aided with NDNAD. Six agencies in the U.K. are approved to provide DNA profiles from offender and crime scene samples to the NDNAD. They are accredited both by the United Kingdom Accreditation Service and the Custodian Unit of the Home Office.

## 22.2 U.S. Combined DNA Index System

### 22.2.1 Brief History

A pilot project started in the United States in 1990 included only 14 state and local agencies. The Congressional DNA Identification Act (1994) authorized the Federal Bureau of Investigation (FBI) to establish a national DNA database. By 1997, 13 STR loci were selected as the core loci for the national database and CODIS was finally formed in 1998. Currently, all 50 states have authorized the collection of samples from convicted felons for DNA databasing. However, state laws vary; all states are required to collect DNA profiles from individuals convicted of felony sex crimes while others authorize collection for additional types of felonies.

### 22.2.2 Infrastructure

CODIS is a database with three hierarchical levels: the Local DNA Index System (LDIS), the State DNA Index System (SDIS), and the National DNA Index System (NDIS). The national tier represents the highest level (Figure 22.1). The DNA profiles from LDIS are loaded to the state level system. NDIS is the central repository of DNA records submitted by states. Communication is mediated using a secured network with encryption.

### 22.2.3 Management

The LDIS is maintained at crime laboratories operated by police departments, sheriff's offices, and local agencies. All forensic DNA records originate at the local level and are transmitted to the state and national levels. Each state maintains a system that allows exchanges and comparisons of DNA profiles with other states. Searches of DNA profiles can be conducted at the national level. NDIS is administered by the FBI and the bureau provides software, installation, user training, and support for all laboratories that participate in CODIS. The quality assurance standards for convicted offender database laboratories were set up by the DNA Advisory Board in 1998.

### 22.2.4 Index

At all three levels of databases, the data collected in CODIS is separated into two indices: (1) the ***forensic index*** containing DNA profiles obtained from evidence collected at crime scenes and (2) the ***offender index*** containing

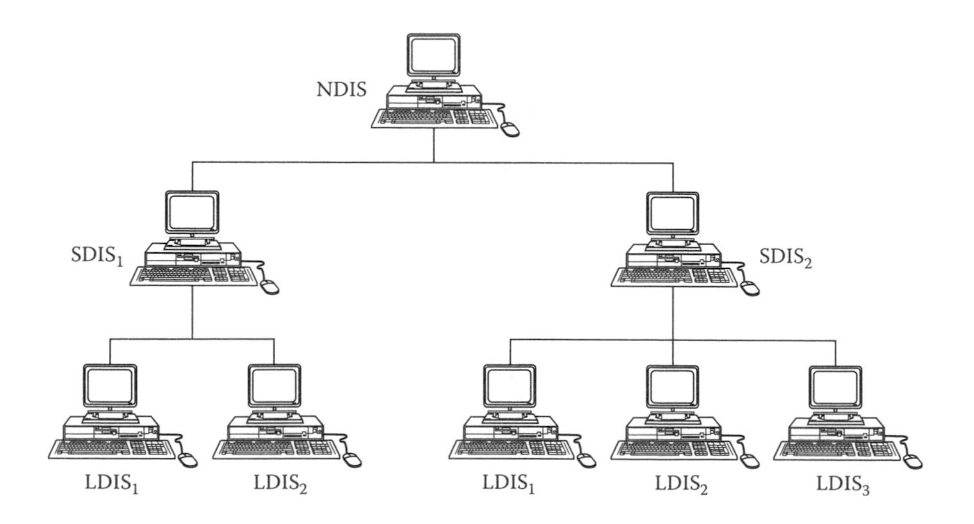


**Figure 22.1** CODIS infrastructure. All DNA profiles originate at LDIS, and then enter SDIS and NDIS. SDIS allows laboratories within states to exchange DNA profiles. NDIS is the highest level of the infrastructure. It allows the participating laboratories to exchange DNA profiles on a national level. (Adapted from www.fbi.gov/hq/lab/codis)

DNA profiles of criminals convicted of sex offenses and several other violent crimes.

### 22.2.5 Database Entries

As of April 2007, the NDIS contained more than 4.5 million DNA profiles of convicted felons and 170,000 forensic profiles. Although the proportion of the population represented on the database is only approximately 1.5%, the NDIS contains the largest number of DNA profiles in absolute numbers. Each profile stored in CODIS includes additional information including specimen identifiers, the sponsoring laboratory that provided the sample, and the DNA profile of the sample. However, it does not include individual criminal history records, dates of birth, social security numbers, or case-related information.

### 22.2.6 DNA Markers

The CODIS software supports the storage and searching of both restriction fragment length polymorphism (RFLP)- and PCR-based DNA profiles (HLA-DQA1, Polymarker, D1S80, and STR). Chapters 17, 18, and 20 discuss RFLP- and PCR-based profiling.

## 22.3 Database Applications for Criminal Investigations

The ultimate goal in the database utilization is assistance in solving crimes. A ***hit*** is a match made from the information provided only by CODIS. ***Investigation-aided*** cases are those assisted by CODIS hits. The number of investigations aided is a useful method for measuring the success of the application of the database (Figure 22.2). As of April 2007, over 48,500 hits made with CODIS assisted in more than 49,400 investigations.

The technological advances and the increase in awareness of the great potential of DNA in solving crimes led to sharp increases in demand for and utilization of databases (Figure 22.3). More jurisdictions are incorporating more felonies into lists of crimes that require DNA profiles and some plan to include all felonies in such databases.

### 22.3.1 Case-to-Offender Searches

Matches of forensic and offender profiles can reveal the identities of perpetrators. In 1998, Leon Dundas, a Florida man, was a suspect in a rape case and refused to provide a blood sample for DNA testing. A year later, Dundas was killed in a drug deal. The investigators were able to obtain a sample of

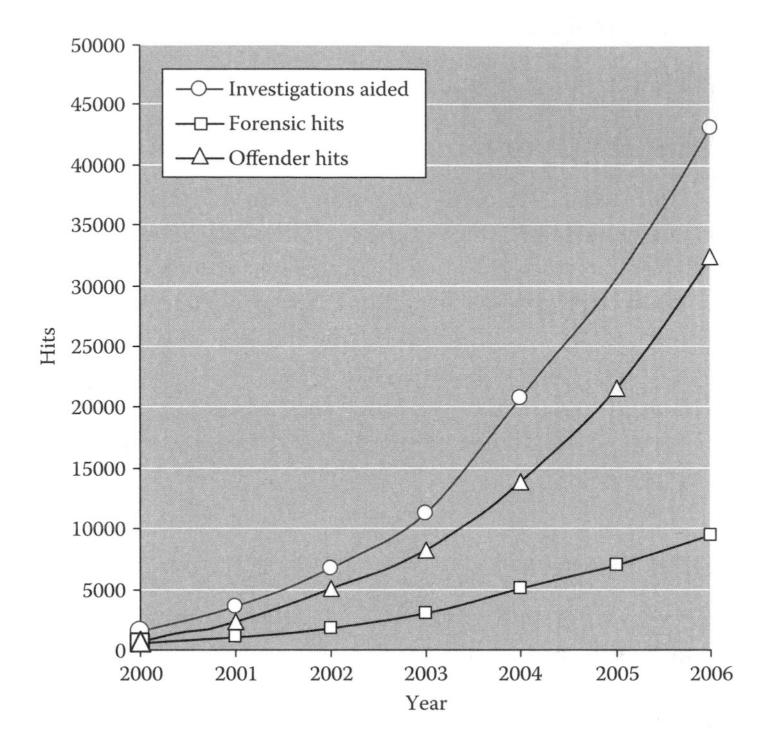


**Figure 22.2** Number of hits and investigations aided from CODIS database. (Adapted from www.fbi.gov/hq/lab/codis)

Dundas' blood from the medical examiner's office. The DNA profile was compared with the forensic index in CODIS. Surprisingly, investigators found that Dundas' profile matched the DNA sample found in the rape case and linked him to three other unsolved rapes in Jacksonville and seven more in Washington, D.C. Without the DNA database, such cases would not be solved.

### 22.3.2 Case-to-Case Searches

Matches of profiles in the forensic index can link crime scenes and aid in identifying serial offenders. Law enforcement agencies in multiple jurisdictions can coordinate their investigations and share leads. In 1996, two young girls were abducted from bus stops in St. Louis. Both girls were raped and DNA samples were collected. Both DNA profiles pointed to the same perpetrator. In 1999, the St. Louis police decided to re-analyze the samples using the new STR technology and sent the profiles to the CODIS database. A match was made to a different rape case to which Dominic Moore, the perpetrator, had already confessed, thus identifying him as the perpetrator of the 1996 rapes.

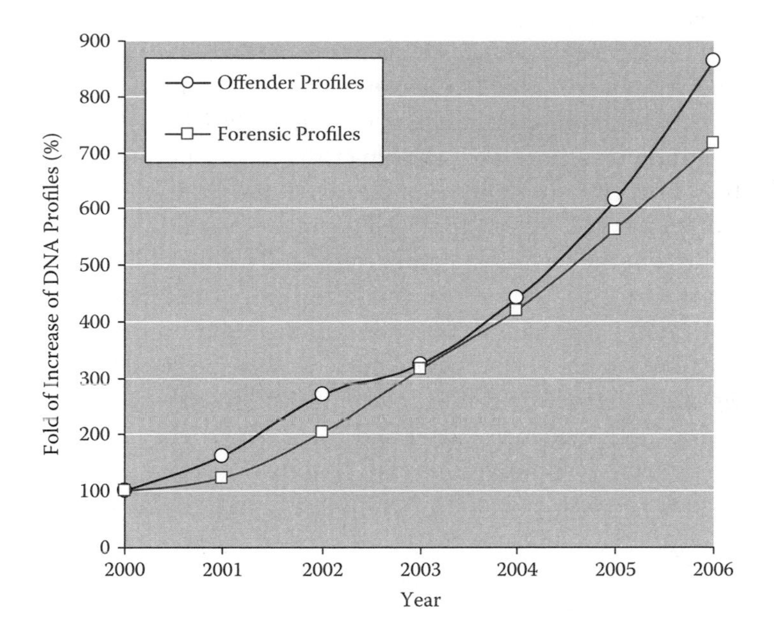


**Figure 22.3** Growth of CODIS database. In 2000, 460,365 offender profiles and 22,484 forensic profiles were entered. (Adapted from www.fbi.gov/hq/lab/codis)

### 22.3.3 Familial Searches

Familial searching, recently introduced in the United Kingdom, is a new method of applying databases in criminal investigations. Familial searches are conducted only for the most serious offenses and in cases when a DNA profile from a crime scene sample has no match in the database. The technique is based on the assumption that close relatives share similar DNA profiles. Thus, databases may be used for familial searches (if a perpetrator's close relative had been convicted of a crime and is listed in the database). These familial searches may lead to identification of a perpetrator who is the source of crime scene evidence.

The first familial search leading to a successful prosecution was conducted in Surrey. In 2003, a truck driver was killed after a brick was thrown through his windshield from a bridge. The perpetrator's DNA profile was obtained from the brick. A search of the NDNAD revealed no match for the perpetrator. Next a familial search of the database was conducted. The system identified a close relative that led police to identify the perpetrator, Craig Harman, who was then convicted of manslaughter.

Policies on familial searching vary among jurisdictions. Familial searches are rare in the U.S. because of concern about the potential impact on individual privacy rights, particularly the potential for these searches to violate the privacy of unrelated people whose genetic profiles happen to resemble those of individuals included in the databases. Federal privacy laws prohibit the FBI from performing familial searches within its database although some states have granted authority to perform familial searches.

In 2003, a perpetrator's DNA sample, left at the scene of a rape and murder committed in 1984, was re-analyzed using forensic DNA techniques. The DNA profile from the scene was compared to the profile of Darryl Hunt who was imprisoned for the crime. The two profiles did not match. The state's database of 40,000 convicted felons was searched and no match was found. The search, however, identified a close relative of the perpetrator. This led the police identify Willard Brown as the perpetrator. Brown was sentenced to life in prison. The DNA evidence exonerated Hunt who spent 18 years in prison for a crime he did not commit. This case demonstrates that DNA databasing can prove guilt and also exonerate the innocent.

## 22.4 Ethical Issues

The use of any database involves a balance of individual civil rights and the interests of the criminal justice systems, which may vary from jurisdiction to jurisdiction. The major ethical issues relate to criteria for sample entry and retention.

### 22.4.1 Entry Criteria

Many countries such as the United States do not allow entry of suspects' profiles into DNA databases. Other jurisdictions such as the United Kingdom allow DNA samples to be taken from individuals suspected of committing recordable offenses that may lead to prison sentences.

The entry criteria of convicted offenders vary by jurisdiction. The database system is projected to include the profiles of minor criminals because statistics show that most offenders found guilty of serious crimes were previously convicted for minor crimes. Broadening the size of the database and including samples from more types of crimes could lead to the assumption that the number of crimes solved would also increase. Although most jurisdictions currently enter only serious offenders into databases, it is possible that one day the databases will include many more offenders and suspects, as well as the general public. One advantage of including an entire population in a database is the ability to identify missing, kidnapped, and abducted individuals in addition to victims of major accidents and mass fatalities. Debates concerning the need to balance the benefits and dangers of developing a broader database will inevitably continue far into the future.

### 22.4.2 Retention of Samples

Austria, Finland, Switzerland, and the United Kingdom retain reference samples that can be retested if new technology becomes available. Conversely, Germany, Netherlands, Norway, and Belgium do not allow retention of samples from offenders after DNA profiles are obtained. Considerable debate has surrounded the retention of samples for possible retesting with future technologies. This would allow retesting of a suspect's sample if required or for purposes of updating with additional loci.

The suggestion was made to use CODIS to obtain more reliable and useful population statistics based on data from a broader population. However, it can be argued from the opposite perspective that the database could reveal private genetic information that could then be misused. The objection is based on concern for protection of privacy rights. If a sample were made available to an unauthorized person, confidential information could be disclosed. Destroying samples would prevent the use of DNA analyses for purposes other than forensic uses.

## Bibliography

### Book Chapters and Review Articles

Asplen, C., and S. A. Lane. 2004. International perspectives on forensic DNA databases. *Forensic Sci Int* 146 Suppl:S119.

Butler, J. M. 2005. Combined DNA index system (CODIS) and the use of DNA databases, in *Forensic DNA Typing*. Burlington, MA: Elsevier.

Corte-Real, F. 2004. Forensic DNA databases. *Forensic Sci Int* 146 Suppl:S143.

Martin, P. D., H. Schmitter, and P. M. Schneider. 2001. A brief history of the formation of DNA databases in forensic science within Europe. *Forensic Sci Int* 119 (2):225.

Schneider, P. M., and P. D. Martin. 2001. Criminal DNA databases: The European situation. *Forensic Sci Int* 119 (2):232.

Werrett, D. J. 1997. The national DNA database. *Forensic Sci Int* 88 (1):33.

### DNA Databases

Harding, H. W., and R. Swanson. 1998. DNA database size. *J Forensic Sci* 43 (1):248.

Linacre, A. 2003. The UK National DNA Database. *Lancet* 361 (9372):1841.

McEwen, J. E. 1995. Forensic DNA data banking by state crime laboratories. *Am J Hum Genet* 56 (6):1487.

McEwen, J. E., and P. R. Reilly. 1994. A review of state legislation on DNA forensic data banking. *Am J Hum Genet* 54 (6):941.

Parson, W., and M. Steinlechner. 2001. Efficient DNA database laboratory strategy for high through-put STR typing of reference samples. *Forensic Sci Int* 122 (1):1.

Peerenboom, E. 1998. Central criminal DNA database created in Germany. *Nat Biotechnol* 16 (6):510.
Walsh, S. J. et al. 2002. The collation of forensic DNA case data into a multi-dimensional intelligence database. *Sci Justice* 42 (4):205.

## Application of Databases

Evett, I. W., J. Scranage, and R. Pinchin. 1992. Efficient retrieval from DNA databases: Based on the second European DNA Profiling Group collaborative experiment. *Forensic Sci Int* 53 (1):45.
Gill, P. et al. 2003. A comparison of adjustment methods to test the robustness of an STR DNA database comprised of 24 European populations. *Forensic Sci Int* 131 (2-3):184.
Harbison, S. A. et al. 2001. The New Zealand DNA databank: Its development and significance as a crime solving tool. *Sci Justice* 41 (1):33.

## Ethical Issues

Balding, D. J. 2002. The DNA database search controversy. *Biometrics* 58 (1):241.
Bereano, P. L. 1990. DNA identification systems: Social policy and civil liberties concerns. *J Int Bioethique* 1 (3):146.
Bieber, F. R., and D. Lazer. 2004. Guilt by association: Should the law be able to use one person's DNA to carry out surveillance on their family? Not without public debate. *New Sci* 184 (2470):20.
Burk, D. L., and J. A. Hess. 1994. Genetic privacy: Constitutional considerations in forensic DNA testing. *George Mason Univ Civ Rights Law J* 5 (1):1.
Guillen, M. et al. 2000. Ethical–legal problems of DNA databases in criminal investigation. *J Med Ethics* 26 (4):266.
Haimes, E. 2006. Social and ethical issues in the use of familial searching in forensic investigations: Insights from family and kinship studies. *J Law Med Ethics* 34 (2):263.
Herkenham, M. D. 2006. Retention of offender DNA samples necessary to ensure and monitor quality of forensic DNA efforts: Appropriate safeguards exist to protect the DNA samples from misuse. *J Law Med Ethics* 34 (2):380.
Lakra, R., and F. Dimitriadis. 2002. Forensic DNA data banks: Considerations for the health sector. *Health Law Can* 22 (3):57.
Lehrman, S. 2006. Partial to crime. Families become suspects as rules on DNA matches relax. *Sci Am* 295 (6):28.
Sankar, P. 1997. The proliferation and risks of government DNA databases. *Am J Public Health* 87 (3):336.
Scheck, B. 1994. DNA data banking: A cautionary tale. *Am J Hum Genet* 54 (6):931.
Shapiro, E. D., and M. L. Weinberg. 1990. DNA data banking: The dangerous erosion of privacy. *Clevel State Law Rev* 38 (3):455.
Shriver, M., T. Frudakis, and B. Budowle. 2005. Getting the science and the ethics right in forensic genetics. *Nat Genet* 37 (5):449.
Simoncelli, T. 2006. Dangerous excursions: The case against expanding forensic DNA databases to innocent persons. *J Law Med Ethics* 34 (2):390.

## Study Questions

1. Which of the following is not part of CODIS?
   (a) A forensic index that contains DNA profiles with crime scene evidence
   (b) An index that contains profiles of convicted offenders
   (c) An index that contains profiles of suspects

2. CODIS is an acronym for:
   (a) Convicted offender DNA index system
   (b) Combined DNA index system
   (c) Codified information system
   (d) Code of detection in identification of suspects

3. Which country's DNA database contains more than 2.5 million profiles?
   (a) United States
   (b) Canada
   (c) United Kingdom
   (d) Germany

4. What are the major concerns about DNA database expansion?

5. What are the advantages of a database containing a high number of DNA profiles in proportion to the population?

6. What are the disadvantages of familial searching using a DNA database?

7. If new DNA loci are included in a database, what problems might be encountered?

8. How would you assess the success of applying a DNA database to criminal investigations?

# Evaluation of Strength of Forensic DNA Profiling Results 23

## 23.1 Basic Principles of Genetics

### 23.1.1 Mendelian Genetics

***Mendel's first law*** is the principle of segregation of alleles. Each pair of alleles segregates from others in the formation of gametes (mature reproductive cells such as spermatozoa and oocytes). Half of the gametes carry one allele, and the other half carry the other allele.

***Mendel's second law*** is the principle of independent assortment of alleles. The segregation of each pair of alleles is independent of the segregation of other pairs during the formation of gametes.

Gametes are formed during a process known as ***meiosis*** in which cells with haploid chromosome numbers are produced by division of cells with diploid chromosome numbers. In meiosis, the chromosome number of human gametes is reduced from 46 (diploid) to 23 (haploid). The fertilization of an egg thus involves a diploid number of chromosomes. A diploid is composed of 22 pairs of autosomes and 2 sex chromosomes (XX in females and XY in males).

Based on Mendelian principles, genes on different chromosomes assort independently of one another in gamete production. Genes residing very closely together on the same chromosome are usually inherited together. Thus, they do not assort independently and are called ***linked genes***. Genes of different loci but distant from each other on the same chromosome tend to become separated during gamete formation. This results from an exchange of segments between a pair of ***homologous*** chromosomes when the chromosomes are paired during early phases of meiotic division. This type of gene exchange event on homologous chromosomes is called ***crossing over*** and it results in the ***recombination*** of genes in a pair of chromosomes (Figure 23.1).

The Mendelian inheritances of genes can often be measured using ***probabilities***. A probability is the ratio of the number of actual occurrences of an event to the number of possible occurrences. Additionally, the probability of two independent events occurring simultaneously is the product of each of their individual probabilities. This is known as the ***product rule of probability***.

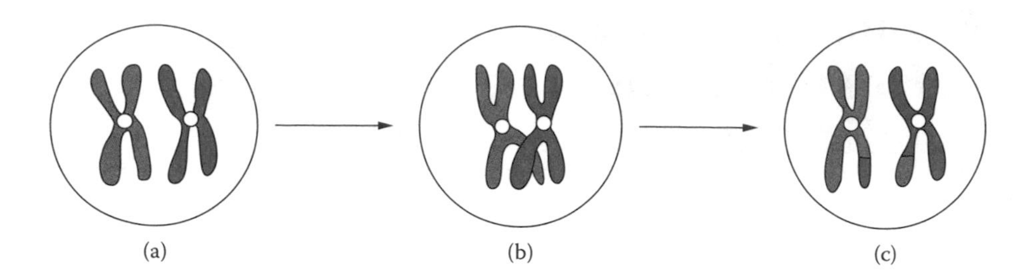


**Figure 23.1** Crossing over hypothesis. (a) Duplicated chromosomes. Each duplicate is called a chromatid, which forms into tetrads. (b) Two chromatids cross each other. (c) Each chromatid breaks at the point of the cross and fuses with a portion of its counterpart.

Mendelian principles apply to the inheritance of loci of the nuclear DNA genome commonly used for forensic testing. Mitochondria contain their own mitochondrial genomes and are inherited maternally. This maternal inheritance of mitochondrial genes does not obey the rules for Mendelian principles and is called nonMendelian inheritance.

### 23.1.2 Population Genetics

Population genetics studies the causes of observed patterns of genetic variations within and among populations.

#### *23.1.2.1 Allele Frequency*

Allele frequency (p) can be calculated directly by counting the number of alleles of one type at a given locus and dividing it by the total number of alleles at that locus in a sampled population. This is called the ***gene counting*** method.

#### *23.1.2.2 Genotype Frequency*

Genotype frequency (P) observed at a given locus can be calculated by dividing the number of individuals with one genotype by the total number of individuals in a sampled population. Each genotype at the locus can be calculated separately. The summation of all genotype frequencies at that locus should equal 1.

#### *23.1.2.3 Heterozygosity*

***Heterozygosity*** is the proportion of alleles, at a given locus, that is heterozygous and is calculated as

$$h = 1 - \sum_i p^2 \qquad \text{(Formula 23.1)}$$

where h is heterozygosity and p is the allelic frequency of the locus for homozygotes. The amount of heterozygosity at a locus in a sampled population is a measure of genetic variation. The higher the heterozygosity, the more variation at a given locus.

#### 23.1.2.4 Hardy-Weinberg Principle

The Hardy-Weinberg (HW) principle, independently discovered by two scholars in the early 1900s, allows predictions of genotype frequencies to be made based on allelic frequencies. However, certain conditions must be met. The population must be large; mate randomly; and lack mutation, migration, and natural selection. If these conditions of the HW principle are met, the population will be in ***equilibrium***, and the following results are expected:

1. The frequencies of the alleles will not change from one generation to the next.
2. Genotype frequencies can be predicted by the allelic frequencies ($p^2$ and $q^2$ for genotype frequencies of homozygotes, and $2pq$ for genotype frequencies of heterozygotes). The sum of the genotype frequencies should equal 1.

$$p^2 + 2pq + q^2 = 1 \qquad \text{(Formula 23.2)}$$

If the observed genotype proportions are different from those expected, one or more of the HW assumptions have been violated.

#### 23.1.2.5 Testing for HW Proportions of Population Database

To determine whether the genotypes of a population in question obey the HW principle, a population database can be constructed. Samples (usually 100 to 200 samples for STR loci) are collected and analyzed at the loci of interest. Allelic frequencies are obtained by using the gene counting method. Table 23.1 shows the allelic frequencies of CODIS loci from a population database. The ***observed genotype frequencies*** at a given locus, as described earlier, are calculated by dividing the number of individuals with one genotype by the total number of individuals in the population sampled. The ***expected genotype frequencies*** are calculated using $p^2$, $2pq$, and $q^2$.

The observed and expected genotype frequencies are then compared using a chi square test. The significance of the differences between observed and expected genotype frequencies can then be determined. The chi square is calculated using the following formula:

$$x^2 = \sum_{i=1}^{n} \frac{(O_i - E_i)^2}{E_i} \qquad \text{(Formula 23.3)}$$

**Table 23.1 Allelic frequencies of 13 CODIS STR Loci**

| Allele | Allelic Frequency (%) | | |
|---|---|---|---|
| | African American | Caucasian | Hispanic |
| D3S1358 | (N=210) | (N=203) | (N=209) |
| <12 | 0.476 | 0.000 | 0.000 |
| 12 | 0.238 | 0.000 | 0.000 |
| 13 | 1.190 | 0.246 | 0.239 |
| 14 | 12.143 | 14.039 | 7.895 |
| 15 | 29.048 | 24.631 | 42.584 |
| 15.2 | 0.000 | 0.000 | 0.000 |
| 16 | 30.714 | 23.153 | 26.555 |
| 17 | 20.000 | 21.182 | 12.679 |
| 18 | 5.476 | 16.256 | 8.373 |
| 19 | 0.476 | 0.493 | 1.435 |
| >19 | 0.238 | 0.000 | 0.239 |
| VWA | (N=180) | (N=196) | (N=203) |
| 11 | 0.278 | 0.000 | 0.246 |
| 13 | 0.556 | 0.510 | 0.000 |
| 14 | 6.667 | 10.204 | 6.158 |
| 15 | 23.611 | 11.224 | 7.635 |
| 16 | 26.944 | 20.153 | 35.961 |
| 17 | 18.333 | 26.276 | 22.167 |
| 18 | 13.611 | 22.194 | 19.458 |
| 19 | 7.222 | 8.418 | 7.143 |
| 20 | 2.778 | 1.020 | 1.232 |
| 21 | 0.000 | 0.000 | 0.000 |
| FGA | (N=180) | (N=196) | (N=203) |
| <18 | 0.278 | 0.000 | 0.000 |
| 18 | 0.833 | 3.061 | 0.246 |
| 18.2 | 0.833 | 0.000 | 0.000 |
| 19 | 5.278 | 5.612 | 7.882 |
| 19.2 | 0.278 | 0.000 | 0.000 |
| 20 | 7.222 | 14.541 | 7.143 |
| 20.2 | 0.000 | 0.255 | 0.246 |
| 21 | 12.500 | 17.347 | 13.054 |
| 21.2 | 0.000 | 0.000 | 0.246 |
| 22 | 22.500 | 18.878 | 17.734 |
| 22.2 | 0.556 | 1.020 | 0.493 |

**Table 23.1 Allelic frequencies of 13 CODIS STR Loci (continued)**

| Allele | Allelic Frequency (%) | | |
|---|---|---|---|
| | African American | Caucasian | Hispanic |
| 22.3 | 0.000 | 0.000 | 0.000 |
| 23 | 12.500 | 15.816 | 14.039 |
| 23.2 | 0.000 | 0.000 | 0.739 |
| 24 | 18.611 | 13.776 | 12.562 |
| 24.2 | 0.000 | 0.000 | 0.000 |
| 24.3 | 0.000 | 0.000 | 0.000 |
| 25 | 10.000 | 6.888 | 13.793 |
| 26 | 3.611 | 1.786 | 8.374 |
| 27 | 2.222 | 1.020 | 3.202 |
| 28 | 1.667 | 0.000 | 0.246 |
| 29 | 0.556 | 0.000 | 0.000 |
| 30 | 0.278 | 0.000 | 0.000 |
| >30 | 0.278 | 0.000 | 0.000 |
| D8S1179 | (N=180) | (N=196) | (N=203) |
| <9 | 0.278 | 1.786 | 0.246 |
| 9 | 0.556 | 1.020 | 0.246 |
| 10 | 2.500 | 10.204 | 9.360 |
| 11 | 3.611 | 5.867 | 6.158 |
| 12 | 10.833 | 14.541 | 12.069 |
| 13 | 22.222 | 33.929 | 32.512 |
| 14 | 33.333 | 20.153 | 24.631 |
| 15 | 21.389 | 10.969 | 11.576 |
| 16 | 4.444 | 1.276 | 2.463 |
| 17 | 0.833 | 0.255 | 0.739 |
| 18 | 0.000 | 0.000 | 0.000 |
| D21S11 | (N=179) | (N=196) | (N=203) |
| 24.2 | 0.279 | 0.510 | 0.246 |
| 24.3 | 0.000 | 0.000 | 0.000 |
| 26 | 0.279 | 0.000 | 0.000 |
| 27 | 6.145 | 4.592 | 0.985 |
| 28 | 21.508 | 16.582 | 6.897 |
| 29 | 18.994 | 18.112 | 20.443 |
| 29.2 | 0.279 | 0.000 | 0.246 |
| 30 | 17.877 | 23.214 | 33.005 |
| 30.2 | 0.838 | 3.827 | 3.202 |

**Table 23.1 Allelic frequencies of 13 CODIS STR Loci (continued)**

| Allele | Allelic Frequency (%) | | |
|---|---|---|---|
| | African American | Caucasian | Hispanic |
| 30.3 | 0.000 | 0.000 | 0.000 |
| 31 | 9.218 | 7.143 | 6.897 |
| 31.2 | 7.542 | 9.949 | 8.621 |
| 32 | 0.838 | 1.531 | 1.232 |
| 32.1 | 0.000 | 0.000 | 0.000 |
| 32.2 | 6.983 | 11.224 | 13.547 |
| 33 | 0.838 | 0.000 | 0.000 |
| 33.2 | 3.352 | 3.061 | 4.187 |
| 34 | 0.838 | 0.000 | 0.000 |
| 34.2 | 0.279 | 0.000 | 0.493 |
| 35 | 2.793 | 0.000 | 0.000 |
| 35.2 | 0.000 | 0.255 | 0.000 |
| 36 | 0.559 | 0.000 | 0.000 |
| >36 | 0.559 | 0.000 | 0.000 |
| D18S51 | (N=180) | (N=196) | (N=203) |
| <11 | 0.556 | 1.276 | 0.493 |
| 11 | 0.556 | 1.276 | 1.232 |
| 12 | 5.833 | 12.755 | 10.591 |
| 13 | 5.556 | 12.245 | 16.995 |
| 13.2 | 0.556 | 0.000 | 0.000 |
| 14 | 6.389 | 17.347 | 16.995 |
| 14.2 | 0.000 | 0.000 | 0.000 |
| 15 | 16.667 | 12.755 | 13.793 |
| 15.2 | 0.000 | 0.000 | 0.000 |
| 16 | 18.889 | 10.714 | 11.576 |
| 17 | 16.389 | 15.561 | 13.793 |
| 18 | 13.056 | 9.184 | 5.172 |
| 19 | 7.778 | 3.571 | 3.695 |
| 20 | 5.556 | 2.551 | 1.724 |
| 21 | 1.111 | 0.510 | 1.970 |
| 21.2 | 0.000 | 0.000 | 0.000 |
| 22 | 0.556 | 0.255 | 0.739 |
| >22 | 0.556 | 0.000 | 1.232 |
| D5S818 | (N=180) | (N=195) | (N=203) |
| 7 | 0.278 | 0.000 | 6.158 |

**Table 23.1 Allelic frequencies of 13 CODIS STR Loci (continued)**

| Allele | Allelic Frequency (%) | | |
|---|---|---|---|
| | African American | Caucasian | Hispanic |
| 8 | 5.000 | 0.000 | 0.246 |
| 9 | 1.389 | 3.077 | 5.419 |
| 10 | 6.389 | 4.872 | 6.650 |
| 11 | 26.111 | 41.026 | 42.118 |
| 12 | 35.556 | 35.385 | 29.064 |
| 13 | 24.444 | 14.615 | 9.606 |
| 14 | 0.556 | 0.769 | 0.493 |
| 15 | 0.000 | 0.256 | 0.246 |
| >15 | 0.278 | 0.000 | 0.000 |
| D13S317 | (N=179) | (N=196) | (N=203) |
| 7 | 0.000 | 0.000 | 0.000 |
| 8 | 3.631 | 9.949 | 6.650 |
| 9 | 2.793 | 7.653 | 21.921 |
| 10 | 5.028 | 5.102 | 10.099 |
| 11 | 23.743 | 31.888 | 20.197 |
| 12 | 48.324 | 30.867 | 21.675 |
| 13 | 12.570 | 10.969 | 13.793 |
| 14 | 3.631 | 3.571 | 5.665 |
| 15 | 0.279 | 0.000 | 0.000 |
| D7S820 | (N=210) | (N=203) | (N=209) |
| 6 | 0.000 | 0.246 | 0.239 |
| 7 | 0.714 | 1.724 | 2.153 |
| 8 | 17.381 | 16.256 | 9.809 |
| 9 | 15.714 | 14.778 | 4.785 |
| 10 | 32.381 | 29.064 | 30.622 |
| 10.1 | 0.000 | 0.000 | 0.000 |
| 11 | 22.381 | 20.197 | 28.947 |
| 11.3 | 0.000 | 0.000 | 0.000 |
| 12 | 9.048 | 14.039 | 19.139 |
| 13 | 1.905 | 2.956 | 3.828 |
| 14 | 0.476 | 0.739 | 0.478 |
| CSF1PO | (N=210) | (N=203) | (N=209) |
| 6 | 0.000 | 0.000 | 0.000 |
| 7 | 4.286 | 0.246 | 0.239 |
| 8 | 8.571 | 0.493 | 0.000 |

**Table 23.1 Allelic frequencies of 13 CODIS STR Loci (continued)**

| Allele | Allelic Frequency (%) | | |
|---|---|---|---|
| | African American | Caucasian | Hispanic |
| 9 | 3.333 | 1.970 | 0.718 |
| 10 | 27.143 | 25.369 | 25.359 |
| 10.3 | 0.000 | 0.246 | 0.000 |
| 11 | 20.476 | 30.049 | 26.555 |
| 12 | 30.000 | 32.512 | 39.234 |
| 13 | 5.476 | 7.143 | 6.459 |
| 14 | 0.714 | 1.478 | 0.957 |
| 15 | 0.000 | 0.493 | 0.478 |
| TPOX | (N=209) | (N=203) | (N=209) |
| 6 | 8.612 | 0.000 | 0.478 |
| 7 | 2.153 | 0.246 | 0.478 |
| 8 | 36.842 | 54.433 | 55.502 |
| 9 | 18.182 | 12.315 | 3.349 |
| 10 | 9.330 | 3.695 | 3.349 |
| 11 | 22.488 | 25.369 | 27.273 |
| 12 | 2.392 | 3.941 | 9.330 |
| 13 | 0.000 | 0.000 | 0.239 |
| TH01 | (N=210) | (N=203) | (N=209) |
| 5 | 0.000 | 0.000 | 0.239 |
| 6 | 10.952 | 22.660 | 23.206 |
| 7 | 44.048 | 17.241 | 33.732 |
| 8 | 18.571 | 12.562 | 8.134 |
| 8.3 | 0.000 | 0.246 | 0.000 |
| 9 | 14.524 | 16.502 | 10.287 |
| 9.3 | 10.476 | 30.542 | 24.163 |
| 10 | 1.429 | 0.246 | 0.239 |
| D16S539 | (N=209) | (N=202) | (N=208) |
| 8 | 3.589 | 1.980 | 1.683 |
| 9 | 19.856 | 10.396 | 7.933 |
| 10 | 11.005 | 6.683 | 17.308 |
| 11 | 29.426 | 27.228 | 31.490 |
| 12 | 18.660 | 33.911 | 28.606 |
| 13 | 16.507 | 16.337 | 10.337 |
| 14 | 0.957 | 3.218 | 2.404 |
| 15 | 0.000 | 0.248 | 0.240 |

*Source:* Adapted from Budowle, B. et al. 1999. *J Forensic Sci* 44:1277.

Where $O_i$ is the ith observed genotype frequency, $E_i$ is the ith expected genotype frequency, and n is the total number of genotypes, the chi square ($x^2$) is calculated as the sum of all genotypes of a given locus.

The chi-square value and the degrees of freedom are then used to determine a p value (do not confuse this p value with the allelic frequency designated p) that can be obtained from a table of p values (such tables can be found at the backs of most statistics text books). If the p value exceeds 0.05, the deviation of expected from observed genotype frequencies is not considered statistically significant. Thus, the observed genotype frequencies fit the expected genotype frequencies predicted by the HW principle.

#### 23.1.2.6 *Probability of Match*

The discriminating power of genetic loci used above can be measured by ***population match probability*** (*P*m). *P*m is defined as the probability of having a matching genotype between two randomly chosen individuals. The lower the *P*m, the less likely a match between two randomly chosen individuals will be. This is calculated as follows:

$$Pm = \sum_i (p^2)^2 + \sum_j (2pq)^2 \qquad \text{(Formula 23.4)}$$

p and q represent the frequencies of two different alleles. *P*m can also be used to compare the discriminating powers of different loci. Tables 23.2 through 23.4 show *P*m values of loci commonly used for forensic applications including SNP, VNTR, and STR.

## 23.2 Statistical Analysis of DNA Profiling Results

It is desirable to evaluate the strength of DNA profiling results, particularly if two DNA profiles match. A DNA profile from crime scene evidence and a profile from a suspect may have the same profiles because the crime scene sample came from the suspect. Another possiblity is that the suspect happens to have the same profile as the evidence found at the crime scene. The significance of a match between DNA profiles can be evaluated by using statistical calculations that determine the rarity of a specific DNA profile in a relevant population. The statistical evaluation of the significance can be included in a case report (Figure 23.2). The guidelines for the DNA profile interpretation such as those issued by the National Research Council, the DNA Advisory Board, and the European DNA Profiling Group can be consulted.

**Table 23.2 Heterozygosity and *Pm* Values of Six SNP Loci**

| Locus | Allele | Heterozygosity | *Pm* |
|---|---|---|---|
| DQA1 | 7 | 0.828 | 0.053 |
| LDLR | 2 | 0.493 | 0.379 |
| GYPA | 2 | 0.498 | 0.376 |
| HBGG | 3 | 0.508 | 0.360 |
| D7S8 | 2 | 0.476 | 0.388 |
| GC | 3 | 0.592 | 0.235 |
| Average | | 0.566 | |
| Product | | | $2.5 \times 10^{-4}$ (1 in 4000) |

*Source:* Office of Justice Programs, National Institute of Justice, U.S. Department of Justice, 2000.

**DEPARTMENT OF FORENSIC BIOLOGY**

**LABORATORY REPORT**

**LAB NO:** IQAS 2003, Lab # 259

**SUSPECTS:** Suspect 1, 2032
Suspect 2, 2033
Suspect 3, 2034

**SUMMARY OF RESULTS:**

Semen* was found on the crime scene sample 2031b, based on the presence of P30 antigen and sperm.

PCR DNA typing was done on crime scene sample 2031b. Results indicate the semen could have come from the suspect 1. This combination of DNA alleles would be expected to be found in approximately:

1 in greater than 1 trillion Blacks**
1 in greater than 1 trillion Caucasians
1 in greater than 1 trillion Hispanics
1 in greater than 1 trillion Asians

The DNA from crime scene sample 2031b could not have come from suspect 2 or suspect 3.

Amylase, a component of saliva, was not found on crime scene sample 2031b.

No semen was found on the control sample 2035.

Crime scene sample 2031a was not analyzed.

**The DNA results in this case do not match any previous PCR (STR) DNA cases to date.**

*Semen has two components: the seminal plasma (which contains the P30 antigen) and spermatozoa. Semen can be identified by detecting either P30 antigen and/or sperm.

** OCME STR database, National Research Council (1996) The Evaluation of Forensic DNA Evidence, Natl. Acad. Press, Washington DC.

**Figure 23.2** Example of laboratory report.

**Table 23.3 Heterozygosity and *Pm* Values of Six VNTR Loci**

| Locus | Bins | Heterozygosity | Pm |
|---|---|---|---|
| D1S7 | 28 | 0.945 | 0.0058 |
| D2S44 | 26 | 0.926 | 0.0103 |
| D4S139 | 19 | 0.899 | 0.0184 |
| D10S28 | 24 | 0.943 | 0.0063 |
| D14S13 | 30 | 0.899 | 0.0172 |
| D17S79 | 19 | 0.799 | 0.0700 |
| Average | | 0.902 | |
| Product | | | $8.26 \times 10^{-12}$ (1 in $1.2 \times 10^{11}$) |

*Source:* Office of Justice Programs, National Institute of Justice, U.S. Department of Justice, 2000.

### 23.2.1 Genotypes

The approaches to performing statistical analysis are (1) calculation of profile probability and (2) use of the likelihood ratio method. Although profile probability is the most commonly used method because of its simplicity, both approaches lead to the same conclusion.

#### *23.2.1.1 Profile Probability*

Profile probability can be calculated based on the following steps:

1. Calculate the locus genotype frequency as follows; p and q are the allelic frequencies observed in the database for a given allele:

   Locus genotype frequency for homozygotes:

   $$P_i = p^2 \qquad \text{(Formula 23.5)}$$

   Locus genotype frequency for heterozygotes:

   $$P_j = 2pq \qquad \text{(Formula 23.6)}$$

2. Profile probability can then be calculated based on the product rule by multiplying all the locus genotype frequencies calculated as above.

The lower the profile probability is, the less likely that an individual chosen at random will have a coincident match with the DNA profile of the evidence sample. Table 23.5 shows calculations of profile probability from a DNA profile.

**Table 23.4 Heterozygosity and *Pm* Values for CODIS Loci**

| Locus | Allele | Caucasian American | | African American | |
|---|---|---|---|---|---|
| | | Heterozygosity | *Pm* | Hetetrozygosity | *Pm* |
| CSF1PO | 11 | 0.734 | 0.112 | 0.781 | 0.081 |
| TPOX | 7 | 0.621 | 0.195 | 0.763 | 0.090 |
| TH01 | 7 | 0.783 | 0.081 | 0.727 | 0.109 |
| vWA | 10 | 0.811 | 0.062 | 0.809 | 0.063 |
| D16S539 | 8 | 0.767 | 0.089 | 0.798 | 0.070 |
| D7S820 | 11 | 0.806 | 0.065 | 0.782 | 0.080 |
| D13S317 | 8 | 0.771 | 0.085 | 0.688 | 0.136 |
| D5S818 | 10 | 0.682 | 0.158 | 0.739 | 0.112 |
| FGA | 19 | 0.860 | 0.036 | 0.863 | 0.033 |
| D3S1358 | 10 | 0.795 | 0.075 | 0.763 | 0.094 |
| D8S1179 | 10 | 0.780 | 0.067 | 0.778 | 0.082 |
| D18S51 | 15 | 0.876 | 0.028 | 0.873 | 0.029 |
| D2S11 | 20 | 0.853 | 0.039 | 0.861 | 0.034 |
| Average | | 0.7812 | | 0.7866 | |
| Product | | | $1.738 \times 10^{-15}$ (1 in $5.753 \times 10^{14}$) | | $1.092 \times 10^{-15}$ (1 in $9.161 \times 10^{14}$) |

*Source:* Office of Justice Programs, National Institute of Justice, U.S. Department of Justice, 2000.

**23.2.1.1.1 Structured Populations** The above calculation of profile probability is based on the assumption that a randomly selected individual is unrelated to a perpetrator. However, it is likely that the individual and the perpetrator are from the same subpopulation (groups within a population) and are thus not completely independent. Mating is more likely to occur within subpopulations than between subpopulations. As a result, the proportion of homozygotes increases and the proportion of heterozygotes decreases in a subpopulation because individuals in a subpopulation appear to be related.

The effect of population structure should be considered and an appropriate correction made to estimate profile probabilities. The correction can be made by using factor θ. Thus, the locus genotype frequency can be calculated as follows:

Locus genotype frequency for homozygotes:

$$P_i = \frac{[2\theta + (1-\theta)p][3\theta + (1-\theta)p]}{(1+\theta)(1+2\theta)} \quad \text{(Formula 23.7)}$$

**Table 23.5 Calculation of Profile Probability of STR Profile with 13 CODIS Loci**

| Locus | Profile | Allelic Frequency* | Formula | Locus Genotype Frequency |
|---|---|---|---|---|
| CSF1PO | 10 | 0.25369 | 2pq | 0.165 |
| | 12 | 0.32512 | | |
| D3S1358 | 14 | 0.14039 | 2pq | 0.0691 |
| | 15 | 0.24631 | | |
| D5S818 | 11 | 0.41026 | p2 | 0.168 |
| | 11 | 0.41026 | | |
| D7S820 | 10 | 0.29064 | 2pq | 0.117 |
| | 11 | 0.20197 | | |
| D8S1179 | 13 | 0.33929 | p2 | 0.115 |
| | 13 | 0.33929 | | |
| D13S317 | 11 | 0.31888 | p2 | 0.102 |
| | 11 | 0.31888 | | |
| D16S539 | 11 | 0.27228 | 2pq | 0.185 |
| | 12 | 0.33911 | | |
| D18S51 | 15 | 0.12755 | 2pq | 0.00911 |
| | 19 | 0.035710 | | |
| D21S11 | 30 | 0.23214 | p2 | 0.0539 |
| | 30 | 0.23214 | | |
| FGA | 23 | 0.15816 | 2pq | 0.0436 |
| | 24 | 0.13776 | | |
| TH01 | 8 | 0.12562 | 2pq | 0.0767 |
| | 9.3 | 0.30542 | | |
| TPOX | 8 | 0.54433 | p2 | 0.296 |
| | 8 | 0.54433 | | |
| vWA | 17 | 0.26276 | 2pq | 0.117 |
| | 18 | 0.22194 | | |
| | | | | Profile probability $=2.76 \times 10^{-14}$ |

* See Table 23.1.

Locus genotype frequency for heterozygotes:

$$P_j = \frac{2[\theta+(1-\theta)p][\theta+(1-\theta)q]}{(1+\theta)(1+2\theta)} \quad \text{(Formula 23.8)}$$

The $\theta$ value is 0.01 for the major U.S. population and 0.03 for Native American populations. Table 23.6 shows calculation of profile probability with subpopulation correction using $\theta = 0.01$. The profile probability is approximately 3 times higher than the value without the correction (Table 23.5). Additional corrections can be calculated for relatives, mixed stains, or database searches using formulas provided by National Research Council's guidelines.

#### *23.2.1.2 Likelihood Ratio*

The ***likelihood ratio*** (LR) method is an alternative for evaluating the strength of a match. The method allows calculation of the probability of the DNA profile under two hypotheses:

Hypotheses 1 ($H_1$) — The evidence and suspect profiles originated from the same source.
Hypotheses 2 ($H_2$) — The evidence and suspect profiles did not originate from the same source (the suspect happens to have the same profile as the evidence).

The LR is the probability under hypothesis $H_1$ divided by the probability under hypothesis $H_2$. Where $Pr_1$ is the probability under hypothesis $H_1$ and $Pr_2$ is the probability under hypothesis $H_2$, this can be expressed as:

$$LR = \frac{Pr_1}{Pr_2} \quad \text{(Formula 23.9)}$$

The greater the likelihood ratio, the greater the $Pr_1$, numerator becomes. The result favors hypothesis $H_1$ (the evidence and suspect profile originated from the same source). $Pr_1$ is equal to 1 (100%) when a match occurs and $Pr_2$ is equal to the profile probability. A LR of 1000 indicates that the evidence is 1000 times as probable if the evidence and suspect profiles originated from the same source.

### 23.2.2 Haplotypes

The ***haplotype*** term was first used to describe very closely linked polymorphic loci. During meiosis, alleles at neighboring loci cosegregate (both alleles segregate as a single allele) because of the close linkage of loci. The term also applies to a genetic region within which recombination is very rare,

**Table 23.6 Calculation of Profile Probability of STR with 13 CODIS Loci and Correction Factor (θ = 0.01)**

| Locus | Profile | Allelic Frequency* | Formula | Locus Genotype Frequency |
|---|---|---|---|---|
| CSF1PO | 10 | 0.25369 | 2[θ+(1-θ)p][θ+(1-θ)q]/ (1+θ)(1+2θ) | 0.168 |
| | 12 | 0.32512 | | |
| D3S1358 | 14 | 0.14039 | 2[θ+(1-θ)p][θ+(1-θ)q]/ (1+θ)(1+2θ) | 0.0734 |
| | 15 | 0.24631 | | |
| D5S818 | 11 | 0.41026 | [2θ+(1-0)p][3θ+(1-θ)p]/ (1+θ)(1+2θ) | 0.180 |
| | 11 | 0.41026 | | |
| D7S820 | 10 | 0.29064 | 2[θ+(1-θ)p][θ+(1-θ)q]/ (1+θ)(1+2θ) | 0.121 |
| | 11 | 0.20197 | | |
| D8S1179 | 13 | 0.33929 | [2θ+(1-θ)p][3θ+(1-θ)p]/ (1+θ)(1+2θ) | 0.126 |
| | 13 | 0.33929 | | |
| D13S317 | 11 | 0.31888 | [2θ+(1-θ)p][3θ+(1-θ)p]/ (1+θ)(1+2θ) | 0.113 |
| | 11 | 0.31888 | | |
| D16S539 | 11 | 0.27228 | 2[θ+(1-θ)p][θ+(1-θ)q]/ (1+θ)(1+2θ) | 0.187 |
| | 12 | 0.33911 | | |
| D18S51 | 15 | 0.12755 | 2[θ+(1-θ)p][θ+(1-θ)q]/ (1+θ)(1+2θ) | 0.012 |
| | 19 | 0.03571 | | |
| D21S11 | 30 | 0.23214 | [2θ+(1-θ)p][3θ+(1-θ)p]/ (1+θ)(1+2θ) | 0.0630 |
| | 30 | 0.23214 | | |
| FGA | 23 | 0.15816 | 2[θ+(1-θ)p][θ+(1-θ)q]/ (1+θ)(1+2θ) | 0.0473 |
| | 24 | 0.13776 | | |
| TH01 | 8 | 0.12562 | 2[θ+(1-θ)p][θ+(1-θ)q]/ (1+θ)(1+2θ) | 0.0815 |
| | 9.3 | 0.30542 | | |
| TPOX | 8 | 0.54433 | [2θ+(1-θ)p][3θ+(1-θ)p]/ (1+θ)(1+2θ) | 0.309 |
| | 8 | 0.54433 | | |

**Table 23.6 Calculation of Profile Probability of STR with 13 CODIS Loci and Correction Factor (θ = 0.01)** (continued)

| Locus | Profile | Allelic Frequency* | Formula | Locus Genotype Frequency |
|---|---|---|---|---|
| VWA | 17 | 0.26276 | $2[\theta+(1-\theta)p][\theta+(1-\theta)q]/(1+\theta)(1+2\theta)$ | 0.120 |
| | 18 | 0.22194 | | |
| | | | | Profile probability = $7.8 \times 10^{-14}$ |

* See Table 23.1.

e.g., within mitochondrial and Y chromosomal DNA. The entire mtDNA sequence can be considered as a single locus or haplotype because of the absence of recombination. Likewise, Y chromosome loci can also be considered haplotypes.

Where recombination is very rare, some allelic combinations occur in populations much more frequently than would be expected. This phenomenon is called ***linkage disequilibrium***. As a result, the HW principle cannot be applied. The two methods for evaluating the strength of a match between haplotypes are (1) mitotype frequency and (2) likelihood ratios. The current most common approach for interpreting mtDNA profiles is mitotype frequency carried out with the gene counting method i.e., calculation of the number of occurrences of a particular sequence or haplotype. The interpretation for Y chromosome profiles is similar to interpretation for mtDNA. The estimation of the frequency of a mitotype can be calculated as shown below.

### *23.2.2.1 Mitotypes Observed in Database*

If a mitotype is observed at least once in a database, Formula 23.10 can be used. $P_{mt}$ is the mitotype frequency, x is the number of observations of the haplotype, and n is the size of the database:

$$P_{mt} = \frac{x+2}{n+2} \qquad \text{(Formula 23.10)}$$

Any sampling error may be addressed by a confidence interval:

$$P_{mt} \pm 1.96\sqrt{\frac{p_{mt}(1-p_{mt})}{n}} \qquad \text{(Formula 23.11)}$$

In this case, $P_{mt}$ is the mitotype frequency and n is the size of the database. The conservative upper bound of the frequency is usually quoted.

#### 23.2.2.2 *Mitotype Not Observed in Database*

If a mitotype has not been observed in a database, Formula 23.12 can be used to calculate the mitotype frequency; α is 0.05 for a 95% confidence interval and n is the size of the database.

$$P_{mt} = 1 - \alpha^{\frac{1}{n}} \quad \text{(Formula 23.12)}$$

## Bibliography

Aitken, C. G. G. 1995. *Statistics and the Evaluation of Evidence for Forensic Scientists.* New York: Wiley.

Bagdonavicius, A. et al. 2002. Western Australian sub-population data for 13 AMPFISTR Profiler Plus and COfiler STR loci. *J Forensic Sci* 47 (5):1149.

Balding, D. J. 1994. The prosecutor's fallacy and DNA evidence. *Crim Law Rev* 11.

_____. 1999. When can a DNA profile be regarded as unique? *Sci Justice* 39:4.

Balding, D. J., and P. Donnelly. 1995. Inferring identity from DNA profile evidence. *Proc Natl Acad Sci USA* 92 (25):11741.

_____. 1996. Evaluating DNA profile evidence when the suspect is identified through a database search. *J Forensic Sci* 41 (4):603.

Balding, D. J., and R. A. Nichols. 1994. DNA profile match probability calculation: How to allow for population stratification, relatedness, database selection and single bands. *Forensic Sci Int* 64 (2-3):125.

Botstein, D. et al. 1980. Construction of a genetic linkage map in man using restriction fragment length polymorphisms. *Am J Hum Genet* 32 (3):314.

Brenner, C. H. 1998. Difficulties in the estimation of ethnic affiliation. *Am J Hum Genet* 62 (6):1558.

Buckleton, J., C. Triggs, and S. Walsh. 2005. *Forensic DNA Evidence Interpretation.* Boca Raton, FL: CRC Press.

Buckleton, J. S., S. Walsh, and S. A. Harbison. 2001. The fallacy of independence testing and the use of the product rule. *Sci Justice* 41 (2):81.

Budowle, B. 1995. The effects of inbreeding on DNA profile frequency estimates using PCR-based loci. *Genetica* 96 (1-2):21.

Budowle, B. et al. 1995. Validation and population studies of the loci LDLR, GYPA, HBGG, D7S8, and Gc (PM loci), and HLA-DQ alpha using a multiplex amplification and typing procedure. *J Forensic Sci* 40 (1):45.

Budowle, B., and R. R. Moretti. 1999. Genotype profiles for six population groups at the 13 CODIS short tandem repeat core loci and other PCR based loci. *Forensic Sci Commun* 1 (2).

Budowle, B. et al. 1999. Population data on the thirteen CODIS core short tandem repeat loci in African Americans, U.S. Caucasians, Hispanics, Bahamians, Jamaicans, and Trinidadians. *J Forensic Sci* 44 (6):1277.

Budowle, B. et al. 2000. Source attribution of a forensic DNA profile. *Forensic Sci Commun* 2 (3).

Budowle, B., D.A. Defenbaugh, and K. M. Keys. 2000. Genetic variation at nine short tandem repeat loci in Chamorros and Filipinos from Guam. *Leg Med (Tokyo)* 2 (1):26.
Budowle, B. et al. 2001. CODIS STR loci data from 41 sample populations. *J Forensic Sci* 46 (3):453.
Butler, J. M. et al. 2003. Allele frequencies for 15 autosomal STR loci on U.S. Caucasian, African American, and Hispanic populations. *J Forensic Sci* 48 (4):908.
Chakraborty, R. 1984. Detection of nonrandom association of alleles from the distribution of the number of heterozygous loci in a sample. *Genetics* 108 (3):719.
_____. 1992. Sample size requirements for addressing the population genetic issues of forensic use of DNA typing. *Hum Biol* 64 (2):141.
Chakraborty, R., and D. N. Stivers. 1996. Paternity exclusion by DNA markers: Effects of paternal mutations. *J Forensic Sci* 41 (4):671.
Chakraborty, R. et al. 1999. The utility of short tandem repeat loci beyond human identification: Implications for development of new DNA typing systems. *Electrophoresis* 20 (8):1682.
Collins, A., and N. E. Morton. 1994. Likelihood ratios for DNA identification. *Proc Natl. Acad Sci USA* 91 (13):5.
Crow, J. F. 1999. Hardy, Weinberg and language impediments. *Genetics* 152 (3):821.
Curran, J. M. et al. 1999. Interpreting DNA mixtures in structured populations. *J Forensic Sci* 44 (5):987.
Edwards, A. et al. 1992. Genetic variation at five trimeric and tetrameric tandem repeat loci in four human population groups. *Genomics* 12 (2):241.
Evett, I., and B. Weir. 1992. *Interpreting DNA Evidence: Statistical Genetics for Forensic Scientists*. Sunderland, MA: Sinauer Associates.
Evett, I. W., and P. Gill. 1991. A discussion of the robustness of methods for assessing the evidential value of DNA single locus profiles in crime investigations. *Electrophoresis* 12 (2-3):226.
Evett, I. W., R. Pinchin, and C. Buffery. 1992. An investigation of the feasibility of inferring ethnic origin from DNA profiles. *J Forensic Sci* 32 (4):301.
Fisher, R. A. 1951. Standard calculations for evaluating a blood-group system. *Heredity* 5 (1):952.
Foreman, L. A. et al. 2003. Interpreting DNA evidence. *Int Stat Rev* 71 (3):13.
Foreman, L. A., and I. W. Evett. 2001. Statistical analyses to support forensic interpretation for a new ten-locus STR profiling system. *Int J Legal Med* 114 (3):147.
Foreman, L. A., A. F. M. Smith, and I. Evett. 1999. Bayesian validation of a quadruplex STR profiling system for identification purposes. *J Forensic Sci* 44:378.
Gaensslen, R. E. 1999. Journal policy on the publication of DNA population genetic data *J Forensic Sci* 44 (4):4.
Gill, P. et al. 2003. A comparison of adjustment methods to test the robustness of an STR DNA database comprised of 24 European populations. *Forensic Sci Int* 131 (2-3):184.
Guo, S. W., and E. A. Thompson. 1992. Performing the exact test of Hardy-Weinberg proportion for multiple alleles. *Biometrics* 48 (2):361.
Hardy, G. H. 1908. Mendelian proportions in a mixed population. *Science* 17:49.
Henderson, J. P. 2002. The use of DNA statistics in criminal trials. *Forensic Sci Int* 128 (3):183.

Holt, C. L. et al. 2000. Practical applications of genotypic surveys for forensic STR testing. *Forensic Sci Int* 112 (2–3):91.

Hosking, L. et al. 2004. Detection of genotyping errors by Hardy-Weinberg equilibrium testing. *Eur J Hum Genet* 12 (5):395.

Lincoln, P., and A. Carracedo. 2000. Publication of population data of human polymorphisms. *Forensic Sci Int* 110 (1):3.

Lins, A. M. et al. 1998. Development and population study of an eight-locus short tandem repeat (STR) multiplex system. *J Forensic Sci* 43 (6):1168.

Morton, N. E. 1992. Genetic structure of forensic populations. *Proc Natl Acad Sci USA* 89 (7):2556.

National Research Council. 1996. *Evaluation of Forensic DNA Evidence*, Press, N.A. Ed. Washington: National Research Council Committee.

Nei, M. 1978. Estimation of average heterozygosity and genetic distance from a small number of individuals. *Genetics* 89 (3):583.

Office of Justice Programs, National Institute of Justice, U.S. Department of Justice. 2000. The future of forensic DNA testing: Predictions of the research and development working group.

Raymond, M., and F. Rousset. 1995. Population genetics software for exact tests and ecumenicism. *J Heredity* 86:12.

Robertson, B., and G. A. Vigneaux. 1992. Expert evidence: Law, practice, and probability. *Oxford J Legal Studies* 12 (1):392.

Roeder, K. 1994. DNA fingerprinting: A review of the controversy. *Stat Sci* 9 (2):222.

Rowold, D. J., and R. J. Herrera. 2005. On human STR sub-population structure. *Forensic Sci Int* 151 (1):59.

Royall, R. 1997. *Statistical Evidence: A Likelihood Paradigm*. New York: Chapman & Hall.

Steinlechner, M. et al. 2001. Population genetics of ten STR loci (AmpFlSTR SGM plus) in Austria. *Int J Legal Med* 114 (4-5):288.

Stockmarr, A. 1999. Likelihood ratios for evaluating DNA evidence when the suspect is found through a database search. *Biometrics* 55 (3):671.

Taroni, F. et al. 2002. Evaluation and presentation of forensic DNA evidence in European laboratories. *Sci Justice* 42 (1):21.

Thompson, W. C., and E. L Schumann. 1987. Interpretation of statistical evidence in criminal trials; the prosecutor's fallacy and the defense attorney's fallacy. *Law Human Behav* 11 (3):11.

Walsh, S. J. et al. 2003. Evidence in support of self-declaration as a sampling method for the formation of sub-population DNA databases. *J Forensic Sci* 48 (5):1091.

Weir, B. S., and W. G. Hill. 2002. Estimating F-statistics. *Annu Rev Genet* 36:721.

## Study Questions

1. The observation of Hardy-Weinberg equilibrium indicates that:
   (a) The loci tested are independent.
   (b) The loci tested are linked.

2. The product rule may be applied to:
   (a) Autosomal STR profiles
   (b) Y chromosomal STR profiles
   (c) Mitochondrial DNA profiles
   (d) Autosomal SNP profiles

3. Which of the following is preferred for human identification?
   (a) Genetic loci with high heterozygosity
   (b) Genetic loci with low heterozygosity

4. The higher the profile probability,
   (a) The more likely a coincidental match of DNA profile will occur
   (b) The less likely a coincidental match of DNA profile will occur

5. Based on genotypes listed in Table 23.6, what would be the profile probability if $\theta = 0.03$ is used?

6. If a mitotype is observed five times in a database (n = 850), what would be the mitotype frequency?

# Quality Assurance and Quality Control 24

***Quality assurance (QA)*** for forensic services requires certain processes to assure confidence that a service will meet laboratory requirements for quality. A QA program must include components that address:

- Continuing education, training, and certification of personnel
- Specification and calibration of equipment and reagents
- Documentation and validation of analytic methods
- Use of appropriate standards and controls
- Sample handling procedures
- Proficiency testing
- Data interpretation and reporting
- Audits (internal and external) and laboratory accreditation
- Corrective actions to address deficiencies and assessments for laboratory competence

Over the years, many guidelines for quality assurance in forensic DNA laboratories have been established. These guidlines will be introduced in this chapter.

***Quality control (QC)*** for forensic services refers to the operational procedures necessary to meet quality requirements. QC procedures may include maintenance of calibration records for equipment and instruments as well as the testing of chemical reagents and supplies used in analysis to ensure reliable results.

## 24.1 United States Quality Standards

DNA profiling methods were first used in criminal investigations in the 1980s. By the early 1990s, emerging forensic DNA techniques had undergone detailed reviews by the ***National Research Council (NRC)*** of the National Academy of Sciences. In 1992, the first published NRC report included recommendations in the areas of (1) technical considerations, (2) statistical interpretation, (3) laboratory standards, (4) data banks and privacy, (5) legal considerations, and (6) societal and ethical issues related to forensic DNA testing. The NRC report attempted to explain the basic scientific principles of forensic DNA technology and made suggestions for applications and improvements. However, the report received negative criticism from both the forensic and legal communities.

In 1996, a second NRC committee was formed "to update and clarify discussion of the principles of population genetics and statistics as they apply to DNA evidence" (*The Evaluation of Forensic DNA Evidence* also known as NRC II report). This report stated:

> The central question that the report addresses is this: What information can a forensic scientist, population geneticist, or statistician provide to assist a judge or jury in drawing inferences from the finding of a match?
>
> To answer this question, the committee reviewed the scientific literature and the legal cases and commentaries on DNA profiling, and it investigated the criticisms that have been voiced about population data, statistics, and laboratory error. Much has been learned since the last report. The technology for DNA profiling as well as the methods for estimating frequencies and related statistics have progressed to the point where the reliability and validity of properly collected and analyzed DNA data should not be in doubt. The new recommendations presented here should pave the way to more effective use of DNA evidence.

The NRC II report consisted of: (1) an introduction describing the 1992 report, changes made subsequent to that report, and the validity and application of DNA typing techniques, (2) assurance of high standards of laboratory performance, (3) population genetics issues, (4) statistical issues, and (5) DNA evidence in the legal system.

In 1995, The ***DNA Advisory Board (DAB)*** was formed as a result of the DNA Identification Act (1994) passed by Congress. The DAB served from 1995 to 2000 to develop guidelines for quality assurance in forensic laboratories. During that time DAB provided two sets of guidelines: (1) Quality Assurance Standards for Forensic DNA Testing Laboratories (1998) and (2) Quality Assurance Standards for Convicted Offender DNA Databasing Laboratories (1999). These standards describe the requirements to ensure quality and integrity of the data and competency of laboratories.

These standards built upon the previous standards set by the ***Scientific Working Group on DNA Analysis Methods (SWGDAM)***. After DAB's assignment ended in 2000, SWGDAM became responsible for providing guidelines to the U.S. forensic community. SWGDAM was established in 1988 by the FBI Laboratory. It was initially called the Technique Working Group on DNA Analysis Methods; the name changed in 1998. SWGDAM is comprised of forensic scientists from DNA laboratories in the United States and Canada. Its purpose is to facilitate forensic DNA community discussions regarding necessary laboratory methods and to share protocols for forensic DNA testing. The FBI sponsors and hosts its meetings and plays an important role in its activities.

SWGDAM established several guidelines including the *Guidelines for a Quality Assurance Program for DNA Analysis* published in 1989, 1991, and 1995. The validation section was revised in 2003. SWGDAM also formed subcommittees to provide guidelines in more specific areas of forensic DNA testing, for example, the *Short Tandem Repeat (STR) Interpretation Guidelines* (2000), *Training Guidelines* (2001), and *Guidelines for Mitochondrial DNA (mtDNA) Nucleotide Sequence Interpretation* (2003). SWGDAM also organized a number of interlaboratory and validation studies for new techniques.

## 24.2 International Quality Standards

In the early stages of forensic DNA testing, the ***International Society for Forensic Genetics (ISFG)*** recognized the potential of DNA testing for criminal investigations and made a number of recommendations related to forensic application of DNA polymorphisms.

ISFG provided various recommendations for forensic DNA testing of STR, mtDNA, and Y chromosome markers for the international community. It formed a working group called the ***European DNA Profiling Group*** (***EDNAP***) in 1991. EDNAP has investigated systems for DNA profiling, and organized a number of collaborative exercises for the evaluation of new methods and published reports of its studies.

The ***European Network of Forensic Science Institutes*** (***ENFSI***) formed its DNA working group about a decade ago to address issues of quality and standards for forensic DNA testing. The ***Interpol European Working Party on DNA Profiling*** (***IEWPDP***) also makes recommendations for applying DNA evidence to criminal investigations in Europe. Based on the EDNAP exercises and recommendations by ENFSI and IEWPDP, the ***European Standard Set*** (***ESS***) for autosomal STR core loci was established. The Standardization of DNA Profiling Techniques in the European Union (STADNAP) group is currently working on the selection of forensic DNA profiling systems, methods for use among European countries, and the maintenance of European population databases.

## 24.3 Laboratory Accreditation

***Accreditation*** is the process used to assess the qualification of a laboratory to meet established standards. The accreditation process evaluates services and performance, in particular, management, operations, personnel, procedures, equipment, physical plant, security, and personnel safety procedures. The accreditation process generally involves several components such as self-evaluation, preparation of supporting documents, on-site inspection and report, and accreditation review reports.

Accreditation is presently a voluntary program for United States and international forensic laboratories seeking accreditation. Accreditation in the U.S. is offered by the ***Laboratory Accreditation Board*** of the ***American Society of Crime Laboratory Directors*** (***ASCLD/LAB***) for forensic laboratories performing casework. The ***National Forensic Science Training Center*** (***NFSTC***) program in Largo, Florida, also provides accreditation. The ***American Association of Blood Banks*** (***AABB***) provides accreditation for laboratories performing DNA parentage testing according to AABB's standards.

The accreditation of a forensic laboratory is granted for five years. To remain in compliance, a laboratory must undergo audits to evaluate its operation according to established guidelines. The areas of operating protocols, instruments and equipment, and personnel training are evaluated based on guidelines. Problems identified during an audit must be documented and actions to resolve the problems must be addressed. Annual internal audit and external audits during alternate years are required under the guidelines.

## 24.4 Laboratory Validation

Validation is the process of demonstrating that a laboratory procedure is robust, reliable, and reproducible. A robust method produces successful results a high percentage of the time. A reliable method produces accurate results that reflect the samples tested. A reproducible method achieves the same or very similar results each time a sample is tested.

## 24.5 Proficiency Testing

Proficiency testing is an important component of quality control and quality assurance. It evaluates a laboratory's performance of DNA analyses according to the laboratory's standard protocols. Proficiency testing also evaluates the performances of individual analysts in terms of quality based on laboratory protocols.

Proficiency tests of DNA analysts must be conducted every 6 months based on DNA Advisory Board Standards. The testing usually involves mock forensic case samples including questioned biological fluid stains and reference samples. The test is assigned to an analyst for processing according to the laboratory procedures. A report must be prepared and is then reviewed. The proficiency test can be administered either as an open or blind test. In the blind test, the analyst is not aware that he or she is being tested. Blind testing is considered a more effective means of evaluating performance.

The tests may be administered internally or by any of a number of external testing organizations. Orchid Cellmark provides the ***International Quality Assessment Scheme*** (***IQAS***) DNA proficiency test. The tests are also offered by ASCLD/LAB providers, namely Collaborative Testing Services, Inc., Quality Forensics, Inc., and the Serological Research Institute. The College of American Pathologists offers external proficiency testing to forensic and paternity testing laboratories. In Europe, the ***German DNA Profiling Group*** (***GEDNAP***) provides proficiency testing for participating European laboratories.

## 24.6 Certification

Certification is a voluntary process that recognizes the attainment of professional qualifications needed for practice in forensic services. The certification is not required, but is desired by some laboratories. In the U.S., the ***American Board of Criminalistics (ABC)*** offers three types of certifications for forensic scientists. A diplomate must pass a general knowledge examination. ABC also requires a bachelor's degree in a natural science and 2 years of experience in a forensic laboratory. To obtain fellow status (higher than diplomate status), an applicant must have two years of experience in his or her specialty and must have met the diplomate requirements in addition to passing a written specialty examination and a proficiency test.

ABC recently added a third certification that is completely separate from the other certification programs. The technical specialist certification for molecular biology was created to recognize the qualifications required for the analysis of biological materials through DNA profiling. An applicant must take a specialist examination containing questions from the general knowledge examination and a subset of questions from the forensic biology fellow examination. This certification also requires a bachelor's degree in a natural science and 2 years of experience along with successful completion of a proficiency examination within 12 months of taking the technical specialist certification examination.

## 24.7 Forensic DNA Analyst Qualifications

DAB's Quality Assurance Standards for Forensic DNA Testing Laboratories require that an examiner or analyst have "at a minimum a BA/BS degree or its equivalent degree in a biology, chemistry or forensic science-related area and must have successfully completed college course work (graduate or undergraduate level) covering the subject areas of biochemistry, genetics and molecular biology (molecular genetics recombinant DNA technology) or other subjects which provide a basic understanding of the foundation of forensic DNA analysis, as well as course work and/or training in statistics and population genetics as it applies to forensic DNA analysis." Additionally, the standards require "a minimum of six (6) months of forensic DNA laboratory experience, including the successful analysis of a range of samples typically encountered in forensic casework prior to independent casework analysis using DNA technology," and must have "successfully completed a qualifying test before beginning independent casework responsibilities."

## Bibliography

### Book Chapters and Review Articles

Butler, J. M. 2005. Laboratory validation, in *Forensic DNA Typing*. Burlington, MA: Elsevier.

Cormier, K., L. Calandro, and D. J. Reeder. 2005. Evolution of quality assurance documents for DNA laboratories. *Forensic Mag* 1.

Presley, L. A. 1999. Evolution of quality standards for forensic DNA analyses in the United States. *Profiles DNA* 10.

Reeder, D. J. 1999. Impact of DNA typing on standards and practice in the forensic community. *Arch Pathol Lab Med* 123 (11):1063.

Schneider, P. M. 2007. Scientific standards for studies in forensic genetics. *Forensic Sci Int* 165 (2–3):238.

### Standards and Guidelines

DNA recommendations: 1992 report concerning recommendations of the DNA Commission of the International Society for Forensic Haemogenetics relating to the use of PCR-based polymorphisms. 1992. *Int J Legal Med* 105 (1):63-4.

Bar, W., B. Brinkmann, B. Budowle, A. Carracedo, P. Gill, P. Lincoln, W. Mayr, and B. Olaisen. 1997. DNA recommendations. Further report of the DNA Commission of the ISFH regarding the use of short tandem repeat systems. International Society for Forensic Haemogenetics. *Int J Legal Med* 110 (4):175-6.

Carracedo, A., W. Bar, P. Lincoln, W. Mayr, N. Morling, B. Olaisen, P. Schneider, B. Budowle, B. Brinkmann, P. Gill, M. Holland, G. Tully, and M. Wilson. 2000. DNA commission of the international society for forensic genetics: Guidelines for mitochondrial DNA typing. *Forensic Sci Int* 110 (2):79-85.

Council, National Research. 1996. *NRC II: The Evaluation of Forensic DNA Evidence.* Washington, D.C.: National Academy Press.

DNA Advisory Board, Federal Bureau of Investigations, US Department of Justice. 1998. Quality Assurance Standards for Forensic DNA Testing Laboratories.

Gill, P., C. Brenner, B. Brinkmann, B. Budowle, A. Carracedo, M. A. Jobling, P. de Knijff, M. Kayser, M. Krawczak, W. R. Mayr, N. Morling, B. Olaisen, V. Pascali, M. Prinz, L. Roewer, P. M. Schneider, A. Sajantila, and C. Tyler-Smith. 2001. DNA commission of the International Society of Forensic Genetics: Recommendations on forensic analysis using Y-chromosome STRs. *Int J Legal Med* 114 (6):305-9.

Gill, P., C. H. Brenner, J. S. Buckleton, A. Carracedo, M. Krawczak, W. R. Mayr, N. Morling, M. Prinz, P. M. Schneider, and B. S. Weir. 2006. DNA commission of the International Society of Forensic Genetics: Recommendations on the interpretation of mixtures. *Forensic Sci Int* 160 (2-3):90-101.

Gusmao, L., J. M. Butler, A. Carracedo, P. Gill, M. Kayser, W. R. Mayr, N. Morling, M. Prinz, L. Roewer, C. Tyler-Smith, and P. M. Schneider. 2006. DNA Commission of the International Society of Forensic Genetics (ISFG): An update of the recommendations on the use of Y-STRs in forensic analysis. *Forensic Sci Int* 157 (2-3):187-97.

Lincoln, P. J. 1997. DNA recommendations—further report of the DNA Commission of the ISFH regarding the use of short tandem repeat systems. *Forensic Sci Int* 87 (3):181-4.

Morling, N., R. W. Allen, A. Carracedo, H. Geada, F. Guidet, C. Hallenberg, W. Martin, W. R. Mayr, B. Olaisen, V. L. Pascali, and P. M. Schneider. 2002. Paternity Testing Commission of the International Society of Forensic Genetics: Recommendations on genetic investigations in paternity cases. *Forensic Sci Int* 129 (3):148-57.

(SWGDAM), Scientific Working Group on DNA Analysis Methods. 2004. Revised Validation Guidelines. *Forensic Sci Comm* 6 (3).

Tully, G., W. Bar, B. Brinkmann, A. Carracedo, P. Gill, N. Morling, W. Parson, and P. Schneider. 2001. Considerations by the European DNA profiling (EDNAP) group on the working practices, nomenclature and interpretation of mitochondrial DNA profiles. *Forensic Sci Int* 124 (1):83-91.

Wilson, M. R., M. W. Allard, K. Monson, K. W. Miller, and B. Budowle. 2002. Recommendations for consistent treatment of length variants in the human mitochondrial DNA control region. *Forensic Sci Int* 129 (1):35-42.

## Interlaboratory Exercises

Andersen, J. et al. 1996. Report on the third EDNAP collaborative STR exercise. *Forensic Sci Int* 78 (2):83-93.

Carracedo, A. et al. 2001. Results of a collaborative study of EDNAP group regarding the reproducibility and robustness of the Y-chromosome STRs DYS19, DYS389 I and II, DYS390 and DYS393 in a PCR pentaplex format. *Forensic Sci Int* 119 (1):28.

Duewer, D. L. et al. 2001. NIST mixed stain studies #1 and #2: Interlaboratory comparison of DNA quantification practice and short tandem repeat multiplex performance with multiple-source samples. *J Forensic Sci* 46 (5):1199.

Gill, P. et al. 1994. Report of the European DNA profiling group (EDNAP)–towards standardisation of short tandem repeat (STR) loci. *Forensic Sci Int* 65 (1):51-9.

Gill, P. et al. 1997. Report of the European DNA profiling group (EDNAP): An investigation of the complex STR loci D21S11 and HUMFIBRA (FGA). *Forensic Sci Int* 86 (1-2):25.

Gill, P. et al. 1998. Report of the European DNA Profiling Group (EDNAP)—an investigation of the hypervariable STR loci ACTBP2, APOAI1 and D11S554 and the compound loci D12S391 and D1S1656. *Forensic Sci Int* 98 (3):193-200.

Kimpton, C. et al. 1995. Report on the second EDNAP collaborative STR exercise. European DNA Profiling Group. *Forensic Sci Int* 71 (2):137-52.

Kline, M. et al. Interlaboratory evaluation of short tandem repeat triplex CTT. *J Forensic Sci* 42 (5):897-906.

Kline, M. C., D. L. Duewer, J. W. Redman, and J. M. Butler. 2003. NIST Mixed Stain Study 3: DNA quantitation accuracy and its influence on short tandem repeat multiplex signal intensity. *Anal Chem* 75 (10):2463-9.

Schneider, P. M. et al. 1999. Results of a collaborative study regarding the standardization of the Y-linked STR system DYS385 by the European DNA Profiling (EDNAP) group. *Forensic Sci Int* 102 (2-3):159-65.

Tully, G. et al. 2004. Results of a collaborative study of EDNAP group regarding mitochondrial DNA heteroplasmy and segregation in hair shafts. *Forensic Sci Int* 140 (1):1.

## Validation Studies

Cotton, E. A. et al. 2000. Validation of the AMPFlSTR SGM plus system for use in forensic casework. *Forensic Sci Int* 112 (2-3):151.

Frank, W. E. et al. 2001. Validation of the AmpFlSTR Profiler Plus PCR amplification kit for use in forensic casework. *J Forensic Sci* 46 (3):642.

Fregeau, C. J., K. L. Bowen, and R. M. Fourney. 1999. Validation of highly polymorphic fluorescent multiplex short tandem repeat systems using two generations of DNA sequencers. *J Forensic Sci* 44 (1):133.

Greenspoon, S. A. et al. 2000. Validation of the PowerPlex 1.1 loci for use in human identification. *J Forensic Sci* 45 (3):677.

Greenspoon, S. A. et al. 2004. Validation and implementation of the PowerPlex 16 BIO System STR multiplex for forensic casework. *J Forensic Sci* 49 (1):71.

Holt, C. L. et al. 2002. TWGDAM validation of AmpFlSTR PCR amplification kits for forensic DNA casework. *J Forensic Sci* 47 (1):66.

Junge, A. et al. 2003. Validation of the multiplex kit genRESMPX-2 for forensic casework analysis. *Int J Legal Med* 117 (6):317.

Krenke, B. E. et al. 2002. Validation of a 16-locus fluorescent multiplex system. *J Forensic Sci* 47 (4):773.

LaFountain, M. J. et al. 2001. TWGDAM validation of the AmpFlSTR Profiler Plus and AmpFlSTR COfiler STR multiplex systems using capillary electrophoresis. *J Forensic Sci* 46 (5):1191.

Levedakou, E. N. et al. 2002. Characterization and validation studies of PowerPlex 2.1, a nine-locus short tandem repeat (STR) multiplex system and penta D monoplex. *J Forensic Sci* 47 (4):757.

Lygo, J. E. et al. 1994. The validation of short tandem repeat (STR) loci for use in forensic casework. *Int J Legal Med* 107 (2):77.

Micka, K. A. et al. 1999. TWGDAM validation of a nine-locus and a four-locus fluorescent STR multiplex system. *J Forensic Sci* 44 (6):1243.

Moretti, T. R. et al. 2001. Validation of STR typing by capillary electrophoresis. *J Forensic Sci* 46 (3):661.

_____. 2001. Validation of short tandem repeats (STRs) for forensic usage: Performance testing of fluorescent multiplex STR systems and analysis of authentic and simulated forensic samples. *J Forensic Sci* 46 (3):647.

Sinha, S. K. et al. 2003. Development and validation of a multiplexed Y-chromosome STR genotyping system, Y-PLEX 6, for forensic casework. *J Forensic Sci* 48 (1):93.

Sinha, S. K. et al. 2003. Development and validation of the Y-PLEX 5, a Y-chromosome STR genotyping system, for forensic casework. *J Forensic Sci* 48 (5):985.

Sparkes, R. et al. 1996. The validation of a 7-locus multiplex STR test for use in forensic casework I. Mixtures, aging, degradation and species studies. *Int J Legal Med* 109 (4):186.

Tomsey, C. S. et al. 2001. Comparison of PowerPlex 16, PowerPlex1.1/2.1, and ABI AmpflSTR Profiler Plus/COfiler for forensic use. *Croat Med J* 42 (3):239.

Wallin, J. M. et al. 1998. TWGDAM validation of the AmpFISTR blue PCR amplification kit for forensic casework analysis. *J Forensic Sci* 43 (4):854.

Wilson, M. R. et al. 1995. Validation of mitochondrial DNA sequencing for forensic casework analysis. *Int J Legal Med* 108 (2):68.

## Proficiency Testing

Peterson, J. L. et al. 2003. The feasibility of external blind DNA proficiency testing. I. Background and findings. *J Forensic Sci* 48 (1):21.

_____. 2003. The feasibility of external blind DNA proficiency testing. II. Experience with actual blind tests. *J Forensic Sci* 48 (1):32.

Rand, S., M. Schurenkamp, and B. Brinkmann. 2002. The GEDNAP (German DNA profiling group) blind trial concept. *Int J Legal Med* 116 (4):199.

Rand, S. et al. 2004. The GEDNAP blind trial concept part II. Trends and developments. *Int J Legal Med* 118 (2):83.

## Study Questions

1. Which of the following is conducted by ASCLD/LAB?
   (a) Laboratory validation
   (b) Proficiency testing
   (c) Laboratory accreditation

2. Which of the following courses are required for qualification of a forensic DNA analyst?
   (a) Biochemistry, genetics, and molecular biology
   (b) Molecular genetics and analytical chemistry
   (c) Analytical chemistry and molecular biology

3. Proficiency tests are based solely on the performance of an analyst.
   (a) True
   (b) False

4. To implement a new laboratory procedure for forensic DNA testing, which of the following is the first step required?
   (a) Proficiency test
   (b) Laboratory accreditation
   (c) Laboratory audit
   (d) Laboratory validation studies

5. What are the minimum qualifications for a forensic DNA analyst?

# Index

## B

## D

## N



## O

## S

## T